M000273323

Engineering and Scientific Computing with Scilab

Engineering and Scientific Computing with Scilab

Claude Gomez (Coordinating Editor)
Carey Bunks
Jean-Philippe Chancelier
François Delebecque
Maurice Goursat
Ramine Nikoukhah
Serge Steer

With Over 100 Illustrations and a CD-ROM

Birkhäuser
Boston • Basel • Berlin

Claude Gomez
(Coordinating Editor)
INRIA
Rocquencourt
78153 Le Chesnay Cedex
France

Carey Bunks
BBN Technologies
Cambridge, MA 02138
USA

Jean-Philippe Chancelier
CERGRENE
École Nationale des Ponts et Chaussées
77455 Marne LaVallée Cedex 02
France

François Delebecque
INRIA
Rocquencourt
78153 Le Chesnay Cedex
France

Maurice Goursat
INRIA
Rocquencourt
78153 Le Chesnay Cedex
France

Ramine Nikoukhah
INRIA
Rocquencourt
78153 Le Chesnay Cedex
France

Serge Steer
INRIA
Rocquencourt
78153 Le Chesnay Cedex
France

Library of Congress Cataloging-in-Publication Data
Engineering and scientific computing with Scilab / Claude Gomez,
   editor.
      p.   cm.
      ISBN 0-8176-4009-6.—ISBN 3-7643-4009-6
      1. Engineering—Computer programs.   2. Science—Computer programs.
   3. Scilab.   I. Gomez, Claude, 1955–
TA345.E5475   1998
620'.00285—DC21
                                                                                     98-16928
                                                                                     CIP

AMS Subject Classifications: 49, 93

Printed on acid-free paper.
© 1999 Birkhäuser Boston

Birkhäuser ®

All rights reserved. This work may not be translated or copied in whole or in part without the written permission of the publisher (Birkhäuser Boston, c/o Springer-Verlag New York, Inc., 175 Fifth Avenue, New York, NY 10010, USA), except for brief excerpts in connection with reviews or scholarly analysis. Use in connection with any form of information storage and retrieval, electronic adaptation, computer software, or by similar or dissimilar methodology now known or hereafter developed is forbidden.
The use of general descriptive names, trade names, trademarks, etc., in this publication, even if the former are not especially identified, is not to be taken as a sign that such names, as understood by the Trade Marks and Merchandise Marks Act, may accordingly be used freely by anyone.

ISBN 0-8176-4009-6
ISBN 3-7643-4009-6

Typeset by the authors in LaTeX.
Printed and bound by Edwards Brothers, Inc., Ann Arbor, MI.
Printed in the United States of America.

9 8 7 6 5 4 3 2 1

Contents

Preface xiii

List of Figures xix

List of Tables xxv

I The Scilab Package 1

1 Introduction 3
   1.1 What Is Scilab? 3
   1.2 Getting Started 7

2 The Scilab Language 17
   2.1 Constants 17
      2.1.1 Real Numbers 17
      2.1.2 Complex Numbers 18
      2.1.3 Character Strings 18
      2.1.4 Special Constants 18
   2.2 Data Types 19
      2.2.1 Matrices of Numbers 19
      2.2.2 Sparse Matrices of Numbers 19
      2.2.3 Matrices of Polynomials 20
      2.2.4 Boolean Matrices 20
      2.2.5 Sparse Boolean Matrices 20
      2.2.6 String Matrices 20
      2.2.7 Lists 21
      2.2.8 Typed Lists 21
      2.2.9 Functions of Rational Matrices 22
      2.2.10 Functions and Libraries 22
   2.3 Scilab Syntax 23

        2.3.1   Variables   23
        2.3.2   Assignments   24
        2.3.3   Expressions   24
        2.3.4   The list and tlist Operations   29
        2.3.5   Flow Control   32
        2.3.6   Functions and Scripts   36
        2.3.7   Commands   42
    2.4 Data-Type-Related Functions   44
        2.4.1   Type Conversion Functions   44
        2.4.2   Type Enquiry Functions   47
    2.5 Overloading   47
        2.5.1   Operator Overloading   49
        2.5.2   Primitive Functions   52
        2.5.3   How to Customize the Display of Variables 53

3   Graphics   55
    3.1 The Media   55
        3.1.1   The Graphics Window   56
        3.1.2   The Driver   58
        3.1.3   Global Handling Commands   58
    3.2 Global Plot Parameters   60
        3.2.1   Graphical Context   61
        3.2.2   Indirect Manipulation of the Graphics Context . . . 63
    3.3 2-D Plotting   64
        3.3.1   Basic Syntax for 2-D Plots   64
        3.3.2   Specialized 2-D Plotting Functions 70
        3.3.3   Captions and Presentation   73
        3.3.4   Plotting Geometric Figures   75
        3.3.5   Some Graphics Functions for Automatic Control . . . 78
        3.3.6   Interactive Graphics Utilities   79
    3.4 3-D Plotting   79
        3.4.1   3-D Plotting   81
        3.4.2   Specialized 3-D Plots and Tools   83
        3.4.3   Mixing 2-D and 3-D Graphics   83
    3.5 Examples   85
        3.5.1   Subwindows   85
        3.5.2   A Set of Figures   86
    3.6 Printing Graphics and Exporting to LATEX 86
        3.6.1   Window to Printer   88
        3.6.2   Creating a Postscript File   88
        3.6.3   Including a Postscript File in LATEX 89

    3.6.4   Scilab, Xfig, and Postscript   92
    3.6.5   Creating Encapsulated Postscript Files   92

4   **A Tour of Some Basic Functions**   95
    4.1   Linear Algebra   95
          4.1.1   QR Factorization   96
          4.1.2   Singular Value Decomposition   98
          4.1.3   Schur Form and Eigenvalues   99
          4.1.4   Block Diagonalization and Eigenvectors   100
          4.1.5   Fine Structure   102
          4.1.6   Subspaces   104
    4.2   Polynomial and Rational Function Manipulation   106
          4.2.1   General Purpose Functions   106
          4.2.2   Matrix Pencils   109
    4.3   Sparse Matrices   111
    4.4   Random Numbers   114
    4.5   Cumulative Distribution Functions and Their Inverses . . .   117

5   **Advanced Programming**   119
    5.1   Functions and Primitives   119
    5.2   The Call Function   121
    5.3   Building Interface Programs   124
    5.4   Accessing "Global" Variables Within a Wrapper   129
          5.4.1   Stack Handling Functions   129
          5.4.2   Functional Arguments   132
    5.5   Intersci   135
          5.5.1   A First Intersci Example   135
          5.5.2   Intersci Descriptor File Syntax   136
    5.6   Dynamic Linking   145
    5.7   Static Linking   147
          5.7.1   Static Linking of an Interface   147
          5.7.2   Functional Argument: Static Linking   148

II **Tools** **149**

6   **Systems and Control Toolbox** **151**
    6.1   Linear Systems   151
          6.1.1   State-Space Representation   151
          6.1.2   Transfer-Matrix Representation   153
    6.2   System Definition   154

	6.2.1	Interconnected Systems	155
	6.2.2	Linear Fractional Transformation (LFT)	159
	6.2.3	Time Discretization	159
6.3	Improper Systems		162
	6.3.1	Scilab Representation	163
	6.3.2	Scilab Implementation	164
6.4	System Operations		166
	6.4.1	Pole-Zero Calculations	167
	6.4.2	Controllability and Pole Placement	168
	6.4.3	Observability and Observers	170
6.5	Control Tools		172
6.6	Classical Control		179
	6.6.1	Frequency Response Plots	183
6.7	State-Space Control		186
	6.7.1	Augmenting the Plant	188
	6.7.2	Standard Problem	190
	6.7.3	LQG Design	191
	6.7.4	Scilab Tools for Controller Design	194
6.8	H_∞ Control		197
6.9	Model Reduction		199
6.10	Identification		203
6.11	Linear Matrix Inequalities		207

7	**Signal Processing**		**209**
7.1	Time and Frequency Representation of Signals		209
	7.1.1	Resampling Signals	210
	7.1.2	The DFT and the FFT	210
	7.1.3	Transfer Function Representation of Signals	212
	7.1.4	State-Space Representation	216
	7.1.5	Changing System Representation	217
	7.1.6	Frequency-Response Evaluation	218
	7.1.7	The Chirp z-Transform	222
7.2	Filtering and Filter Design		224
	7.2.1	Filtering	224
	7.2.2	Finite Impulse Response Filter Design	226
	7.2.3	Infinite Impulse Response Filter Design	234
7.3	Spectral Estimation		238
	7.3.1	The Modified Periodogram Method	241
	7.3.2	The Correlation Method	243

8   Simulation and Optimization Tools **247**
  8.1   Models 247
  8.2   Integrating ODEs 248
         8.2.1   Calling **ode** 250
         8.2.2   Choosing Between Methods 261
         8.2.3   ODE Integration with Stopping Times 261
         8.2.4   Sampled Systems 263
  8.3   Integrating DAEs 265
         8.3.1   Implicit Linear ODEs 265
         8.3.2   General DAEs 267
         8.3.3   DAEs with Stopping Time 279
  8.4   Solving Optimization Problems 280
         8.4.1   Quadratic Optimization 281
         8.4.2   General Optimization 284
         8.4.3   Solving Systems of Equations 291

9   SCICOS — A Dynamical System Builder and Simulator **293**
  9.1   Hybrid System Formalism 293
  9.2   Getting Started 295
         9.2.1   Constructing a Simple Model 295
         9.2.2   Model Simulation 299
         9.2.3   Symbolic Parameters and "Context" 301
         9.2.4   Use of Super Block 303
         9.2.5   Simulation Outside the Scicos Environment 305
  9.3   Basic Concepts 307
         9.3.1   Basic Blocks 307
         9.3.2   Inheritance and Time Dependence 311
         9.3.3   Synchronization 312
  9.4   Block Construction 313
         9.4.1   **Super Block** 314
         9.4.2   **Scifunc** Block 314
         9.4.3   **GENERIC** Block 315
         9.4.4   **Fortran** Block and **C** Blocks 315
         9.4.5   *Interfacing* Function 315
         9.4.6   *Computational* Function 323
  9.5   Example 331
  9.6   Palettes 332
         9.6.1   Existing Palettes 332
         9.6.2   Constructing New Palettes 336

10 Symbolic/Numeric Environment **339**
   10.1 Introduction 339
   10.2 Generating Optimized Fortran Code with Maple 342
   10.3 Maple to Scilab Interface 349
   10.4 First Example: Simulation of a Rolling Wheel 354
   10.5 Second Example: Control of an n-Link Pendulum 358
      10.5.1 Simulation of the n-Link Pendulum 360
      10.5.2 Control of the n-Link Pendulum 363

11 Graph and Network Toolbox: Metanet **367**
   11.1 What Is a Graph? 368
   11.2 Representation of Graphs 369
      11.2.1 Standard Tail/Head Representation 369
      11.2.2 Other Representations 371
      11.2.3 Graphs and Sparse Matrices 378
   11.3 Creating and Loading Graphs 378
      11.3.1 Creating Graphs 378
      11.3.2 Loading and Saving Graphs 380
      11.3.3 Using the Metanet Window 380
   11.4 Generating Graphs and Networks 385
   11.5 Graph and Network Computations 387
      11.5.1 Getting Information About Graphs 387
      11.5.2 Paths and Nodes 388
      11.5.3 Modifying Graphs 388
      11.5.4 Creating New Graphs From Old Ones 389
      11.5.5 Graph Problem Solving 389
      11.5.6 Network Flows 390
      11.5.7 The Pipe Network Problem 397
      11.5.8 Other Computations 398
   11.6 Examples Using Metanet 398
      11.6.1 Routing in the Paris Metro 399
      11.6.2 Praxitele Transportation System 400

III Applications **411**

12 Modal Identification of a Mechanical Structure **413**
   12.1 Modeling the System 413
   12.2 Modeling the Excitation 415
      12.2.1 Decomposition of the Unknown Input 415
      12.2.2 Contribution of the Colored Noise 416

12.2.3 Contribution of the Harmonics 418
12.2.4 The Final Discrete-State Model 420
12.3 State-Space Representation and an ARMA Model 421
12.4 Modal Identification 423
12.4.1 Instrumental Variable Method 423
12.4.2 Balanced Realization Method 429
12.5 Numerical Experiments 431
12.5.1 Basic Computations 431
12.5.2 Some Plots of Results 434

13 Control of Hydraulic Equipment in a River Valley **441**
13.1 Introduction 441
13.2 Description of a Managed River Valley 442
13.2.1 Hydraulic Equipment in a River Valley 442
13.2.2 Power Production 442
13.2.3 Structural Analysis 442
13.2.4 Controller Structure 444
13.2.5 Central Hydraulic Supervision Station 444
13.2.6 Local Controllers 446
13.3 Race Modeling 447
13.3.1 Physical Description 447
13.3.2 Mathematical Model 448
13.3.3 Race Numerical Simulation 449
13.4 Choice of Observation 452
13.4.1 Volume Observer 454
13.4.2 Level Observer 455
13.5 Control of a Race 457
13.5.1 Race Dynamics Identification 457
13.5.2 Local Control Synthesis 460
13.5.3 Series Anticipations Design 461
13.5.4 Parallel Anticipation Design 463
13.5.5 Feedback Controller Design 465
13.6 Metalido Overview 466
13.6.1 Graphical User Interface 466
13.6.2 Scicos 467
13.6.3 Data Structures 477

Bibliography **481**

Index **487**

Preface

Overview

Scilab is a scientific software package that provides a powerful open computing environment for engineering and scientific applications. Distributed freely via the Internet since 1994, Scilab is currently being used in educational and industrial environments around the world.

This book contains all the information needed to master Scilab: how to use it interactively as a super calculator, how to write programs, how to develop complex applications, and more. The authors, Carey Bunks (BBN[1]), Jean-Philippe Chancelier (ENPC[2]), François Delebecque, Claude Gomez, Maurice Goursat, Ramine Nikoukhah, and Serge Steer (INRIA[3]), have not only been involved in the development of Scilab, but have used it for teaching and industrial applications for many years. A CD-ROM, containing the entire Scilab source code as well as a set of precompiled binary executables for a variety of computing platforms, is included with this book.

The objective here is to give a thorough description of Scilab's use, including how to master its environment and programming language, the use of the integrated graphics, the incorporation of user-provided functions, and a tour of the numerous application toolboxes. The purpose is to provide students and professionals with an introduction to Scilab and its use in engineering and scientific problem solving. The numerous practical examples serve as a framework that can be used as a basis for developing other applications.

The philosophy of this book is to go beyond simple software documentation. Thus, a strong tutorial language that seeks to explain the subject matter of many engineering and scientific disciplines has been adopted. This book will be of interest to all scientists who need to perform numer-

---

[1]BBN Technologies (a subsidiary of GTE), Cambridge, MA, USA.
[2]École Nationale des Ponts et Chaussées, Marne la Vallée, France.
[3]Institut National de Recherche en Informatique et en Automatique, Rocquencourt, France.

ical computations, especially those of simulation, optimization, signal processing, and control. The associated numerical facilities in conjunction with the open computing environment, the object-oriented programming tools, the Scilab interpreter, and the integrated graphics make for a powerful scientific computing tool.

Organization and Features

This book has four primary objectives. The first is to give a description of the bulk of Scilab's existing functions, including the integrated graphics facilities. Although an explanation of the full functionality of Scilab would require a much longer book than this one, the covered topics should provide a rich understanding of Scilab's fundamental structure and use. This leads to the second objective, which is to provide a basis for self-instruction and discovery within Scilab by explaining the use of its on-line documentation facilities.

It would be naive to believe that any finite set of preprogrammed functions could satisfy all scientific computing needs. Thus, the third objective of this book is to describe the open and extensible computing environment that Scilab provides and that allows the integration of user-defined functions written in Fortran, C, or in the Scilab programming language.

Finally, this book provides a description of several complex real-world problems that have been solved using Scilab. The objective here is that readers be able to use these examples as a framework to develop solutions for their own problems.

This book, consisting of thirteen chapters, is organized into three parts. The first part of the book gives an overview of Scilab and contains five chapters:

- Chapter 1 presents a brief introduction to Scilab.

- Chapter 2 describes Scilab's most basic elements: the data types and syntax of the Scilab language. This chapter also discusses some especially useful functions for data-type manipulation.

- Chapter 3 introduces graphics in Scilab. The chapter begins with a discussion of the global graphic structure and context. This is followed by a description of high-level 2-D and 3-D plotting capabilities. The last section of this chapter is devoted to how to export graphics for plotting and for incorporation into other documents.

- Chapter 4 describes a selected set of general-purpose functions that can be used as building-block tools. These functions also serve as an introduction to many related tools that can be found using the on-line help.

- Chapter 5 describes how users can extend the functionality of Scilab by linking in custom Fortran and C programs and by so doing creating new functional Scilab primitives. Furthermore, this chapter discusses the creation of abstract data types and operator overloading.

The second part of this book presents the use of Scilab for various scientific and engineering problems. This part can serve as a tutorial introduction to several engineering fields such as signal processing, control, simulation, and optimization. Each of these topics is supported by a Scilab toolbox, which is a rich collection of functions and primitives relevant to the subject. The objective here is to introduce the student to the topic and to aid the professional in making effective use of Scilab in the application area. This part of the book contains six chapters:

- Chapter 6 describes the main Scilab functions for system analysis and control. It begins with the various system representations that can be used in Scilab. Tools for analyzing system performance (controllability, observability, etc.) are discussed. This is followed by the presentation of the solution to a standard control problem using both LQG and H_∞ methods. This chapter ends with an examination of some identification and LMI problems.

- Chapter 7 illustrates the use of Scilab in a signal-processing context. It discusses the core signal-processing tools, which include discussions on signal representation, FIR and IIR filter design, and spectral estimation.

- Chapter 8 is devoted to simulation and optimization tools. It begins by introducing various models used for simulation and optimization problems. This is followed by a description of Scilab's numerical solver for differential equations. This solver is effective for both ODE and DAE systems. Finally, Scilab's nonlinear optimization tools are described and are illustrated with examples. Particular emphasis is given in this chapter to mechanical problems that can be described using DAEs.

- Chapter 9 describes Scicos (the Scilab Connected Object Simulator), which is an interactive graphical user interface for modeling and simulation of hybrid dynamical systems. This chapter provides a step-by-step guide for the construction of dynamical models in Scicos using predefined as well as user-designed graphical blocks. The various functionalities of Scicos are illustrated with examples.

- Chapter 10 describes how to link symbolic with numeric computations by making use of the Maple-to-Scilab interface.[4] This interface allows the user to make use of optimized Scilab, C, or Fortran code symbolically generated by Maple. Two working examples, the simulation of a rolling wheel, and the control of a 3-link pendulum, are given.

- Chapter 11 describes Metanet, a toolbox for graphs and network flow computations. The way graphs are represented in Metanet and the corresponding underlying data structures are described. The main functionalities are presented, in particular the special graphical interface window used for displaying and modifying graphs. Emphasis is put on the types of problems Metanet can solve. Two examples are presented; in particular, the Praxitele transportation system is described in great detail.[5]

The third and final part of the book presents two applications, each presented in a single chapter:

- Chapter 12 describes an application concerning the identification of the vibrational modal behavior of a mechanical structure. The material covered in this chapter assumes a basic knowledge of signal processing and the theory of vibration.

- Chapter 13 describes a real industrial application built with Scilab for E.D.F., the French electric company. It addresses the problem of control for hydraulic equipment in a river valley. This application makes extensive use of many of Scilab's tools.

CD-ROM and Computational Aspects

This book comes with a CD-ROM containing the complete source distribution for Scilab. The CD also contains a large number of precompiled binary

---

[4]Maple is a separate, commercial computer algebra software package and is not included in Scilab.

[5]Praxitele is a transportation system based on a fleet of electric cars available in a self-service mode and in a limited area.

distributions available for a great many computing platforms. Scilab runs on most Unix workstations. Examples are Linux PCs and PPCs, SUN Solaris, SUN OS, SGI, Dec Alpha (OSF and Linux), and IBM RS6000s. Scilab also runs on Windows 95/98 and NT computers. For most platforms the binary distributions will run right out of the box. Installation procedures are provided on the CD for both binary and source versions of Scilab. All this material can also be obtained freely from the Internet at the Web site

```
http://www-rocq.inria.fr/scilab
```

The CD-ROM also contains all the Scilab code used in the example sessions found in this book. These examples are provided in LaTeX and HTML files, which should allow the user to easily test and experiment with them in their own Scilab sessions. An HTML file, including a tour of a few Scilab standard demos together with a demonstration of Scilab's full-color graphic capabilities, is also included on the CD.

It should be noted that all the examples in the book were tested on a DEC Alpha 300 running a Unix OSF1 operating system.

Acknowledgments

Scilab makes use of much public-domain software and as such contains the contributions of many authors. It is not possible to cite all of them here. However, many of these contributions are cited in the ACKNOWLEDGMENTS file in the Scilab distribution. The source for other contributions can be determined by direct examination of the source files distributed with Scilab.

List of Figures

1.1   Scilab's graphical user interface   7
1.2   Examples of Scilab plots   14
1.3   Scilab's help GUI   15

3.1   First example of plotting   65
3.2   Different 2-D plotting styles   67
3.3   Black and white (mark) plotting styles   69
3.4   Box, captions, and ticks   70
3.5   Vector field in the plane   71
3.6   Gray plot using a true gray colormap   73
3.7   Grid, title eraser, and comments   74
3.8   Geometric graphics and comments   77
3.9   Some specialized plots for automatic control   80
3.10   Presentation of plots   81
3.11   2-D and 3-D plot   84
3.12   Use of **xsetech**   85
3.13   Group of figures   87
3.14   Blatexpr2 example   91
3.15   Encapsulated Postscript using Xfig   93

4.1   2-D mesh   112
4.2   Nonzero entries in A   113
4.3   Reduced bandwidth matrix   114
4.4   Cumulative distribution for beta distribution   118

6.1   Interconnection of linear systems   156
6.2   Interconnection of linear systems   160
6.3   Discretization and Tustin transform   162
6.4   Conversion functions   164
6.5   Observer output tracks x_2   172
6.6   Step and impulse responses for $H(s) = (1-s)/(s^2 + 0.2s + 1)$ 175

6.7   Step and impulse responses for $H(s) = (1 - s)/(s^2 + 0.2s + 1)$ 177
6.8   A mixed continuous discrete-time system 178
6.9   DC motor: Open loop step response 180
6.10 DC motor: PID controls 181
6.11 DC motor: root locus 182
6.12 DC motor: effect of lag compensator 183
6.13 DC motor: Bode plot 184
6.14 Nyquist plot of the system **Sys** 185
6.15 Zoomed Nyquist plot 185
6.16 Feedback control system 186
6.17 Output sensitivity functions 187
6.18 Input sensitivity functions 187
6.19 Formulation as a standard problem 190
6.20 Standard problem 191
6.21 Matching ideal time response 193
6.22 Observer-based compensator 195
6.23 Nyquist plot and M-circle 199
6.24 Step response 200
6.25 Magnitude responses 202
6.26 Impulse response identification **arl2** and **imrep2ss** 204
6.27 Matching magnitude with stable system 205
6.28 Arma simulation 207

7.1   Resampling a signal using **intdec** 211
7.2   Computing the DFT using **fft** 213
7.3   Bode plot obtained using **freq** 215
7.4   Impulse response obtained with **rtitr** 216
7.5   Group delay computed with **group** 222
7.6   The chirp z-transform samples the z-transform on spirals . 223
7.7   Filtering using **flts** 227
7.8   Examples using **wfir** 229
7.9   Comparing **eqfir** with **wfir** 232
7.10 Design of a multiband filter using **eqfir** 233
7.11 Design of a triangular magnitude response with **remezb** . . 234
7.12 Design of IIR filters using **iir** 237
7.13 Design of IIR filters using **eqiir** 239
7.14 Example of spectral estimation using **pspect** 244
7.15 Example of spectral estimation using **cspect** 246

8.1   RLC circuit simulation 252
8.2   Simple ode plot 252

8.3   Ode with surface crossing 263
8.4   ODE with sampling 265
8.5   3-D trajectory of the pendulum 275
8.6   Minimum variance distribution of samples 289
8.7   Criterion variation (percentage) around the solution 290

9.1   Continuous and discrete-time signals 294
9.2   A hybrid signal and its activation time set 295
9.3   Scicos main window 296
9.4   Choice of palettes 297
9.5   Inputs/Outputs palette 297
9.6   Blocks copied from the Inputs/Outputs palette 298
9.7   Complete model 298
9.8   **Clock**'s dialogue panel 299
9.9   Simulation result 300
9.10 **MScope** original dialogue box 300
9.11 Symbolic expression as parameter 302
9.12 Context is used to give numerical values to symbolic expressions 302
9.13 Use of symbolic expressions in block parameter definition . 303
9.14 Super block in the diagram 304
9.15 Super block content 304
9.16 Complete diagram with super block 305
9.17 **MScope** updated dialogue box 306
9.18 Constructing an event clock using feedback on a delay block 310
9.19 Activation links: split and addition 312
9.20 Super block defining a 2-frequency clock 313
9.21 Context of the diagram 332
9.22 Linear system with a hybrid observer 333
9.23 Model of the system. The two outputs are **y** and **x**. 333
9.24 Model of the hybrid observer. The two inputs are **u** and **y**. 334
9.25 Linear system dialogue box 334
9.26 **Gain** block dialogue box 335
9.27 Simulation result 335
9.28 Super block realizing the Saint Venant function. The parameter g is defined in the context of the super block. 337

10.1 User model of the interface between Scilab and Maple . . . 342
10.2 Wheel description 355
10.3 Snapshot of wheel animation 359
10.4 3-link pendulum description 359

10.5 Snapshot of 3-link pendulum animation 364
10.6 Time evolution of the controlled pendulum 366

11.1 Definition of edges and arcs 369
11.2 Small directed graph 370
11.3 Adjacency list representation of graphs 372
11.4 Graph `toto` 376
11.5 Graph `toto` after bandwidth reduction of its node–node incidence matrix 376
11.6 Chained list representation of graphs 377
11.7 Smallest directed graph 379
11.8 Small undirected graph 379
11.9 Directed graph 379
11.10 Directed graph in Metanet window 382
11.11 Highlighted arcs and nodes in Metanet window 383
11.12 Generated network 387
11.13 Maximum-capacity path from source node to sink node . . 391
11.14 Maximum flow from source node to sink node 393
11.15 Maximum flow from source node to sink node 396
11.16 A travel in the Paris metro 400
11.17 The elements of a Praxitele line 402
11.18 Simulation results of the sizing problem 408
11.19 Number of praxicars of praxiparks 1 and 2 409
11.20 Number of praxicars in service vehicles 1 and 2 410

12.1 Spectrum of output component 1 427
12.2 Stabilization diagram 428
12.3 Selection with a window focus 429
12.4 Output of a sensor 434
12.5 Spectrum of previous signal 435
12.6 Stabilization diagram 436
12.7 Selected eigenfrequencies 437
12.8 Evolution of a selected eigenfrequency 438
12.9 MAC criterion for the previous eigenfrequency 439
12.10 Evolution of the damping coefficient 439
12.11 MAC criterion 440

13.1 Plants along the Rhine River 443
13.2 Race sequence 444
13.3 Example of central hydraulic supervision structure 445
13.4 Local controller typical structure 446

13.5 Typical race diagrams 447
13.6 Branches interconnection representation using Metanet . . . 448
13.7 3-D view of a branch 450
13.8 Steady-state water level for various volumes V_1, V_2, V_3 . . . 451
13.9 Steady-state water level for various flows 452
13.10 Transient behavior 453
13.11 Step response at a given point 453
13.12 Daily flow modulation 454
13.13 Water surfaces for different steady states with constant volume 455
13.14 Individual $(h_i - h_r)$ and observer $(C_m = y - h_r)$ evolution vs. steady-state flow 456
13.15 J vs. position and various steady-state flows 456
13.16 Water surfaces for different steady flows with constant level 457
13.17 Control diagram of a race 458
13.18 Step responses with one measurement point in the middle of the race for various flows 459
13.19 Typical step responses for three-point composite measure . 459
13.20 Impulse response 460
13.21 Fourth-order identification of the impulse response 461
13.22 Series anticipation diagram 462
13.23 Series anticipation simulation 463
13.24 Series anticipation simulation for big flow changes 464
13.25 Diagram of parallel anticipation actions due to top input flow 464
13.26 Diagram of two degrees of freedom I.M.C. 465
13.27 Two degree of freedom I.M.C. action 466
13.28 Metalido main menu 467
13.29 Metalido observation determination menu 468
13.30 Steady-state water surface in Beauchastel main branch . . . 468
13.31 Identification submenu 469
13.32 An example of Metalido dialogues 469
13.33 Simple Scicos simulation diagram 471
13.34 Scicos main diagram 472
13.35 Scicos regulator main diagram 474
13.36 Scicos regulator diagram 475
13.37 Scicos compensator diagram 476
13.38 Scicos race diagram 478
13.39 Scicos observation 479

List of Tables

2.1	Type numbers	47
2.2	Type names	48
2.3	Overloading operation codes	49
2.4	Overloading type codes	50
5.1	**addinter** examples	129
5.2	**call** examples	131
5.3	Description of a pair of a Scilab function and a programming language subroutine (various fields are separated by blanks or tabs)	137
5.4	Description of a variable element of a list	142
5.5	Intersci examples	144
8.1	Coding ODE in Fortran or C	256
9.1	Tasks of *Computational* function and their corresponding flags	324
9.2	Different types of *Computational* functions. Type 0 is obsolete.	324
9.3	Arguments of *Computational* functions of type 1. I: input, O: output.	326
9.4	Arguments of *Computational* functions of type 2. I: input, O: output.	328
9.5	Arguments of *Computational* functions of type 3. I: input, O: output.	330

Part I

The Scilab Package

Chapter 1

Introduction

1.1 What Is Scilab?

Scilab is a powerful interactive programming environment that greatly facilitates the task of numerical computation and data analysis. Scilab's main features are:

- A high-level programming language

- an interpreter

- a data environment

- data abstraction

- transparent data types

- a large collection of built-in primitive functions

- a suite of Scilab coded function toolboxes

- integrated graphics

In the following paragraphs each of these elements is described in some detail.

The Programming Language Scilab is fully endowed with a powerful programming language whose syntax is easy to use and has all the elements of a high-level programming language. The elements include assignment, operators, functions, conditionals, and loops. For reasons that will become clearer in what follows, a program written in Scilab normally takes many fewer lines of code than one written in C or in FORTRAN.

The Interpreter When Scilab is executed, a user prompt within a graphical user interface appears. The prompt is Scilab's interpreter waiting for the user to type a command, which is then immediately evaluated. An interpreted programming language for data analysis is much more powerful than one that requires the construction of complete programs followed by compilation before the data begins to be manipulated. An interpreter in conjunction with a data environment provides immediate access to data, which is easier and less time-consuming than having to plan out a series of computations that lead to some final result (an activity that supposes a clear idea of the computational objectives to be achieved). With an interpreter, the user can operate on data a command at a time, which leads to better understanding of a potential processing sequence and allows for rapid experimentation with parameters.

The Data Environment Scilab's interpreter works hand in hand with the data environment. As data is created or imported into Scilab it is directly available in the data environment. This means that individual values or groups of values in the data can be examined simply by typing the variable names. Powerful data extraction tools exist that aid in the examination of specific data values. Also, data can be easily displayed using the integrated plotting capabilities of Scilab.

A Rich Collection of Transparent Data Types Scilab works with a rich collection of data types. Scilab is fundamentally matrix based. However, Scilab's matrices can contain many interesting data types. For example, matrices can contain real and complex numbers, and in Scilab they can also contain variable-length strings, polynomials, rationals, and Boolean data types. Scalars and vectors are just simpler types of matrices. The sparse matrix is a special type in Scilab that aids in accelerating many numerical matrix-based computations.

Scilab's most interesting data types are lists and typed lists. These powerful data types are the basis of Scilab's ability to perform data abstraction.

Functions, which are either compiled or uncompiled programs, are also data types. Since functions are Scilab data types, it is possible to operate on them and pass them as input and output arguments to other functions.

In contrast with most high-level programming languages, Scilab's data types have transparent properties. The data knows its own type, its dimensions, and how much memory it uses, which relieves the programmer from having to know it. This feature greatly diminishes the kinds of programming bugs that can easily creep into high-level programming languages such as C or FORTRAN, since in Scilab it is impossible, for example, to incorrectly combine different data types or inadvertently overrun the memory of an array.

Data Abstraction   One of the most powerful aspects of Scilab is its capabilities to perform data abstraction. Data abstraction is implemented using typed lists, which allow for the aggregation of heterogeneous data types and the labeling of the new abstract data type. Data abstraction, in conjunction with operator overloading, allows the user to define new data types, which can be combined with each other or with other data types using any of Scilab's built-in operators. Furthermore, functions can easily be constructed to handle abstract data types transparently.

Scilab is delivered with three built-in abstract data types: rationals, linear systems, and graphs. Furthermore, Scilab users can easily define new abstract data types, implement operator and function overloading for these new types, and develop Scilab functions that will perform other typical operations for them.

Primitive Functions   The most fundamental operations in Scilab, called primitive functions, are implemented in C and FORTRAN. The use of primitives is as transparent as the manipulation of Scilab data types. Furthermore, the user can create new primitives by writing C or FORTRAN programs, which are then made accessible to the user in Scilab through a variety of interfacing methods.

Numerous Toolboxes of Functions   Since Scilab has its own programming language, it is useful to combine collections of Scilab commands into programs called functions. These are referred to as Scilab coded functions to differentiate them from primitive functions. The creation of functions adds a great deal of functionality to Scilab. A large collection of useful function toolboxes has already been created, among which there are the following:

- Linear Algebra Toolbox

- ARMA Toolbox

- Classical Control Toolbox

- Polynomial Toolbox

- Hybrid System Simulation Toolbox (Scicos)

- Communication Toolbox

- Parallel Toolbox (using PVM)

- Graphs and Networks Toolbox (Metanet)

- Nonlinear Optimization and Simulation Toolbox

- Robust Control Toolbox

- Signal Processing Toolbox

- Hidden Markov Models Toolbox

- Fuzzy Logic Toolbox

- Fractal Analysis Toolbox

Toolboxes are just collections of functions that have some common application, and consequently, new toolboxes can be created by any Scilab user.

Integrated Graphics   Finally, integrated graphics are one of Scilab's most useful tools, since they allow the user to examine data interactively. Data understanding is a very important aspect of any scientific or engineering activity, and the capability to rapidly examine large collections of data and present them graphically is very powerful. Just as in the case of Scilab primitive and coded functions, the plotting tools act directly on data in the environment.

Scilab can manage multiple plot windows simultaneously. Furthermore, Scilab plots in 2-D and 3-D, performs contouring and parametric plots, and has a variety of other useful plotting modes.

Scilab is free software, which is distributed with all its source code, and many thousands of copies have already been distributed via the Internet throughout the world. Relevant updates, news, and contributed packages can be obtained starting from Scilab's home page on the web:

    http://www-rocq.inria.fr/scilab/

Also, there is a newsgroup for discussions on usage and development in Scilab. The newsgroup can be found on

    comp.soft-sys.math.scilab

Postings to the newsgroup are also archived at Scilab's home page.

1.2 Getting Started

To illustrate some of Scilab's main points as well as to provide the novice user with a starting point, an annotated sample session is presented in this section.

When Scilab is first called, the graphical user interface illustrated in Figure 1.1 appears. The window consists of three main areas. The first area,

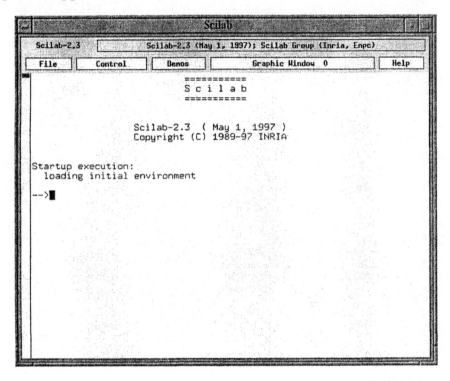

Figure 1.1: *Scilab's graphical user interface*

showing the Scilab startup banner and the --> prompt, is used for typing commands and viewing the results of operations. The second area is a collection of pull-down menus labeled **File, Control, Demos, GraphicWindow,** and **Help**. The third area is the scroll bar, which appears at the left-hand side of the window. Typing the following command at the prompt (followed by a carriage return)

```
-->sin(%pi/2)+sqrt(4)*%i
 ans =
    1. + 2.i
```

illustrates several things. First, typical operations such as addition, multiplication, division, and subtraction use their usual symbols **+** , ***** , **/** , and **-** . Also, functions such as the square root and the sines operate as they would in other typical programming languages. The example also illustrates that Scilab contains some special constants such as π, which is represented in Scilab by **%pi**, as well as the imaginary number $\sqrt{-1}$, which is represented by **%i**. Finally, Scilab's interpreter immediately evaluates the expression and, since no assignment has been declared, gives the value to the standard output variable **ans**.

As a second example,

```
-->a=7;b=3
 b  =
    3.

-->c=a*b
 c  =
    21.
```

Here two assignment statements are made at the same prompt. The first, **a=7**, is terminated in a semicolon, which prevents the result from being printed, whereas the second assignment, **b=3**, is displayed, since it does not use the semicolon. At the second prompt the result of the product of the two previous assignments is in turn assigned to the variable **c**. This illustrates how Scilab interprets operations and how it can make use of variables that were previously defined and now live in the data environment. The command

```
-->who
your variables are...

    c           b           a           home        pwd
    startup     ierr        scicos_pal              TMPDIR
    fraclablib              soundlib    blockslib
    scicoslib               xdesslib    utillib     tdcslib
    siglib      s2flib      roblib      percentlib
    optlib      metalib     elemlib     commlib     polylib
    autolib     armalib     alglib      SCI         %F
    %T          %z          %s          %nan        %inf
    old         newstacksize            $           %t
    %f          %eps        %io         %i          %e
    %pi
    using    4259 elements   out of 1000000.
             and      45 variables out of      1023
```

can be used to see all the variables in the environment. As shown here, the variables **a**, **b**, **c**, and **ans** can be seen, as well as a number of predefined or preloaded variables and function toolboxes. Furthermore, it is indicated that the environment has room for 1,000,000 numbers,[1] of which 4259 are used, and has room for 1023 variable names, of which 45 are used. The next example illustrates some simple matrix operations:

```
-->A=[1,2,3;4,5,6]
 A   =
!   1.     2.     3. !
!   4.     5.     6. !

-->v=1:0.5:2
 v   =
!   1.     1.5    2. !

-->A*v
    !--error    10
inconsistent multiplication

-->A*v'
 ans   =
!    10.  !
!    23.5 !
```

At the first prompt a 2×3 matrix is defined by separating column entries by commas and row entries by semicolons. The second prompt illustrates the use of the **:** syntax for creating vectors. A vector in Scilab is a sequence of numbers starting with the value of the left-hand number and incrementing by the second number up to the value of the right-hand number. If the format **n:m** is used, then the middle number is presumed to be unity. At the third prompt, a matrix–vector product is attempted, which gives rise to a Scilab error, since the vector does not have the appropriate dimensions for the operation. This is resolved by transposing the vector at the following prompt using **'**, the transpose operator.

Numerous tools for creating matrices are available in Scilab:

```
-->B=rand(3,3);

-->size(B)
 ans   =
!    3.     3. !
```

_____

[1]This number can be made bigger by using the function **stacksize**. See the on-line help for more details.

```
-->B*inv(B)
 ans   =
!   1.      0.      0. !
!   0.      1.      0. !
!   0.      0.      1. !
```

Here a 3×3 matrix is created from numbers randomly distributed in $[0, 1)$. The size is verified by the size function and the product of the matrix with its inverse is illustrated by the resulting identity matrix, as shown at the third prompt.

Some other functions that are useful for creating matrices are **eye**, **ones**, and **zeros**, which create an identity matrix, a matrix of ones, and a matrix of zeros, respectively. The final prompt also illustrates that standard matrix functions for computing inverses, such as the determinant, rank, eigenvalues, and eigenvectors, also exist in Scilab.[2]

Another important feature in Scilab is the ability to easily concatenate matrices as well as to extract parts of matrices:

```
-->C=[ones(3,3),zeros(3,2);rand(2,3),eye(2,2)]
 C   =
!   1.              1.              1.              0.      0. !
!   1.              1.              1.              0.      0. !
!   1.              1.              1.              0.      0. !
!      .0683740        .6623569        .1985144     1.      0. !
!      .5608486        .7263507        .5442573     0.      1. !

-->C(1:2:5,[1,5])
 ans   =
!   1.              0. !
!   1.              0. !
!      .5608486     1. !
```

At the first prompt in this example a larger matrix is composed of four smaller matrices by concatenating them together using the same syntax with these matrices as what would be used with scalars. Of course, the dimensions of the smaller matrices are suitable for concatenation. The extraction feature is illustrated at the second prompt. Here the desired parts of the matrix **c** are the elements at the intersection of the first, third, and fifth rows and the first and fifth columns. The extraction is specified by using the **:** syntax in the first argument to the matrix and specifying the vector of elements **[1,5]** for the second argument.

---

[2]See the Linear Algebra toolbox for more details.

Scilab can manipulate and operate on strings. The following example illustrates this:

```
-->D=['This','is';'a','mat'+'rix']
 D  =
!This   is         !
!                  !
!a      matrix   !
```

Notice that the lower right-hand element of the matrix is created by concatenating two strings with the + operator. Scilab contains many operators for strings.

Scilab can create and manipulate polynomials and rational functions:

```
-->p=poly([1,2],'w')
 p  =
                 2
    2 - 3w + w

-->q=poly([-1,2,3],'w','c')
 q  =
                 2
  - 1 + 2w + 3w

-->roots(q)
  ans  =
!     .3333333 !
! - 1.         !

-->r=(p+q)/(p*q)
  r  =
                     2
        1 - w + 4w
    -------------------
              2    3    4
   - 2 + 7w - w - 7w + 3w
```

The first prompt shows how a second-degree polynomial in w with zeros at 1 and 2 is created. The second prompt shows a second-degree polynomial created with the coefficients $-1, 2$, and 3. The roots of q can be computed using Scilab's root primitive. The final prompt shows that normal algebra can be performed with polynomials to create other polynomials and rational functions. Furthermore, matrices in Scilab can contain both polynomials and rational functions, and all normal matrix operations can be performed with them.

Lists in Scilab are illustrated by the following example:

```
-->L=list('Scilab',[1,2;3,4],poly([1,2],'w'))
 L  =
       L(1)

 Scilab

       L(2)

!   1.      2. !
!   3.      4. !

       L(3)

                2
    2 - 3w + w

-->L(2)(2,2)
 ans  =
     4.
```

At the first prompt a list is defined that contains a string, a matrix, and a polynomial. Lists in Scilab support sophisticated concatenation and extraction. The second prompt shows how the (2,2) element of the matrix, which is the second element of the list, can be extracted.

Lists play an important role in Scilab for data abstraction. In fact, the rational data type illustrated above is really a typed list, as illustrated here:

```
-->r=(p+q)/(p*q)
 r  =
                        2
            1 - w + 4w
    --------------------
                2    3    4
    - 2 + 7w - w - 7w + 3w

-->length(r)
 ans  =
     4.

-->r(1)
 ans  =
!r   num   den   dt   !
```

```
-->r(2)
 ans  =
                  2
    1 - w + 4w

-->r(3)
 ans  =
                  2    3    4
   - 2 + 7w - w - 7w + 3w

-->r(4)
 ans  =
       []
```

At the first prompt above, the variable **r**, which was previously created, is displayed as a rational function. However, examining its length reveals that it has four elements. This is because it is in reality a typed list. The first element of this list is a matrix of strings that identify it as a rational function with a numerator and denominator and potentially an associated sampling interval.[3] The vector of strings describes the type of each of the remaining elements in the list, which explains the name "typed list." The second element is the numerator polynomial, the third element is the denominator polynomial, and the fourth element is an empty matrix indicating that this rational has no associated sampling interval.

One way of creating programs in Scilab is with a text editor. However, it is also possible to create a program with the **deff** command:

```
-->deff('[y]=foo(x)','y=x^2;')

-->foo(2)
 ans  =
      4.
```

Here a function named **foo** is created that squares its input. The syntax of **deff** takes two input arguments, both of which are strings. The first string defines the calling syntax of the new function, and the second argument is a vector of strings that gives the sequence of Scilab statements needed to create the output from the input.

Scilab has a host of commands that allow the user to read and write data as well as to load in new functions (using **getf**) and to run files that contain collections of Scilab commands. All of these commands are available from the menu item **File Operations** contained in the pull-down

---

[3]This is important when the rational function represents a linear time-invariant system.

menu labeled **File** at the top of the Scilab graphical user interface (see Figure 1.1).

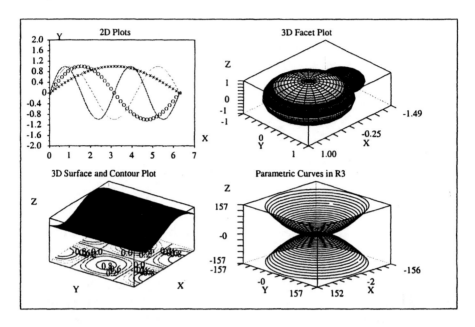

Figure 1.2: *Examples of Scilab plots*

Some of Scilab's integrated plotting functions are illustrated in Figure 1.2. The figure illustrates four plots. The upper left-hand corner illustrates some of Scilab's 2-D plotting capabilities. The upper right-hand corner shows a 3-D facet plot, and the lower left-hand corner shows a 3-D surface plot with contour lines rendered in a plane below the surface. Finally, the lower right-hand corner shows a 3-D parametric plot. These plots and more can be studied in more detail by clicking on the graphics demos contained in the **Demos** button on the user interface shown in Figure 1.1.

Scilab has a powerful on-line help facility, which is illustrated in Figure 1.3. As can be seen in the figure, the graphical user interface for help is in four parts. The uppermost part has a scrollbar and a list of function names each with a brief description. Clicking on a line makes a window appear with a more complete description of the function as well as its calling syntax. The part just below this is the list of installed toolboxes and function categories. Clicking on a line in this part displays the associated functions in the uppermost part. By default, the functions in the uppermost part are those associated with Scilab programming. The lowermost part is an entry box where the user can type a keyword. Typing a keyword followed by a carriage return makes the names of all the Scilab functions that have that word in their help description appear in the uppermost part of the window. This is called the **Apropos** function. Clicking on the **Done** button closes the on-line help GUI.

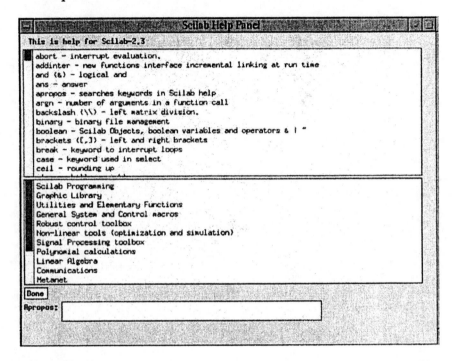

Figure 1.3: *Scilab's help GUI*

Finally, Scilab has numerous demos that illustrate the depth and range of its capabilities. These can be run by clicking on the **Demos** button at the top of Scilab's graphical user interface, seen in Figure 1.1.

Chapter 2

The Scilab Language

Scilab is an interpreted high-performance language with control-flow statements. This, combined with functions, high-level data types, and object-oriented programming features, allows for the rapid creation of experimental programs as well as that of complete applications packages. Scilab interprets and evaluates statements typed in its main window. Simple statements are assignments such as **variable=expression**. Here the computed value of the expression is stored in the specified variable. Expression values are recursively computed from referred variables or constant values using operators and function calls. This chapter describes the basic elements of the Scilab language. The ensuing sections will describe Scilab constants, data types, and syntax. The chapter finishes with a description of some special data-type-related functions.

2.1 Constants

The Scilab language recognizes only two types of constants: real numbers and character strings. Some special constants are predefined in Scilab.

2.1.1 Real Numbers

Scilab follows conventional decimal notation, with representation of numbers having an optional decimal point, leading sign, and power-of-ten factor. For example, some valid real number syntaxes are **5, 732., -78, 0.00128, -6.0356 1.345e3, -190e5, 1.976e-10, 785.e-2, 3.976e-102**. Note that no distinction is made between integers and real numbers.

17

2.1.2   Complex Numbers

Complex numbers are made by combining real numbers with the predefined constant `%i`, which denotes $\sqrt{-1}$:

```
5+3*%i, -0.367*%i.
```

2.1.3   Character Strings

The syntax for character strings is any nonempty sequence of characters enclosed between single (`'`) or double (`"`) quotes, as in the following examples:

```
'This is a Scilab string'
"and this is another one"
```
Single and double quotes must be doubled to be included in strings:

```
'It''s true.'
'""Scilab"" is a language'
```

2.1.4   Special Constants

Commonly used constants are stored in predefined variables such as:

- `%pi`: approximate value of π.

- `%e`: base of the natural logarithm.

- `%inf`: representing ∞.

- `%nan`: representing "not a number."

- `%s`: the polynomial with a single root at 0 and with **s** as formal variable name.

- `%z`: the polynomial with a single root at 0 and with **z** as formal variable name.

- `%t` or `%T`: logical value for "true."

- `%f` or `%F`: logical value for "false."

2.2   Data Types

Scilab supports several data types. Primarily, these are full-storage two-dimensional matrices with constants, Booleans, polynomials, strings, and rational functions (i.e., quotients of polynomials) as entries. Other basic data types are lists, typed-lists, functions, and libraries. There is also a special sparse matrix type with constant and Boolean entries. All data types may be constructed using the constants in expressions or in function calls. In particular, matrices can be formed by entering an explicit list of entries.

Data types in Scilab do not require type declarations or dimensioning. When Scilab encounters a new variable name, it automatically allocates the appropriate amount of storage for it and infers its type. If the variable already exists, the appropriate changes are made to its type, and if necessary, new storage is allocated. All matrix types (vectors and scalars just being special cases of matrices with a single row or column) are displayed in a matrix tabular form.

2.2.1   Matrices of Numbers

The data type matrix of numbers is for the representation of all full-storage two-dimensional matrices with real or complex entries. Real and complex matrices differ only in their internal storage (real matrices store only real parts of the entries).

[1 2 3] creates a row vector with three entries, 1, 2, and 3.

[0.1;-1] creates a column vector with two entries, 0.1 and -1.

[0.1 -0.1 %e ;1 2 3] creates a two by three matrix, which is displayed as

```
!   0.1   - 0.1    2.7182818 !
!   1.    2.       3.        !
```

2.2.2   Sparse Matrices of Numbers

The sparse matrix data type can be used to efficiently represent two-dimensional matrices that contain only a few nonzero entries that are real, complex, or Boolean. Only the nonzero entries are stored. The sparse matrix data type allows for the compact representation of large matrices with only a few nonzero entries. This data type may be constructed using the **sparse** function as described on page 45. The expression

```
sparse([1 10;3 5],[8.4,-9.1],[10,10])}
```

creates a **10** by **10** matrix with the two nonzero entries **8.4** and **-9.1** located at the positions **(1,10)** and **(3,5)**.

Many possible internal representations may be used to store sparse matrices. In Scilab we have chosen the following representation: an integer vector that contains the number of nonzero entries for each line, a second integer vector that contains the column index of each nonzero entry, and finally the floating-point vector of the nonzero entry values.

2.2.3   Matrices of Polynomials

The data type matrix of polynomials is for the representation of two-dimensional matrices of polynomials in a single formal variable. Each matrix entry is a polynomial with real or complex coefficients. Each polynomial entry is defined by its coefficients stored in a full form (zero coefficients are stored), and each polynomial has its own degree. If a single polynomial has complex coefficients, the associated matrix polynomials data structure stores all polynomials in the complex form. As an example, the evaluation of **[%s+1 %s^3]** results in a row vector with two entries, $s + 1$ and s^3, and is displayed as follows:

```
!                3 !
!     1 + s       s !
```

2.2.4   Boolean Matrices

The Boolean matrix data type is for the representation of full two-dimensional matrices of logical values. An example is **[%T;%F;%F]**, which yields the three-entry column vector displayed as

```
! T !
! F !
! F !
```

2.2.5   Sparse Boolean Matrices

Sparse Boolean matrices are similar to full Boolean matrices. However, only the nonzero entries are stored. This data type may be constructed using the **sparse** function (see page 45).

2.2.6   String Matrices

The string matrix data type is for the representation of two-dimensional matrices of character strings. Each matrix entry is a string whose length is independent of the length of any other entry. As an example,

```
    ['speed'  '10m/s'  ;'position'  '12.5m']
```
results in a two by two matrix displayed as

```
!speed        10m/s   !
!                     !
!position    12.5m    !
```

2.2.7 Lists

Lists can be used to group a set of heterogeneous objects into a single data object. A list in Scilab is quite similar to the "struct" definition of the C programming language and allows for the construction of object-oriented data behavior in Scilab. The entries in a list may contain other lists. This means that `list` is a recursive data structure. Note that data objects entered into lists are just copies of the original variable values, so that the latter may be changed without modifying the list value. A list can be constructed using the `list` function (see page 46). An example is
```
x=list('tax',['bread';'car';'book'],[3;20.8;5]),
```
which results in the ordered structure that is displayed as

```
x(1)

tax

x(2)

!bread !
!      !
!car   !
!      !
!book  !

x(3)

!   3.  !
!  20.8 !
!   5.  !
```

2.2.8 Typed Lists

Typed lists are similar to lists in that they allow for the grouping of heterogeneous objects. However, typed lists also permit the specification of structure class and field names. The class name and field names are given in a string vector in the first entry of the typed list:

```
x=tlist('tax',['bread';'car';'book'],[3;20.8;5])
```
                                    ← defines **x** as a **tax** class structure;

```
x=tlist(['Chars';
         'Names';
         'Ages';
         'Sizes'],..
         ['Peter';
         'John';
         'Franck'],..
         [25;40;18],[1.85;1.76;1.70])
```

defines **x** as a **Chars** class structure with component names **Names, Ages,** and **Sizes.**

```
x('Names')(1)
```
                                    ← returns the first character names

```
x('Ages')(find(x('Names')=='Peter'))
```
                                        ← returns Peter's age

Typed lists are particularly valuable, since they can be used to define new abstract data objects in Scilab. They are also used to allow for operator over-loading on the newly defined data types (see page 47).

2.2.9   Functions of Rational Matrices

The data type matrix of rational functions allows for the representation of full, univariate, two-dimensional matrices of rational functions. Each matrix entry represents the rational fraction of two polynomials with either real or complex coefficients. If one of the polynomial entries has complex coefficients, then all the associated rational entries are stored in complex form. As an example, **[1 %s]/(%s+1)** defines a two-entry rational row vector, which is displayed as

```
!    1          s     !
!   -----      -----  !
!   1 + s      1 + s  !
```

Note that rational matrices are in fact a particular class of a **tlist** data structure.

2.2.10   Functions and Libraries

In addition to the above-described data types, the Scilab language also treats functions and libraries as special data types. Scilab coded functions (see page 37) are stored in a special data type. They can be manipulated like any other data structure, and in particular, they can be passed as arguments to other functions.

Libraries are collections of Scilab coded functions. This data type is created using the **lib** function and is used by the Scilab function search mechanism. For example, the **elemlib** library is displayed as

```
Functions files location :SCIDIR/macros/elem/

and        atanm       atanhm      atanh       asinhm
asinh      asinm       asin        acoshm      acosh
acosm      acos        calerf      cosm        cothm
coth       cotg        coshm       cosh        erfcx
erfc       erf         fix         GLoad       inttrap
intsplin   integrate               intc        intl
interpln   logm        log10       modulo      null
or         pertrans    sprand      spzeros     speye
smooth     sqrtm       sinm        signm       sinhm
sinh       toeplitz    tanhm       tanh        tanm
tan
```

2.3 Scilab Syntax

Scilab's syntax is easy to use and to understand. The syntax consists of variables, assignment, expressions, list and typed list operations, flow control, functions and scripts, and commands.

2.3.1 Variables

As already noted, Scilab is an interpreted language. Variables are dynamically created through the assignment operator, = . Thus a defined variable always has a value.

Names of Variables

A variable name consists of a letter or the symbol **%** followed by any number of letters, digits, and special characters in (**#** _ **$** !). Only the first 24 characters are taken into account. Upper-case and lower-case letters are distinguished. **Car_Speed, a, A,%pi, x$, my_first_function**, are valid variable names.

Variable Management

In Scilab it is not necessary to declare variable types or specify dimensions, since these quantities are automatically set to those of the assigned value. Thus the type and dimensions of new variables are automatically created. Also, a new assignment will change a variable's type and definition:

```
-->a=1
 a =
    1.

-->a=['simple','string']
 a =
!simple  string !
```

2.3.2 Assignments

Assignments have the following syntax:

- **var = expr**, where **var** is a variable name (see page 23) and **expr** a valid Scilab expression (see page 24).

- **[v1,v2,...,vn] = expr**, where **v1**, ..., **vn** are variable names and **expr** a multivalued expression.

- **expr,** which gives rise to a default variable name **ans**, which is the evaluated value of **expr**.

2.3.3 Expressions

Scilab expressions consist of arithmetic, relational, and logical operations. There are also string and special matrix operations.

Arithmetic Operators

Scilab uses the familiar arithmetic operators **+** , **-** , ***** , **/** , **^** , **'** with the usual precedence rules. These operators are extended to matrix objects. For example, the operator **+** represents scalar addition between scalars or matrices.

There are also some matrix-specific arithmetic operators such as **.*** , **./** , **.^** , **.'** , **.*.** , **:** . The interpretation of these Scilab operators depends on the operand types and sizes.

- **+** Addition, or unary plus. **A+B** adds matrices **A** and **B**. **A** and **B** must have the same size, unless one is a scalar. A scalar can be added to a matrix of any size. A scalar added to an empty matrix yields the scalar value.

- **–** Subtraction, or unary minus. **A-B** subtracts **B** from **A**. Matrices **A** and **B** must have the same size, unless one is a scalar. A scalar can be subtracted from a matrix of any size.

- ***** Matrix multiplication. **A*B** is the matrix product of the matrices **A** and **B**. For nonscalar **A** and **B**, the number of columns of **A** must equal the number of rows of **B**. A scalar can multiply a matrix of any size.

- **.*** Entrywise multiplication. **A.*B** is the entry-by-entry product of the entries of the matrices **A** and **B**. Matrices **A** and **B** must have the same size, unless one of them is a scalar.

- **/** Matrix right division. If **B** is a square matrix, then **A/B** is the solution (if it exists) of the linear equation **X*B=A** for the unknown **X**. If **B** is a rectangular matrix, then **A/B** is a minimum squared-error solution of the linear equation **X*B=A** for the unknown **X**.

- **** Matrix left division. **B\A** is equivalent to **A'/B'**.

- **./** Entrywise right division: **A./B** is the matrix with the entries **A(i,j)/B(i,j)**. The matrices **A** and **B** must have the same size, unless one of them is a scalar.

- **.** Entrywise left division: **A.\B** is the matrix with entries **B(i,j)/A(i,j)**. The matrices **A** and **B** must have the same size, unless one of them is a scalar.

- **^** or ****** Power: If **p** is a scalar and **X** a square matrix, then **X^p** is the matrix **X** to the power **p**; if **X** is a vector, then **X^p** is equivalent to **X.^p**. If **X** is a scalar and **p** is a matrix, then **X^P** is **X** raised to the matrix power. Note that **X^p**, where **X** and **p** are both matrices, is not implemented.

- **.^** Entrywise power: **A.^B** is the matrix with entries **A(i,j)** to the **B(i,j)** power. The matrices **A** and **B** must have the same size, unless one of them is a scalar.

- **.'** Matrix transpose: **A.'** is the transpose of **A**.

- **'** Matrix conjugate transpose: **A'** is the complex conjugate transpose of **A**.

- **.*.** Matrix Kronecker product: **A.*.B** is the matrix formed by the blocks **A(i,j)*B**. The Kronecker product may be written as follows with concatenation operations:

  ```
  A.*.B=[A(1,1)*B,...,A(1,n)*B;...;A(m,1)*B,...,A(m,n)*B]
  ```

- **./.** Matrix Kronecker right quotient: **A./.B** is the matrix with blocks **A(i,j)./B**.

- **.\.** Matrix Kronecker left quotient: **A.\.B** is the matrix with blocks **A(i,j).\B**.

These arithmetic operators work with most types of numerical matrices: matrices of real or complex numbers, matrices of polynomials, matrices of rational functions. They also work for linear systems state space representations in a sense equivalent to that of transfer function operators.

Relational Operators

The relational operators are **<** , **<=** , **>** , **>=** , **==** , **~=** , and **<>** .

The relational comparisons **A<B, A<=B, A>B, A>=B**, work entry by entry between two matrices of real numbers. Matrices **A** and **B** must have the same size, unless one of them is a scalar. These operators return a Boolean matrix of the corresponding size, with entries set to **%t** where the relation is true, and entries set to **%f** where it is not.

The relational operators **A==B**, and **A~=B** or **A<>B**, perform entry by entry comparisons if **A** and **B** are matrices with the same size, or if one of them is a scalar. In this case they return a Boolean matrix of the same size with entries set to **%t** where the relation is true, and entries set to **%f** where it is not.

In all other cases relational operators perform global variable comparisons and return a scalar Boolean. In this way, any pair of variable types may be checked for equality as shown in the following examples:

```
[1 2 3]>=2  evaluates to  [%f,%t,%t]
[1;2;3]<[2;1;4]  evaluates to  [%t,%f,%t]
[1;2;3]==[1,2,3]  evaluates to  %f
[1;2;3]=='a string'  evaluates to  %f
```

Logical Operators

The logical operators are **&** , **|** , and **~** .

The operations **A&B** and **A|B** respectively perform a logical entrywise AND and OR. Matrices **A** and **B** must have the same size, unless one of them is a scalar. The result is a Boolean matrix of the corresponding size. The operation **~A** complements the entries of the Boolean matrix **A**. The operations **&** , **|** , and **~** are defined only for matrices with logical-type entries and matrices of numbers.

String Operator

String concatenation is obtained by using the **+** operator. To concatenate **A** and **B** both must be strings, or matrices of strings of the same dimension:

```
'test'+'1'  evaluates to 'test1'
```
and
```
['speed:';'Position:']+['10m/s';'12.5m']
```
evaluates to
```
['speed:10m/s';'Position:12.5m']
```

Special Matrix Operators

• Brackets: Brackets **[]** are used to form vectors and matrices from scalar- or matrix-valued subexpressions. Within brackets, commas (**,**) or blanks separate column entries within a row. Newlines or semicolons (**;**) separate row entries within a column. Brackets may be nested, but within a pair of brackets matrices must be entered row by row: **[1;2,3;4]** is not valid, and **[a,b;c]** is interpreted as **[[a,b];c]** (where this works only if **[a,b]** has the same column dimension as **c**). Empty brackets **[]** define an empty matrix (0 rows and 0 columns). As examples:

```
[1 %s+1 %i]  is a row vector with three entries
[11 12; 21 22;31 32]  is a 3-by-2 matrix
[[11;21;31],[12;22;32]]  is the same 3-by-2 matrix
a=[];for k=1:10,a=[a,k^2];end  build a row vector
```

Note: Within matrix brackets arithmetic operators have higher priority than blanks. Thus, **[1 -2]** and **[1- 2]** yield two different matrices. The first evaluates to **[1,-2]** and the second one to **[1 - 2]**, which is just **-1**.

• Colon: The colon **:** is used to define vectors with regularly spaced entries:

```
1:3  produces the row vector [1 2 3]
6.5:-2:1  produces the row vector [6.5 4.5 2.5]
```

The syntax for the `:` operator is `V1:V2` or `V1:Step:V2`, where `V1`, `V2`, and `Step` are real scalars. For the first syntax, `Step` defaults to 1.

• Parentheses: Parentheses `()` are used to enclose subscripts for vector and matrix insertion and extraction operations. Extraction is for selecting a subset of elements from a matrix, and insertion is for defining or replacing a subset of elements in a matrix.

 - Extraction syntax: `variable(subscripts)`
   Here the `subscripts` must not exceed the dimensions of `variable`.

 - Insertion syntax: `variable(subscripts)=expression`
   The dimensions of the submatrix determined by `subscripts` must correspond to the dimensions of `expression`.

   For example, `a(1:2)=[5,5]` is a valid syntax, but `a(1:3)=[5,5]` is not. This rule holds true unless `expression` is a scalar (for example, `a(1:2)=5`) or the empty matrix (for example `a(1:2)=[]`).

In the above, `subscripts` is a comma-separated sequence of scalars, vectors, or implicit vectors (see below) with real positive values.

If `v` and `I` are vectors, then `v(I)` is the subvector formed with the entries of `v` selected by the indices contained in `I`. Some examples are:

`V(1:2)` is the submatrix formed by first two entries of `V`
`V(:)` reshapes `V` as a column vector
`V(5:-1:1)` gets the first five entries of `V` in reverse order

If `M` is a matrix and `I` and `J` are vectors, then `M(I,J)` is the submatrix of `M` formed with the rows selected by the indexes contained in `I` and the columns selected by the indices contained in `J`. `M(1:2,1)` is the submatrix formed by first two rows of the first column of `M`.

The implicit vector (`:`) selects all the indices relative to the corresponding dimension. Some examples are:

`M(:,1:2)` is the submatrix formed by first two columns of `M`
`M(:)` is the column vector formed with all the columns of `M`

A very useful device for both insertion and extraction is the final index symbol `$`. This symbol can be used to designate the final index in a matrix without the dimensions of the matrix actually being known. Some examples of this useful feature are

`M(1,$)` is the last entry of the first row of `M`
`V(1:2:$)` is equivalent to `V(1:2:size(V,'*'))`
`M(5:$,$/2)` is equivalent to `M(5:size(M,1),size(M,2)/2)`

Some additional important syntactical constructions for insertions are illustrated here:

- `v(i) = []` deletes entry(ies) `i` of vector `v`.

- `M(i,:) = []` deletes row(s) `i` of matrix `M`.

- `M(:,j) = []` deletes column(s) `j` of matrix `M`.

- `M(k) = []` reshapes matrix `M` into a column vector and deletes the entries indicated by `k` (note that `k` could be a vector of indices).

Expression Evaluation Rules

For rules on precedence Scilab operators are classed in seven groups (from highest to lowest priority):
```
(' , .')
(^)
( * , / , \ , .* , ./ , .\ , .*. , ./. , .\. )
( + , - )
( == , > , < , <= , >= , ~= )
( & )
( | )
```
Within each group precedence is from left to right. As an example of precedence:

`1+2^2==5&%t` is parsed as follows: `((1+(2^2))==5)&%t`

Parentheses may be used, in the usual way, to change the default evaluation order.

2.3.4   The list and tlist Operations

The `list` and `tlist` (see page 21) data types may be seen as tree data structures. Thus, for this type of object, the insertion and extraction operations are applied to the leaves or subtrees of the tree. The syntaxes are:

Extraction

```
[...]=l<path>
l<path>(subscripts)
```

Here `l` is the name of the variable containing the list, or typed list and `<path>` stands for a sequence of parenthesized subscripts, for instance `(s1)` or `(s1)(s2)(s3)`. All the parenthesized expressions in `<path>` except for the last must evaluate to a scalar. The last parenthesized expression in `<path>` may evaluate to a vector whose dimension equals the number of left-hand-side arguments.

For lists the subscript arguments must have positive integer values. For typed lists the subscripts may also evaluate to character strings that correspond to the typed list entry names. As an example:

```
-->l=list(10,list(30,40,50),60);                    ← a nested list definition

-->l(1)                                             ← extract first component
 ans  =
    10.

-->[a,b]=l(2)([1 2])                                ← extract two leaves
 b  =
    40.
 a  =
    30.

-->l=tlist(['foo','A','B','C'],..                   ← a tlist
         10,list(30,40,50),60);
-->l('A')                                           ← extract A  component
 ans  =
    10.

-->[a,b]=l('B')([1 $])
 b  =
    50.
 a  =
    30.
```

If the `<path>` points to a leaf of the tree that contains a matrix, then `l<path>(subscripts)` can be used to extract the matrix or a submatrix as shown below:

```
-->l=list(10,list(30,[40,50,60]));

-->l(2)(2)(1,2:3)
 ans  =
!   50.    60. !

-->l=tlist(['foo','A','B'],10,list(30,[40,50,60]));
```

```
-->1('B')($)([$:-1:1])
 ans  =
!   60.    50.     40. !
```

In this example the first $ refers to the last element of the B sublist, and the second $ to the last element of the vector [40,50,60].

Note that the $ symbol can also be used to denote the final element (or subtree) of a list or typed list.

Insertion

The syntax for a list or typed list insertion is as follows:

```
l<path> = expression
l<path>(subscripts) = expression
```

Here 1 is a list, or typed list and <path> is as described above in extraction. However, here all parenthesized subscripts in <path> must evaluate to scalars. Normally, the parenthesized entries must be positive scalars for a list. However, for typed lists they may also be character chains corresponding to the typed list's field names. If the parenthesized value is zero, then the indices of all elements of the list are incremented, and the insertion is then made at the first index. The following example should clarify this point:

```
-->1=list([1 2]);
```

```
-->1(1)=10                          ← changes first entry value
 1  =
          1(1)
     10.
```

```
-->1(0)=-1;                          ← add an entry on the left
```

```
-->1($+1)='foo'                      ← add an entry on the right
 1  =
          1(1)
  - 1.
          1(2)
     10.
          1(3)
 foo
```

```
-->1=tlist(['foo','A'],[%s+1,3])
```

```
-->1('A')=1('A')^2
```

```
l  =
        l(1)
!foo   A  !
        l(2)
!                 2          !
!    1 + 2s + s        9   !
```

If `<path>` is a path that leads to a leaf containing a matrix, then the `l<path>(subscripts)` syntax may be used to insert a matrix or submatrix, as shown in the following example:

```
-->l=list([1 2;3 4]);

-->l(1)(1,:)=[5 6]
 l  =
        l(1)
!    5.    6. !
!    3.    4. !
```

Other Operations

As already noted, the **tlist** data type can be used to define new abstract data types. It is also possible to extend the definition of all Scilab operators and basic primitives to this new data type (i.e., operator overloading). The definition of operators on abstract data types made using typed lists can be constructed with Scilab functions (see page 47).

As an example of operator overloading, the function **%CharsfChars** below defines the column concatenation of two variables of type **Chars** (see page 22):

```
function r=%CharsfChars(a,b)
  r=tlist(a(1),[a('Names');b('Names')],[a('Ages');..
                b('Ages')], [a('Sizes');b('Sizes')])
```

2.3.5   Flow Control

The Scilab programming language has five flow control constructs:

- **if** statements

- **select** and **case** statements

- **for** loops

- **while** loops

- **break** statements

if

The **if** statement evaluates a logical expression and executes a group of statements when the expression is true. The syntax is

```
if expr1 then statements,
elseif expr2 then statements,
....
else statements,
end
```

The variables **expri** are expressions with numeric or Boolean values. If **expri** are matrix-valued, the condition is true only if all matrix entries are true. For example, the following instructions display nothing:

```
v=[1 0 3]
if v>1 then
  disp('first case ')
elseif ~(v>1) then
  disp('second case ')
end
```

The optional **elseif** and **else** keywords provide for the execution of alternative groups of statements. An **end** keyword, which matches the **if**, terminates the last group of statements. The line structure given above is not significant, but the **then** keyword must be on the same line as the corresponding **if** or **elseif**. This is shown in the two following examples:

```
if err>0 then x_message('an error has occurred'),end
```

```
if a>5 then
  x=a+1
  y='high'
elseif a<1                   ← then may be replaced by a comma or an EOL
  x=5
  y='low'
else
  x=0
  y='intermediate'
end
```

select/case

The **select** statement syntax is a means of conditionally executing code. In particular, **select** executes one set of statements selected from an arbitrary number of alternatives. Each alternative is called a case. The optional **else** provides an alternative group of statements if none of the cases are selected. The syntax is

```
select expr,
case expr1 then statements,
case expr2 then statements,
...
case exprn then statements,
else statements,
end
```

The expressions **expr** and **expr1** through **exprn** are any valid Scilab expressions. A **case** condition is executed if all the elements of **expr** are equal to those of **expri**. In the event that more than one case evaluates to true, only the instructions associated to the first case are executed. As an example, the following can be used to select the visualization mode for a dynamic system's frequency response:

```
h=syslin('c',(%s^2+18*(%s+100)/((%s^2+6.06*(%s+102.01)))));
viewing_modes=['bode','nyquist','black'];
view=x_choose(viewing_modes,'Choose a visualization mode')
select view
case 1 then
  bode(h)
case 2 then
  nyquist(h)
case 3 then
  black(h)
end
```

for

The **for** construct is used to repeat statements a specific number of times. The syntax is **for variable = expression, statements, end**.

If **expression** is a matrix or a row vector, **variable** takes as values the entries of each column of the matrix. In the example

```
for variable=n1:step:n2,...,end
```

the **variable** takes as successive values n1, n1+step, n1+2*step ... .

If **expression** is a list, **variable** takes as values the successive components of the list. Examples below illustrate the use of **for**:

• Creates a 5-by-5 Hilbert matrix.

```
-->n=5;
```

```
-->for i = 1:n,
-->   for j = 1:n,
-->     a(i,j) = 1/(i+j-1);
-->   end
-->end
```

• Successively set **e** to the unit 3-vectors:

```
-->for e=eye(3,3),e,end
 e  =
!    1. !
!    0. !
!    0. !
```

```
 e  =
!    0. !
!    1. !
!    0. !
 e  =
!    0. !
!    0. !
!    1. !
```

• Successively displays the values of the **x** list entries:

```
-->x=list(1,2,'example');
```

```
-->for l=x; l,end
 l  =
     1.
 l  =
     2.
 l  =
 example
```

In using **for** loops it should be kept in mind that it is computationally much more efficient to make use of Scilab's matrix capabilities whenever possible, since instructions within loops are interpreted. For example, the n-by-n Hilbert matrix may be efficiently created using the following statements:

```
i=(1:n)';j=1:n;
a=ones(n,n) ./ (i*ones(1,n) + ones(n,1)*j - 1)
```

while

The **while** construct repeats statements while a condition is true. The syntax is:

while expr, statements, else statements, end

If present, the **else** statements are executed when the condition becomes false. The expression **expr** has numeric or Boolean values, the condition is true if the expression's value is true or nonzero. The following example computes the biggest floating point value such that **1+eps==1**:

```
eps = 1;
while eps+1 > 1
    eps = eps/2;
end
```

When **expr** is matrix-valued, the condition is true only if all matrix entries are true or nonzero.

break

The **break** statement permits an early exit from a **for** or **while** loop. In nested loops, **break** exits only from the innermost loop. As an example:

```
h=syslin('c',(%s^2+18*(%s+100)/((%s^2+6.06*(%s+102.01))))));
viewing_modes=['bode','nyquist','black'];
while %t ,
  view=x_choose(viewing_modes,'Choose a visualization mode')
  select view
  case 0 then
    break                                    ← exit while loop
  case 1 then
    bode(h)
  case 2 then
    nyquist(h)
  case 3 then
    black(h)
  end
end
```

2.3.6 Functions and Scripts

Scilab is a powerful programming language as well as an interactive computational environment. In addition, Scilab's functionality is easily extensible through the creation of user applications. These applications can be constructed using scripts and functions.

Scripts

Scripts are just sequences of Scilab statements. They do not accept input arguments or return output arguments. Scripts can operate on existing data in the workspace, or they can create new data on which to operate. Although scripts do not return output arguments, any variables that they create remain in the workspace, to be used in subsequent computations. Script statements may be written into files and executed using

exec script_file_name or **exec('script_file_name')**.

Any file name may be used, but it is the Scilab group's convention to use .sce as the filename extension.

It is also possible to store script statements in a scilab vector of strings and to execute them using **execstr(vector_of_strings)**, as illustrated in the following example:

```
-->to_do=['for i = 1:n,'
          '  for j = 1:n,'
          '    a(i,j) = 1/(i+j-1);'
          '  end'
          'end'];

-->n = 4;execstr(to_do)

-->a,
 a  =
!   1.            0.5            0.3333333     0.25         !
!   0.5           0.3333333      0.25          0.2          !
!   0.3333333     0.25           0.2           0.1666667    !
!   0.25          0.2            0.1666667     0.1428571    !
```

Functions

Unlike scripts, functions can accept input arguments and return one or several output arguments. Functions may be hard coded in the C or Fortran languages (see Section 5.3). In this case they are called "primitives" or "primitive functions." They can also be written in the Scilab programming language, in which case they are called "Scilab coded functions" or simply "functions."

Scilab coded functions are, like scripts, sequences of Scilab statements, but associated with a calling syntax description that gives the function name and the formal input and output variable names. Scilab coded function definitions may be constructed in a file. The first line of the function definition must begin with the **function** keyword followed by the calling syntax definition. As an example:

```
function [xmin,xmax]=bounds(x)
 xmin=min(x)
 xmax=max(x)
```

defines a function named **bounds** that computes the minimum and maximum values of a vector.

A file may contain several function definitions:

```
function [xmin,xmax]=bounds(x)
xmin=min(x)
xmax=max(x)

function m=mod(a,b)
m=a-int(a./b).*b
```

Here the **bounds** function is defined that a second function called **mod** that computes the integer modulus of two vectors.

New function files may be loaded into the Scilab environment using the **getf** function. The syntax **getf file_name** or **getf('file_name')** are both valid. Any file name can contain Scilab functions to be loaded. However, it is the Scilab group's convention to use **.sci** as the filename extension. Also, it is possible to organize a set of functions into Scilab libraries using the **lib** function. Libraries are useful for the automatic loading of referenced functions. For libraries, the name of the main function, as defined in the first line of the function file, should be the same as the name of the file containing the library (without the **.sci** extension).

It is also possible to define a Scilab coded function through the use of the **deff** function:

```
deff(function_call_syntax,deff_str)
```

where **function_call_syntax** is a character string containing the calling syntax definition, and **deff_str** is a vector of strings containing the sequence of Scilab statements. For example, the following instruction is another way to define the **mod** function:

```
deff('m=mod(a,b)','m=a-int(a./b).*b')
```

Loading a Scilab coded function into Scilab creates a variable in the environment whose name is that given in the function syntax. For example, the previous statement produces a variable **mod** in the current workspace.

Function type variables can, like any other Scilab type, be used as an argument of other functions, as shown in the example below:

```
-->deff('[y]=my_fun(x)',['y=x*sin(30*x)';
                    'y=y/sqrt(1-((x/(2*%pi))^2))')
-->intg(0,2*%pi,my_fun)
```

This example computes $\int_0^{2\pi} \frac{x\sin(30x)}{\sqrt{1-(x/(2\pi))^2}} dx.$

Calling Functions

The syntax used for calling functions can take three forms:

```
func()
func(input_arguments)
[output_arguments]=func(input_arguments)
```

Here **func** is the function name, **input_arguments** is a comma-separated sequence of expressions, and **output_arguments** is a comma-separated sequence of variable names. As an example:

```
[xmin,xmax]=bounds(rand(3,3))
```

If **func** returns only one output argument, the brackets **[ ]** may be omitted and **func(input_arguments)** may be used as a factor in an expression, as the **mod** function below:

```
find(mod(v,2)==0)
```

Functions can be invoked with fewer input or output parameters than specified in the formal argument list. Depending on the construction of the function, its behavior may change as a function of the number of inputs and outputs, as shown with the definition of the function **f** used in the following example:

```
function [y1,y2]=f(x1,x2)
y1=x1+x2
y2=x1-x2
```

• The first call to **f** is made with all the arguments supplied

```
-->[y1,y2]=f(2,1)
y2  =
      1.
y1  =
      3.
```

- The second call is made with only one output argument

```
-->x=f(2,1)
 x  =
      3.
```

Here only the leftmost formal argument of the function is returned.
- The next call is made with only one input argument:

```
-->f(1)
  y1=x1+x2;
            !--error      4
  undefined variable : x2
  at line        2 of function f
```

Here x2 is referred to but was not defined in the calling list. The attempt at execution produces an error.
- Finally, the call to the function can be made with named input arguments:

```
-->[y1,y2]=f(x2=1,x1=2)
 y2  =
      3.
 y1  =
      1.
```

In some cases it is of use to be able to create functions that can accept an unknown number of input arguments or return an unknown number of output arguments. For that you can use the keywords **varargin** for input and **varargout** for output. These keywords must appear respectively at the end of the input argument sequence of the function or at the end of the output argument sequence of the function. Then when the function is called, the input arguments corresponding to keyword **varargin** are stored in a list named **varargin** within the function. The same mechanism applies for **varargout**.

For example, suppose we define the following function:

```
function mydisp(comment,varargin)
write(%io(2),comment)
for k=1:size(varargin)
  write(%io(2),'===============================')
  disp(varargin(k))
end
```

We can use it in the following way:

```
-->mydisp('Hello','arg2','arg3')
Hello
================================

 arg2
================================

 arg3
```

Scilab coded functions can have a variable number of inputs and outputs where nevertheless the number of possibilities is finite and known in advance. The exact number of input and output arguments can be obtained using the function `[lhs,rhs] = argn()`. For example:

```
function [v,ksat]=stat(v,level)
[lhs,rhs]=argn()
if rhs<2 then level=1
if lhs==1 then
  v(find(v>level))=level
else
   ksat=find(v>level)
   v(ksat)=level
end
```

This example shows a function **stat** that accepts an optional input argument **level** and returns an optional output argument **ksat**. This type of construction is particularly useful for creating functions that can provide default values when not all the arguments are specified by the user.

Global and Local Variables

Variables introduced in the body of a function, as well as those specified in a function's syntax are all local to the scope of the function. This means that assignments of values to these variables are created and destroyed on entry and exit of the function. Thus, a variable **x** can exist in the global environment and another variable **x** can be created within the environment of a function, and the values of these two variables have no relationship to each other.

On the other hand, if a variable used in a Scilab coded function expression is not defined (and is not among the input parameters), then it takes the value of a variable having the same name in the calling environment. The following example illustrates these concepts:

```
function x=test()
y=y+1
x=y^2
```

Here no input arguments are defined, so for **test** to work it must find a value for **y** in the calling environment:

```
-->y=2;x=test(),y
 x   =
     9.
 y   =
     2.
```

As can be seen, the value of **y** in the calling environment is unchanged even though the value within the function environment was changed.

Recursivity

Scilab coded functions may be called recursively. The following example illustrates this:

```
function x=fact(n)
if n==1 then
   x=1
else
   x=n*fact(n-1)
end
```

This defines a function **fact** that computes $x = n!$. The function recursively calls itself. Note that although the function **fact** is syntactically correct, it is an inefficient way to code the factorial function, and **prod(1:n)** is a much more computationally efficient method, especially when **n** is large.

2.3.7   Commands

Commands Syntax

Commands are statements with the following syntax:

```
command_name args
command_name
```

Here **args** is a sequence of groups of characters separated by blanks. The command ends on a newline or the first comma (**,**) or semicolon (**;**). Examples:

```
clear                                    ← clears all defined variables
clear a b                                ← removes variables  a  and  b
clear a b,a=1                            ← the command clear a b followed by
                                           the assignment a=1
disp 'Scilab is a cute language'
```

Commands and Functions

Any function that has only character string arguments and at most one output argument may be called in a command-like form. For example:

```
getf('SCI/macros/auto/bode.sci')
```

is equivalent to

```
getf SCI/macros/auto/bode.sci
```

Unlike primitive functions, Scilab coded functions are variables. A function name not followed by an input argument list refers to the value of the variable. For this reason it is not possible to call Scilab coded functions without an input argument in a command-like form. It is necessary to add () after the variable name to effectively perform a function call and not just a variable evaluation. The following example clarifies this point:

```
-->deff('prompt_me()','disp(''Hello to you'')')

-->prompt_me                            ← display function syntax
prompt_me   =
[]=prompt_me()

-->prompt_me()                          ← function evaluation
 Hello to you
```

Useful Commands

help The **help** command gives help on defined functions and on many keywords. Without argument it gives help on **help**.

apropos The **apropos** command searches for help relative to specified keywords.

clear The **clear** command without argument clears all local variables (predefined variables are not removed). When the command is followed by a blank-separated sequence of variable names, it clears (if possible) all the specified variables.

who The **who** command displays a list of all defined variables in the environment.

pause The **pause** command suspends current execution, protects defined variables, and prompts for statements or commands in a new local environment.

return The **return** command (or equivalently **resume**) is used to exit a local environment and returns to the calling environment.

exit The **exit** command is used to exit Scilab.

abort The **abort** command terminates current execution, exits all local environments, and returns control in the top-level environment.

2.4   Data-Type-Related Functions

2.4.1   Type Conversion Functions

poly

The function **poly** is used to create polynomials.

For **A** a matrix, **poly(A, 'x')** is the characteristic polynomial of **A**, i.e., the determinant of **x*eye-A**, where **x** is the symbolic variable name.

For **v** a vector, **poly(v,'x' [,'roots'])** is the polynomial whose roots are the values of **v**, and again **x** is the symbolic variable name.

The expression **poly(v,'x','coeff')** is the polynomial whose coefficients are the values of **v**, where **v(1)** is the lowest-order coefficient and **v($)** is the highest-order coefficient. As an example:

```
-->poly([1 2;-2 1],'z')
 ans  =
                 2
   5 - 2z + z

-->poly([1+2*%i,1-2*%i],'x','roots')
 ans  =
                 2
   5 - 2x + x

-->poly([1 2 3],'z','coeff')
 ans  =
```

$$1 + 2z + 3z^2$$

coeff

Let **Mp** stand for a matrix of polynomials in the symbolic variable **s**. Then **C=coeff(Mp)** returns the matrix **C** that contains the coefficients of the matrix of polynomials **Mp**. The matrix **C** is partitioned as **C=[M0,M1,...,Mn]**, where **Mi** is the ith-degree coefficient of **Mp**. In this way **C** can be viewed as the coefficients of a polynomial with matrix coefficients: $Mp=\sum_{i=0}^{n} s^i Mi$. As an example:

```
-->coeff([1+%s^2;5*%s^3])
 ans =
!   1.      0.      1.      0. !
!   0.      0.      0.      5. !
```

sparse

The **sparse** function is used to build sparse matrices. The call **sparse(X)** converts the full-storage matrix **X** to sparse form by eliminating any zero-valued entries. The call **sp=sparse(ij, v[,mn])** builds an **mn(1)**-by-**mn(2)** sparse matrix with **sp(ij(k,1),ij(k,2))=v(k)**. The matrices **ij** and **v** must have the same column dimension. If the optional **mn** parameter is not given, the **sp** matrix dimensions are the maximum value of **ij(:,1)** and **ij(:,2)**. The next examples illustrate different uses of the **sparse** function:

```
-->A=[5 6;0 0];
```

```
-->sparse(A)                              ← full to sparse
 ans =
(    2,     2) sparse matrix
(    1,     1)           5.
(    1,     2)           6.
```

```
-->ij=[1 1;1 2];
```

```
-->v=[%t;%t];
```

```
-->sparse(ij,v,[2,2])          ← Boolean sparse from "true" entries
 ans =
(    2,     2) sparse matrix
(    1,     1)     T
(    1,     2)     T
```

full

The **full** function converts sparse matrices to full storage. For example, **X=full(sp)** converts the sparse matrix **sp** into its full-storage representation. If **sp** is already full, then **X** equals **sp**.

list

The **list** function is used to create data objects of type **list** (see page 21). As an example, **list(a1,...,an)** creates a **list** whose entries are the arbitrary Scilab objects **a1, ..., an**.

tlist

The **tlist** function is used to create **tlist** type data structures (see page 21). The call **tlist(T,a1,...,an)** creates a variable of type **tlist** whose entries are the arbitrary Scilab objects **a1, ..., an**, where **T** is a string or vector of strings giving the **tlist** class name and component names.

str2code

Scilab uses a specific encoding scheme to store characters. The digits 0–9 are represented by the integer codes 0–9. The lower-case letters a–z are represented by the integer codes 10–35, and upper-case letters A–Z by the negative codes (-10)–(-35). The call **c=str2code(str)** returns the Scilab integer codes associated with each character of the string **str**. Example:

```
-->str2code('Scilab')
 ans  =
!  -  28.  !
!      12.  !
!      18.  !
!      21.  !
!      10.  !
!      11.  !
```

code2str

The function **code2str** is the inverse of **str2code**. Given a vector of Scilab integers with values in the interval $[-63, 63]$, **code2str** returns the associated code characters. Example:

```
-->code2str([-28,12,18,21,10,11])
  ans  =
  Scilab
```

1	real or complex constant matrix
2	polynomial matrix
4	Boolean matrix
5	sparse matrix
6	sparse Boolean matrix
10	matrix of character strings
11	not compiled function
13	compiled function
14	function library
15	list
16	tlist
128	pointer
129	size implicit relative vector

Table 2.1: *Type numbers*

2.4.2 Type Enquiry Functions

type

The function **type(x)** returns an integer that is the code for the data object type of the variable **x**. The codes for the various Scilab types are listed in Table 2.1.

typeof

The function **typeof(object)** returns a string indicating the name of the type associated with the variable **object**. Table 2.2 describes the type names.

2.5 Overloading

As already seen, it is possible to create new abstract data types in Scilab using **tlist** (see Section 2.2.8). Moreover, it is possible to extend the scope of primitive functions (i.e., internal hard-coded functions) and operators to these new data types. Furthermore, this can be done without having to resort to changing the low-level internal Scilab code. Such extensions, called overloading, are not limited only to abstract data types obtained using **tlist**. It is also possible to extend the scope of a primitive function or operator to other previously undefined arguments. In general, the error message obtained when trying to use an invalid argument to a primitive

'constant'	real or complex constant matrix
'polynomial'	polynomial matrix
'function'	function
'string'	m character string matrix
'Boolean'	Boolean matrix
'list'	list
'rational'	rational matrix (transfer matrix)
'state-space'	state-space model
'sparse'	sparse matrix
'Boolean sparse'	Boolean sparse matrix
'size implicit'	size implicit vector
'pointer'	sparse LU factorization pointer
<class>	tlist in class <class>

Table 2.2: *Type names*

function or operator gives a hint as to what function needs to be implemented in order to permit an extension:

```
-->a=['1','2','3'];
```

```
-->sum(a)                             ← sum is not defined for a matrix of strings
        !--error      4
undefined variable : %c_sum           ← function %c_sum must be
                                      implemented to extend sum to matrices of strings
```

In the preceding example, the %c_sum function can be a Scilab function or a new Scilab primitive (see Chapter 5 to learn how to do this).

Overloading cannot be used to change Scilab's built-in functions. For example, sum(x) is well-defined for real or complex matrices, and this default behavior cannot be changed with overloading. However, it is possible (but certainly a bit dangerous) to completely erase the primitive sum and to replace it with an alternatively defined function:

```
-->sum([1,2,5]);
  ans   =
      8.
```

```
-->sum_code=funptr('sum');            ← internal code for sum
```

```
-->clearfun('sum');                   ← removing sum primitive
```

```
-->deff('[y]=sum(x)','y=prod(x)');          ← a new fancy sum

-->sum([1,2,5])
 ans  =
    10.

-->newfun('old_sum',sum_code);              ← restoring sum primitive
                                               under a new name

-->old_sum([1,2,5])
 ans  =
     8.
```

2.5.1   Operator Overloading

This subsection is devoted to the naming conventions needed for implementing operator overloading. Each operator is associated to a code (listed in Table 2.3), as is each basic data type (listed in Table 2.4). Note that the code for the **tlist** type uses the **tlist** name field entry. The following describes the rules for mixing operators and argument code in order to overload a function name.

'	t	+	a	-	s
*	m	/	r	\	l
^	p	.*	x	./	d
.\.	q	.*.	k	./.	y
.	z	:	b	.*	u
./	v	.\	w	[a,b]	c
[a;b]	f	() extraction	e	() insertion	i
==	o	<>	n	|	g
&	h	.^	j	~	5
.'	0	<	1	>	2
<=	3	>=	4		

Table 2.3: *Overloading operation codes*

- For unary operators such as - , ' , and .' , the name of the overloading function must be **%<arg_type>_<op_code>**. Here **<arg_type>** is the string code associated to the operand type, and **<op_code>** is a string code associated to the operator name.

string	c
polynomial	p
function	m
constant	s
list	l
tlist	<tlist_type>
Boolean	b
sparse	sp
Boolean sparse	spb

Table 2.4: *Overloading type codes*

The function `%<arg_type>_<op_code>` has one input argument and one output argument. For example, the unary minus operator for rational matrices can be implemented as follows:

```
function f=%r_s(f)
    f('num')=-f('num')                    ← change sign of the numerator
```

Indeed, in Scilab rational matrices are of type `tlist`. The field **num** stores the rational matrix numerator and field **den** the rational matrix denominator.

- For binary operators, the name of the overloading function must be `%<arg1_type>_<op_code>_<arg2_type>`, where `<arg1_type>` and `<arg2_type>` are the string codes associated to the types of each of the function's operands, and `<op_code>` is a string associated to the type of operator.

- **Extraction** Extraction, which corresponds to expressions such as `x(i1,..,in)`, may be overloaded by constructing a function with the name `%<obj_type>_e`, where `<obj_type>` is the type code for the object from which entries are to be extracted. Thus the evaluation of `x(i1,...,in)` results in the call `%<x_type>_e(i1,...,in,x)` (where `x` is the last argument). For example, submatrix extraction for rational matrices can be implemented as follows:

```
function f=%r_e(i,j,f)
    f('num')=f('num')(i,j)                ← form new numerator
    f('den')=f('den')(i,j)                ← form new denominator
```

Note that, for **tlist** object extraction with only one argument, **x(i1)** is alway seen as **tlist** component extraction and thus cannot be over-loaded:

```
-->f=1/%s;                          ← a simple rational fraction

-->f(3)
 ans  =
    s                               ← f(3)   extracts the denominator
```

- **Insertion** The insertion operation, which corresponds to instructions such as **x(i1,..,in)=y**, may be overloaded by constructing a func-tion with the name **%<target_type>_i_<source_type>**, where **<source_type>** is the type code of object **y**, and **<target_type>** is the type code of object **x**. A scilab evaluation of **x(i1,...,in)=y** will result in the following call **%<x_type>_i_<y_type>(i1,...,in,x,y)**.

  For example, submatrix insertion for rational matrices **r(i,j)=n** can be implemented as follows:

```
function r=%r_i_s(i,j,r,n)
   d=ones(n);                ← form denominator d associated with n
   n(i,j)=r('num'),               ← create result numerator
   d(i,j)=r('den'),               ← create result denominator
   r('num')=n;r('den')=d;         ← update data structure
```

  Note that for **tlist** object insertion with only one argument, **x(i1)** is alway seen as **tlist** component insertion and thus cannot be over-loaded.

- Row and column concatenation are binary operators, and they follow the general rules of binary operators. The associated **<op_code>** is **c** for row concatenation and **f** for column concatenation. Remember that complex row and column concatenation expressions are just com-positions of simple row and column concatenations. For example, the expression **[a,b,c;d,e,f]** is parsed as **[[[[a,b],c];[[d,e],f]]]**.

A Small Example   This example implements a new data type called **func**. It is used to make symbolic computations on mathematical functions, and the following function allow the definition of variables of the **func** type:

```
function [f]=func(var,expr)        ← new data type func constructor
f=tlist(['func','var','expr'],var,expr);
```

Let's define some overloading functions:

```
function %func_p(f)                      ← print function for func data type
write(%io(2),f('var')+'->'+f('expr'));

function [f]=%func_a_func(f1,f2)                              ← f1+f2
f=tlist(f1(1),f1('var'),f1('expr')+'+'+..
      strsubst(f2('expr'),f2('var'),f1('var')));

function f=%func_b_func(f1,f2)                              ← f1:f2
f=tlist(f1(1),f2('var'),strsubst(f1('expr'),f1('var'),..
      f2('expr')));
```

It is then possible to perform simple symbolic computations:

```
-->f1=func('x','sin(x)')
 f1  =
x->sin(x)                    ← %func_p performs the display for func objects



-->f2=func('x','log(x)')
  f2  =
x->log(x)

-->f=f1+f2
 f  =
x->sin(x)+log(x)                        ←      the + operator for func objects

-->g=f1:f2                              ← the : operator for func objects
 g  =
x->sin(log(x))                    ← implemented as a composition operator
```

2.5.2   Primitive Functions

Primitive functions may be overloaded for new data types by defining a function whose name is formed as follow %<arg_type>_<function_name>, where <arg_type> is the string code of one of the input arguments and <function_name> is the function name.

Let us now give a small example that implements the **sum** primitive for polynomial matrices:

```
function a=%p_sum(a,flag)
  [m,n]=size(a);
  if flag==1|flag=='r' then a=ones(1,m)*a;
  else if flag==2|flag=='c' then a=a*ones(n,1);
```

```
else a=ones(1,m)*a*ones(n,1);
end
```

For functions with more than one argument, one question remains: Which argument must be used to give the `<arg_type>` code? No systematic rule can be used to determine which is the most "significant" argument of a given function. But as pointed out in the introduction, the answer is coded in the error message returned by the function:

```
sum([%t %f %t],'r')                    ←    try sum for a Boolean vector
             !--error    4
undefined variable : %b_sum      ←    requested overloading function
```

2.5.3   How to Customize the Display of Variables

Normally, the display of `tlist` data is like that of `list` data. However, the display of new objects defined by `tlist` structures may be overloaded to have a more pleasing and intuitive output. The overloading function must have the following syntax:

```
function %<tlist_type>_p(x)
```

where `<tlist_type>` is the name field entry of the `tlist`.

The following function illustrates the overloading of the display function for the `chars` data type, which was defined on page 22:

```
function %Chars_p(a)
nchars=size(a(2),'*')                          ← number of characters
txt=['Name            ','Age ','Size ';
    '--------------','----','-----']
for k=1:nchars
  txt=[txt;
    a('Names')(k),string(a('Ages')(k)),string(a('Sizes')(k))]
end
disp(txt)
```

When the function `%Chars_p` is loaded into the Scilab environment, the result of the Scilab statement

```
-->x=tlist(['Chars';'Names'; 'Ages'; 'Sizes'],..
        ['Peter'; 'John'; 'Franck'],..
        [25;40;18],[1.85;1.76;1.70])
```

will be displayed as

```
 x   =
!Name           Age   Size   !
!                             !
!--------------  ----  ----- !
!                             !
!Peter           25    1.85   !
!                             !
!John            40    1.76   !
!                             !
!Franck          18    1.7    !
```

Chapter 3

Graphics

This chapter introduces graphics in Scilab. Scilab contains numerous integrated graphics functions that can be used for the illustration and visualization of ongoing work and final results. Scilab also contains several interactive graphical user interfaces that can be used for the presentation of specific objects and systems as well as for the usual visualization of numerical results. Many high-level 2-D and 3-D graphics functions are at the disposal of the user; low-level functions are also available. In short, no matter how complex or sophisticated the graphical task, it can be performed in Scilab.

In the following sections a description of the global manipulation of the graphical media (i.e., windows and software drivers) is given. This is followed by an explanation of the graphics context. We then describe high-level 2-D and 3-D plotting capabilities. Finally, the last section of this chapter is devoted to exporting graphical output, for example, for plotting to printers.

3.1 The Media

In Scilab the term "the media" refers to two different objects. The first one is the graphics window used by Scilab for on-screen plotting. The second object is the graphics driver, which is the software support for plots. The driver is the interface between the graphics commands and the physical support on which the graphics are to appear.

We note here that for graphics there are some small differences between the X Window under Unix and the Windows 95 and NT versions of Scilab. The functionalities are identical. However, there are some differences in the use of some interactive window buttons.

3.1.1   The Graphics Window

Scilab contains several types of windows, including those for user interface control, message windows, and graphics windows. In this section the last of these is described.

In Scilab graphics windows are labeled **ScilabGraphicN**, where **N** is the graphics window number. Several graphics windows can be open simultaneously, but at any time only one window is active. The presentation of the active window and its behavior can be controlled by graphics commands as well as by control buttons found directly on the graphics window itself and in the main Scilab window.

The default working mode for graphics is the superposition of graphs. This mode cannot be changed, and the window must be cleared if a new plot is to displayed in the same graphics window. The command to clear a window is **xbasc(N)**. The command **xbasc()** clears the current graphics window, and the **File** menu button on a graphics window also contains a selection **Clear**, which does the same (see below).

Some basic graphics window manipulations can be performed with the mouse. On the main Scilab window the button **Graphics Window N** can be used to manage graphics windows. The value of **N** denotes the number of the active window, and this graphics window can be set (created if it does not exist), raised, or deleted. The **+** and **-** buttons allow for the incremental selection of graphics windows, as does the **Select** button in the **File** menu on each graphics window. The execution of a plotting command automatically creates a window if necessary.

High-level 2-D and 3-D functions need the definition of some parameters defining a *graphical environment*: These parameters have default values that can be changed if necessary; every graphics window has its specific context, so the same plotting command can give different results on different windows.

There are 4 buttons on each graphics window:

- **3D Rot.**: for applying a rotation with the mouse to a 3-D plot. This button is inhibited for a 2-D plot. To facilitate the operation (i.e., rotation to specific angles) the current angle is displayed at the top of the window.

- **2D Zoom**: zooming on a 2-D plot. This command can be recursively invoked. For a 3-D plot this button is not disabled, but it has no effect.

- **UnZoom**: return to the initial plot size (not to the plot corresponding to the previous zoom in case of multiple zooms).

These 3 buttons, each affecting the plot in the window, are not always available for use. It will be seen later that there are different choices for the underlying device driver, and the zoom and rotation functions need access to the record of the prior plotting commands to function correctly (see Section 3.1.2).

- **File**: this button contains a menu of different commands, which are described below,

  - The first menu entry, **Clear**, as described above erases the graphics in the window as well as the stored record of the graphics.

  - The second entry, **Select**, makes the graphics window the active one.

  - The command **Print** opens a selection panel for printing.

    In the Unix environment, the list of printers can be defined by the environment variable **PRINTERS** or directly at the beginning of the shell script **SCIDIR/bin/scilab**. Under Windows 95/NT, the choice of the printer is driven by the system.

  - The **Export** command opens a panel selection for outputting a graphic to a file of a specified format (Postscript, Postscript-LaTeX, Postscript No Preamble, and Xfig are available). The export can be in color or black and white, and the orientation can be landscape or portrait. The **Export** button is perhaps the most expeditious way to obtain a graphic that can be imported into another document. Later, more specific commands will be described for the export of graphics. For the moment it is useful to know that exporting Postscript is useful for paper and slide hardcopy, Postscript-LaTeX can be used for the insertion of graphics into a LaTeX document, and Xfig output yields a file that can be interpreted and modified by the Xfig program (an interactive vector drawing program in the Unix environment).

  - The **Save** command directly saves the plot to a file with a specified name. This file can be later reloaded and replotted in Scilab. This is useful for the superposition of a graphics that is to be shared among several other plots.

  - The **Load** command is for loading saved plots.

  - The **Close** command, as described above, clears the graphics window as well as the windows graphical context.

3.1.2 The Driver

The default for graphical output is the **ScilabGraphic0** window. However, there are different graphics devices in Scilab that can be used to send graphics to windows or files (and, subsequently, the printer). Also, for a graphics window there are several different driver possibilities. Thus, the interface from the output of a plotting command and the physical support one performed by the driver. When it is desired to change the physical support or display mode, the driver can be selected using the command **driver**.

The different drivers are:

- **X11**: graphics driver for output to the screen of the computer. This driver works for X Window and Windows 95/NT. The plot is directly displayed in the active window.

- **Rec**: the same driver as in the previous item but all the graphics commands are recorded (i.e., stored). This is needed for the zoom and rotate buttons to operate. This driver is the default.

  As a simple illustration of the difference between these two drivers, type **driver('X11'); plot2d()** and then resize the graphics window with the mouse: The plot disappears since it was not recorded. In contrast, for the commands **driver('Rec'); plot2d()** a resizing of the window does not destroy the plot. The "record" functionality allows the graphics to be redrawn. If these commands are followed by **xclear(); plot3d()** and the window is subsequently resized, the previous *two* plots appear, since they have both been recorded by the driver **Rec**.

- **Pos**: graphics driver for Postscript printers; this driver is one of the ways to obtain hardcopy of a plot.

- **Fig**: graphics driver for importation into the Xfig system (an interactive vector graphics program in the Unix environment).

In the first two cases above the "implicit" output device is a graphics window. The last two cases are for export to files. More details will be given later on Postscript and Xfig output formats.

3.1.3 Global Handling Commands

Fundamental graphics commands consist of operations like the opening of a window, deleting a plot, resizing a window, or waiting for a mouse click.

The following list gives the set of fundamental Scilab graphics commands:

- **driver('name')**: selects a graphics driver; the input parameter is the string that is the name of one of the drivers defined above.

  The next commands are specific of the **X11** or **Rec** drivers:

- **xclear(v)**: clears one or more graphics windows; does not affect the graphical context of these windows. The input parameter is a vector containing the numbers of the windows to be cleared.

- **xbasc(v)**: clears one or more graphics windows and erases the recorded graphics; does not affect the graphical context of the window.

- **xpause(ms)**: a pause in milliseconds; useful for animation effects when the direct plotting is too fast for the desired visual output.

- **xselect()**: raises the current graphics window and makes it current (equivalent to the **Select** button in the **File** menu of the graphics window and the **Raise** button of the main Scilab window).

- **xclick()**: waits for a mouse click; this command returns the number of the window where the click occurs, the position of the click, and the number of the button of the mouse used for the click. All these elements are useful for the development of interactive graphical tools.

- **xbasr(i)**: redraws the plot of the graphics window numbered **i**.

- **xdel(i)**: deletes the window **i** (equivalent to the **Close** button).

- **xinit('name')**: initializes a graphics device; this command is necessary for Postscript and Xfig drivers, since it opens a file where the graphics operations are recorded. This command simply opens a graphics window for the **X11** or **Rec** drivers.

- **xend()**: closes a graphics session (and the associated device) that was previously opened with **xinit**.

- **xgetmouse()**: returns the current position of the mouse.

- **xselect()**: raises the active graphics window.

- **xsave('fname',i)**: saves the graph of the window **i** in the file designated by the string **'fname'**. This save is machine-independent.

- **xload**: loads a graph previously saved in a file by the command **xsave**. This load is machine-independent.

- **xinfo('string')**: adds a string in the title bar above graphics window buttons. Note that during the use of the **3D Rot.** button the angle position overwrites the previous title in the bar, and the **xinfo** command needs to be reexecuted.

- **winsid()**: returns the window numbers of the open graphics windows.

The usual default driver, **Rec**, is the one that is most commonly used. There are special commands that avoid the need to change drivers. This is useful when a plot has been interactively created as the result of many graphics commands and it is desired to output the plot, for example, to a Postscript or Xfig format file. This can be accomplished using the **xbasimp** and **xs2fig** commands (which is equivalent to choosing these options from the **File/Export** button in the graphics window). The following is an example:

```
driver('Pos')
xinit('foo.ps')
plot(1:10)
xend()

driver('Rec')
plot(1:10)
xbasimp(0,'foo1.ps')
```

With the preceding commands two identical Postscript files are obtained: **foo.ps** and **foo1.ps.0** (the appended 0 is the number of the active window where the plot was created). The difference is that the second Postscript file was obtained without invoking the Postscript driver.

Superposition of plots is the default. To clear a preceding plot to make way for a new one, two commands can be used: **xbasc(window-number)**, which clears the window and erases the recorded Scilab graphics commands associated with the window **window-number**, and **xclear**, which simply clears the window but preserves the recorded commands.

Finally, in the Unix environment, we note that the environment variable **DISPLAY** can be used to specify the specific X11 display where graphics windows are to be displayed. As an alternative, the **xinit** function can be used to specify the desired display device.

3.2 Global Plot Parameters

If **mycurv** is a vector or an array of data, the command **plot(mycurv)** plots the data. The plot is constructed within an environment that defines the characteristics, i.e., colors, axis tick marks, etc., that were chosen automatically (by default). All the graphics characteristics can be chosen by the user, and this section discusses these global parameters, explains how to set and query their values, and describes some special functions that indirectly manipulate the global parameters in order to achieve special effects such as iso-scaling.

3.2.1 Graphical Context

Some graphics parameters are implicitly controlled by the graphics context (e.g., line thickness), and others are directly controlled through graphics call arguments specified in plotting commands. The graphics context consists of default values that can be queried with the command **xget** and changed with the command **xset**, which can be used interactively by calling it without arguments (i.e., **xset()**), which then opens a graphical user interface called the **Scilab Toggles Panel**. With this panel the values of various graphics context variables can be changed interactively with the mouse.

There are 19 different graphics context parameters that can be set with **xset**. The following is a description of a subset of the total collection of parameters (see the online help for more details) that can be set with the function **xset**:

- **xset("default")**: returns to the default graphics context. It is very often used to clear any preceding graphics context. *In particular, it is better to use it before trying the examples given in this chapter.*

- **xset("font",fontid,fontsize)**: selects the current font and its size.

- **xset("mark",markid,marksize)**: sets the current mark and mark size.

- **xset("use color",flag)**: selects color plotting when **flag=1** and gray-scale plotting when **flag=0**.

- **xset("clipping",x,y,w,h)**: sets dimensions of the clipping zone, which specifies where plotted values can be seen. The values **x**, **y**, **w**, **h** specify the **(x,y)** origin and the width **w** and height **h** of the plot.

- **xset("colormap",cmap)**: sets the colormap using an $m \times 3$ matrix, where **m** is the number of colors and color **i** is specified as the 3-tuple **cmap[i,1]**, **cmap[i,2]**, **cmap[i,3]** corresponding to the saturation of red, green, and blue, respectively. The values of **cmap** must be in $[0, 1]$.

- `xset("wdim",width,height)`: sets the current graphics window's width and height, where `width` and `height` are in units of pixels. This command has no effect for the Postscript and the Xfig drivers.

- `xset("window",window-number)`: sets the specified window number to be the active window. The window number is created if it does not already exist.

- `xset("wpos",x,y)`: sets the position of the upper left point of the graphics window, where `x` and `y` are in units of pixels.

- `xset("pixmap",1)`: the plot is drawn to a pixmap that can be displayed on the screen with the command `xset("wshow")`. The pixmap is cleared with the command `xset("wwpc")` or with the usual command `xbasc()`. This driver is particularly useful for graphics animation, since the graphics operations are split into two parts: a plot creation phase (which is often the most time-consuming) and the display phase.

- `xset("alufunction",n)`: the logical function for drawing. This parameter specifies the way the next graphics object is combined with already displayed graphics objects. There are 16 logical functions including OR, XOR, AND (see the online help for details). The following examples (which the user should try) should give an idea of what the `xset('alufunction',n))` is useful for:

```
plot3d()
plot3d()
xset('alufunction',7)
xset('window',0)
plot3d()
xset('default')
plot3d()
xset('alufunction',6)
xset('window',0)
plot3d()
xclear()
plot3d()
plot3d()
```

In the preceding discussion it has been seen that choices exist for fonts. These choices can be extended with the command `xlfont`, which is used to load a new family of fonts from the X Window Manager (some knowledge of X11 fonts is needed).

The following example session (which the user should try) illustrates some of the **xset** commands:

```
x=0:0.1:2*%pi;
plot2d2('gnn',[x;x;x]',[sin(x);sin(2*x);sin(3*x)]')
xbasc();xset("font",2,3)
plot2d2('gnn',[x;x;x]',[sin(x);sin(2*x);sin(3*x)]')
xbasc();xset("wdim",200,200)
plot2d2('gnn',[x;x;x]',[sin(x);sin(2*x);sin(3*x)]')
xbasc();xset("thickness",4)
plot2d2('gnn',[x;x;x]',[sin(x);sin(2*x);sin(3*x)]')
xset("pixmap",1);xbasc()
plot2d2('gnn',[x;x;x]',[sin(x);sin(2*x);sin(3*x)]')
xset("wshow")
xset("pixmap",0);
xset('default');
xset("wdim",610,455)
```

3.2.2 Indirect Manipulation of the Graphics Context

In the previous section direct manipulation of graphics context parameters was discussed using the function **xset**. Scilab also contains a number of functions that manipulate the graphics context to achieve certain results without having to make direct changes using **xset**.

Scilab functions that indirectly set the graphics context are listed below:

- **isoview**: yields an isometric scaling without changing the graphics window's size. This is useful, for example, to plot a circle in a standard graphics window or hardcopy output where the circle appears as a circle and not as an ellipse. The following example (which the user should try) clarifies the use of isoview:

  ```
  t=(0:0.1:2*%pi)';
  plot2d(sin(t),cos(t))
  xbasc()
  isoview(-1,1,-1,1)
  plot2d(sin(t),cos(t),-1,'001');
  ```

- **square**: yields an isometric scale by changing the graphics window's size (which is in contrast to **isoview**, which does not change the graphics window's size).

- **scaling**: scales data.

- **rotate**: rotates data.

    scaling and **rotate** execute an affine transform and a geometric rotation of a $2 \times N$ matrix, respectively. The $2 \times N$ matrix represents the **(x,y)** values of a set of points.

- **xclip**: sets a clipping rectangle. This function can be very useful for superimposing one graph on another where the two graphs are not of the same dimensions. Without any extra computation a plot can be restricted to appear only inside a specified rectangle. This function is also useful for plotting in the margins of the main graphics window.

- **xsetech**: specify a subwindow of the current graphics window as the current graphics surface. This is accomplished by setting the graphics scale within the window. This command is useful for plotting several graphs in the same window. The command **xgetech** queries the current graphics scale. The following example (which the user should try) illustrates the use of **xsetech**:

```
t=(0:0.1:2*%pi)';
xsetech([0.,0.,0.6,0.3],[-1,1,-1,1])
plot2d(sin(t),cos(t))
xsetech([0.5,0.3,0.4,0.6],[-1,1,-1,1])
plot2d(sin(t),cos(t))
```

3.3   2-D Plotting

In Scilab there are 5 distinct commands for displaying 2-D plots. This section describes each one and illustrates them with examples. These descriptions are accompanied with a variety of other commands that are used for annotating plots. Some high-level, specialized 2-D plotting functions are also described. The ensuing subsection presents a set of commonly used low-level functions for geometric figures. This is followed by a description of an ensemble of specialized plotting tools for the domain of automatic control. The description of some miscellaneous commands ends this section.

3.3.1   Basic Syntax for 2-D Plots

The simplest 2-D plot is **plot(x,y)** or **plot(y)**. This yields the plot of **y** as a function of **x**, where **x** and **y** are two vectors. If **x** is not given, it is replaced

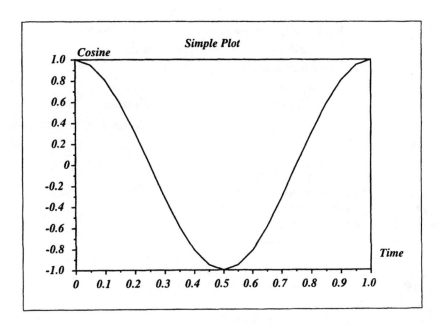

Figure 3.1: *First example of plotting*

by the default vector $(1, \ldots, \text{size}(y))$. If **y** is a matrix, its rows are plotted. There are optional arguments (see the on-line help for more details).

A first example is given by the following commands, and the result is illustrated in Figure 3.1:

```
t=(0:0.05:1)';
ct=cos(2*%pi*t);
plot(t,ct)                                          ← plot the cosine
xset("font",5,4);xset("thickness",3)                ← xset() opens
           the toggle panel and some parameters can be changed with mouse clicks
                                  given by commands for the demo here
plot(t,ct,'Time','Cosine','Simple Plot')            ← plot with captions
                                  for the axis and a title for the plot;
                     if a caption is empty, the argument ' ' is needed
```

Now click on a color of the **xset** toggle panel and do the previous plot again to get the title in the chosen color; afterwords, you can return to the default context by issuing the command **xset('default')**. The **plot()** command is actually a shorthand for the generic syntax for a 2-D multiple plot, which is

   plot2di(str,x,y,[style,strf,leg,rect,nax])

where **i** in **plot2di** can take the values **i=missing,1,2,3,4**. The different values of **i** signify:

   i=missing: piecewise linear plotting
   i=1: as above but with an option for logarithmic scales
   i=2: piecewise constant plotting
   i=3: plot vertical bars (useful for plotting sample data)
   i=4: plot piecewise linear arrows (useful for plotting vector fields)

The first 4 modes are illustrated by the following examples, which are displayed in Figure 3.2:

```
t=(1:0.1:8)';xset("font",2,3)
xsetech([0.,0.,0.5,0.5],[-1,1,-1,1])
plot2d([t t],[1.5+0.2*sin(t) 2+cos(t)])
xtitle('Plot2d')
titlepage('Piecewise linear')

xsetech([0.5,0.,0.5,0.5],[-1,1,-1,1])
plot2d1('oll',t,[1.5+0.2*sin(t) 2+cos(t)])
xtitle('Plot2d1')
titlepage('Logarithmic scale(s)')

xsetech([0.,0.5,0.5,0.5],[-1,1,-1,1])
plot2d2('onn',t,[1.5+0.2*sin(t) 2+cos(t)])
xtitle('Plot2d2')
titlepage('Piecewise constant')

xsetech([0.5,0.5,0.5,0.5],[-1,1,-1,1])
plot2d3('onn',t,[1.5+0.2*sin(t) 2+cos(t)])
xtitle('Plot2d3')
titlepage('Vertical bar plot')
```

The commands **xtitle** and **titlepage** are described in Section 3.3.3.
  The syntax of the call to the 2-D plotting functions consists in specifying the values of a set of arguments, which are described here:

- **str**: is a string (not part of the syntax for **plot2d()**) of length three that takes the form **"abc"**, where the values of **a**, **b**, and **c** may be:

  a=e means that the **x** values are not to be used (however, the user must give dummy values for **x**).

  a=o means that only one set of **x** values will be used for all the curves. Here **x** is a vector and **y** is a matrix.

  a=g means that for each column of **y** there must be a separate column for **x**.

  b=n displays a plot on a linear (i.e., normal) scale on the x-axis.

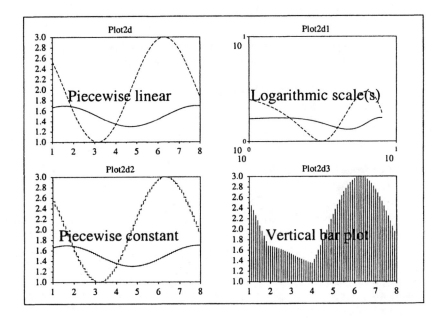

Figure 3.2: *Different 2-D plotting styles*

b=1 displays a plot on a logarithmic scale on the x-axis.

c=n displays a plot on a linear (i.e., normal) scale on the y-axis.

c=1 displays a plot on a logarithmic scale on the y-axis.

- **x,y**: two matrices of the same size **[nl,nc]** where **nc** designates the number of curves and **nl** the number of points for each curve. As described above, if **a=o**, then **x** can be a vector instead of a matrix. In the event that both **x** and **y** are vectors, then they can be either (both) column or row vectors (i.e., **plot2d(t',cos(t)')** and **plot2d(t,cos(t))** are equivalent).

- **style**: a real vector of size **(1,nc)** used to describe the drawing style for each of the curves. The style of curve **j** is given by **style(j)**. If **style(j)** is nonnegative, it specifies the color number from the colormap to be used to plot curve **j**. If it is negative, it specifies the number of the mark to be used for curve **j**. This is shown in the following example:

```
x=0:0.1:2*%pi;
u=[-0.8+sin(x);-0.6+sin(x);-0.4+sin(x);..
   -0.2+sin(x);sin(x)];
```

```
u=[u;0.2+sin(x);0.4+sin(x);0.6+sin(x);0.8+sin(x)]';
plot2d1('onn',x',u,[9,8,7,6,5,4,3,2,1,0],..
        "011"," ",[0,-2,2*%pi,3],[2,10,2,10])
x=0:0.2:2*%pi;
v=[1.4+sin(x);1.8+sin(x)]';
xset("mark",1,5)
plot2d1('onn',x',v,[7,8],"011"," ",..
        [0,-2,2*%pi,3],[2,10,2,10])

xbasc();xset('default')
x=0:0.1:2*%pi;
u=[-0.8+sin(x);-0.6+sin(x);-0.4+sin(x);..
   -0.2+sin(x);sin(x)];
u=[u;0.2+sin(x);0.4+sin(x);0.6+sin(x);0.8+sin(x)]';
plot2d1('onn',x',u,[-9,-8,-7,-6,-5,-4,-3,-2,-1,0],..
        "011"," ",[0,-2,2*%pi,3],[2,10,2,10])
x=0:0.2:2*%pi;
v=[1.4+sin(x);1.8+sin(x)]';
xset("mark",1,5)
plot2d1('onn',x',v,[-7,-8],"011"," ",..          ← Figure 3.3
        [0,-2,2*%pi,3],[2,10,2,10])
```

The second half (mark drawing styles) of the above example is illustrated in Figure 3.3.

- **strf** is a string of length three that takes the form **"xyz"**, where the values of **x**, **y**, and **z** may be:

  x=0 means that captions are not displayed.

  x=1 means that captions are displayed.

  y=0 means that the current frame boundaries are used.

  y=1 means that the argument **rect** is used to specify the frame boundaries of the plot, where **rect** is a 4-vector containing the min and max values of the frame in **x** and **y**.

  y=2 means that the boundaries of the plot are computed from the min and max values of **x** and **y**.

  y=3 is the same as **y=1** but with iso-scaling.

  y=4 is the same as **y=2** but with iso-scaling.

  y=5 is the same as **y=1** but with pretty graduations.

  y=6 is the same as **y=2** but with pretty graduations.

  z=1 means that axes are drawn and the tick marks can be specified by the argument **nax** (see below).

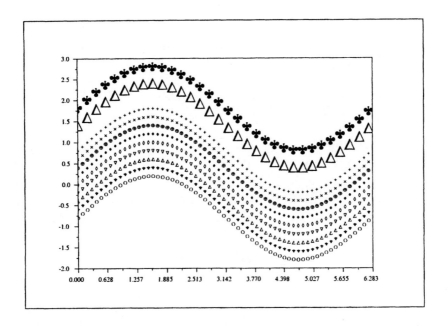

Figure 3.3: *Black and white (mark) plotting styles*

z=2 means that the plot is surrounded by a box without tick marks.

z=k where **k** is anything but **1** or **2** means that no frame and no tick marks are drawn around the plot.

- **leg**: this parameter gives the legend for each of the drawn curves. The format of **leg** is a string of the form

  'legend_1@legend_2@...@legend_N'

  where **N=nc** (the number of curves). The use of **leg** produces a legend in the lower margin of the plot with a short segment of colored line or a mark corresponding to the curve associated with the legend.

- **rect**: this parameter specifies the frame boundaries of the plot, where **rect** is a 4-vector containing the min and max values of the frame in **x** and **y**. This parameter takes the form **rect=[xmin,ymin,xmax,ymax]**.

- **nax**: is a vector **[nx,Nx,ny,Ny]** where **nx** and **ny** are the minor graduations on the **x** and **y** axes, respectively, and **Nx** and **Ny** are the major graduations of these axes.

The following example illustrates how the captions, axis boundaries, and axis tick marks can be controlled:

```
x=-%pi:0.1:%pi;
y1=sin(x);y2=cos(x);y3=x;y4=1.5*sin(5*x);
X=[x;x;x;x]; Y=[y1;y2;y3;y4];
plot2d1("gnn",X',Y',[-1 -2 -3 -4]','"111",..
        "sin(x)@cos(x)@x@1.5 sin(5x)",..
        [-3,-3,3,2],[2,20,5,5])
```

In the above example four curves are defined, and each one is designated by a legend. Furthermore, the visible part of the plot is clipped to the zone $-3 \leq x \leq 3$ and $-3 \leq y \leq 2$. Finally, the x-axis graduations are specified to have 20 major tick marks and 2 minor ones, whereas the y-axis graduations consist of 5 major and 5 minor tick marks. The result is illustrated in Figure 3.4.

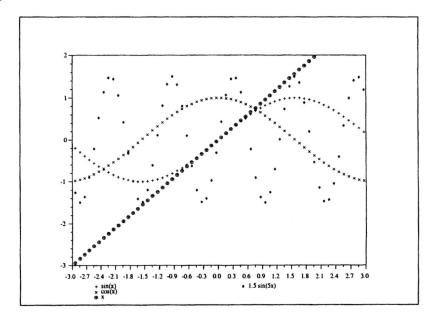

Figure 3.4: *Box, captions, and ticks*

A demo of each of the 2-D plotting commands can be obtained by simply typing the name of the command without arguments (i.e., **plot2d()**, **plot2d1()**, **plot2d2()**, **plot2d2()**, and **plot2d3()**).

3.3.2 Specialized 2-D Plotting Functions

Scilab contains some specialized plotting functions. These are described here:

- **champ**: produces a vector field in $\mathbb{R}^2$ with variable-length arrows (the function **champ1** can be used to obtain a colored vector field). As an example of **champ**:

```
x=[-1:0.1:1];y=x;u=ones(x);
fx=x.*.u';fy=u.*.y';
champ(x,y,fx,fy)
xset("font",2,3)
xtitle(['Vector field plot';'(with champ command)'])
```

The command **xtitle** is described in Section 3.3.3. The result of the above example is illustrated in Figure 3.5. You can do the same example with colors (with a large stacksize, see Section 1.2):

```
x=[-1:0.004:1];y=x;u=ones(x);
fx=x.*.u';fy=u.*.y';
champ1(x,y,fx,fy)
```

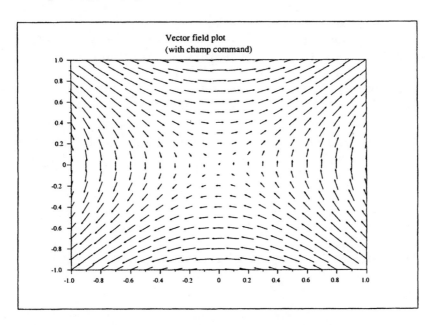

Figure 3.5: *Vector field in the plane*

- **fchamp**: produces a vector field in $\mathbb{R}^2$ exactly like **champ** but where the field is defined by a Scilab function.

- **fplot2d**: produces a 2-D plot just like **plot2d** but where the curve is defined by a Scilab function.

- **grayplot**: 2-D pixelated grayscale (using the current colormap) of a surface where the surface values are defined by a matrix. The function **Sgrayplot** is like **grayplot**, but here the grayscale values are smoothed using the function **fec** (see its on-line help). Since the current colormap is used, the name **Sgrayplot** is somewhat misleading, since the result is not a grayscale plot unless the colormap is set with **xset('colormap',c)**, where **cmap=(0:N)'*[1,1,1]/N** produces **N**+1 values from a true grayscale.

- **fgrayplot**: produces a grayscale plot where the surface is defined by a Scilab function. The function **Sfgrayplot** gives a smoothed rendition of the plot obtained by **fgrayplot**.

  The following example (see Figure 3.6) illustrates the usage of the function **fgrayplot**:

```
R=[1:256]/256;RGB=[R' R' R'];
xset('colormap',RGB);
deff('[z]=surf(x,y)','z=-((abs(x)-1)**2+(abs(y)-1)**2)')
fgrayplot(-1.8:0.02:1.8,-1.8:0.02:1.8,surf,"111",..
          [-2,-2,2,2])
xset('font',2,3)
xtitle(["Grayplot";"(with fgrayplot command)"])
```

  The command **xtitle** is described in Section 3.3.3. The same plot could be done with one given color:

```
R=[1:256]/256;
G=0.1*ones(R);
RGB=[R' G' G'];
xset('colormap',RGB)
fgrayplot(-1.8:0.02:1.8,-1.8:0.02:1.8,surf,"111",..
          [-2,-2,2,2])
```

  Notice that in the above example the Scilab function **surf** is used as the argument to **fgrayplot**. The first part of the example illustrates the construction of a true grayscale plot (see Figure 3.6), whereas the second part creates a colorscale plot.

- **errbar**: creates a plot with error bars.

- **histplot**: plot the histogram of a data set for a given number of (equally spaced) classes.

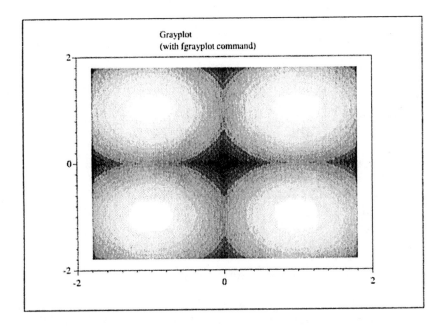

Figure 3.6: *Gray plot using a true gray colormap*

3.3.3   Captions and Presentation

A large variety of commands is available to transform a rough plot into a more attractive and illustrative one. Some of these commands are described here.

The most useful commands for improving graphics output are:

- **xgrid**: adds a grid on a 2-D graphic. If a passed argument is specified, it gives the value of the color in the colormap to make the grid lines (the default is the first color in the colormap).

- **xtitle**: adds a title above the plot and labels to the x and y axes.

- **titlepage**: creates a graphics title at the plot center.

- **plotframe**: pre-creates a graphics frame with specified scaling, grids, captions, and subwindows.

- **graduate**: a simple tool for computing pretty axis graduations before plotting with any of the **plot2di** or **plotframe** commands.

The following example illustrates some of the above commands:

```
titlepage("Titlepage")
x=-%pi:0.1:%pi;
y1=sin(x);y2=cos(x);y3=x;y4=1.5*sin(5*x);
X=[x;x;x;x]; Y=[y1;y2;y3;y4];
plot2d1("gnn",X',Y',[-1 -2 -3 -4]','"111",..
        "caption1@caption2@caption3@caption4",..
        [-3,-3,3,2],[2,20,2,5])
xgrid()
xclea(-2.7,1.5,1.5,1.5)
xstring(0.6,-0.65,"(with titlepage command)")
xstring(-2.6,.7,["xstring command after";..
        "xclea command"],0,1)
xset("font",4,2);
xtitle(["General Title";"(with xtitle command)"],..
        "x-axis title",..
        "y-axis title (with xtitle command)")
```

The commands **xclea** and **xstring** are defined in next section. Notice that in the above example a title is created in the center of the plot window. An additional title is placed at the top of the plot, and x and y axis labels are created. Also illustrated is how a grid may be placed over the plot. The result of the above example is illustrated in Figure 3.7.

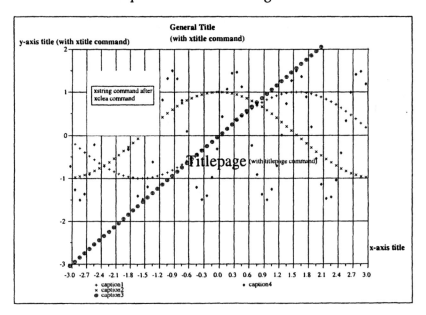

Figure 3.7: *Grid, title eraser, and comments*

It has been shown how to control the tick marks on the axes, choose the frame size for a plot, and add a grid using **plotframe**. A useful feature is that this operation can be performed once and then used for a sequence of different plots. For example, if several curves of different domains and ranges are to be superimposed, then **plotframe** can be used to specify an appropriate frame with graduations corresponding to whole numbers, and the ensuing curves are then plotted within this framework. Using **plotframe,** this can also be done for subwindows. The way that ensuing plot commands are superimposed without changing the **plotframe** settings is by setting the parameter **strf** to the value **'000'**. The following example illustrates how **plotframe** can be used to specify a frame, tick marks, a grid, a title, and axis labels:

```
rect=[-%pi,-1,%pi,1]
ticks=[2,10,4,10]
plotframe(rect,ticks,[%t,%t],..
['Plot with grids and automatic bounds','angle','velocity'])
```

3.3.4 Plotting Geometric Figures

Scilab provides a set of functions for drawing filled or empty geometric figures and for writing text (numbers or characters).

Polygon Plotting

The following Scilab functions provide a variety of geometric figures:

- **xsegs**: draws a set of unconnected segments.

- **xrect**: draws a single rectangle.

- **xfrect**: draws a single filled rectangle.

- **xrects**: fills or draws a set of rectangles.

- **xpoly**: draws a suite of connected straight lines (a polyline).

- **xrpoly**: draws a regular polygon with a specified number of sides.

- **xpolys**: draws a set of polylines.

- **xfpoly**: draws a filled polygon.

- **xfpolys**: draws a set of filled polygons.

- **xarrows**: draws a set of unconnected arrows.

- **xclea**: erases a rectangle in a graphics window.

Elliptical Plotting

Scilab also offers a number of functions for plotting simple curves:

- **xarc**: draws an ellipse.

- **xarcs**: draws a set of ellipses.

- **xfarc**: draws a filled ellipse.

- **xfarcs**: draws a set of filled ellipses.

Displaying Text in a Plot

Scilab provides a number of functions for displaying text in a plot:

- **xstring**: draws a string or a matrix of strings.

- **xstring1**: computes a rectangle that surrounds a string.

- **xstringb**: draws a string in a specified box.

- **xnumb**: draws a set of numbers.

The following example illustrates the use of several commands from the above list:

```
plot2d(0,0,0,"012"," ",[-2,0,12,12]);
xrect(0,1,3,1)
xfrect(3.1,1,3,1)
xstring(0.5,0.5,"xrect(0,1,3,1)")
xv=[0 1 2 3 4];
yv=[2.5 1.5 1.8 1.3 2.5];
xpoly(xv,yv,"lines",1)
xstring(0.5,2.,"xpoly(xv,yv,""lines"",1)")
xa=[5 6 6 7 7 8 8 9 9 5];
ya=[2.5 1.5 1.5 1.8 1.8 1.3 1.3 2.5 2.5 2.5];
xarrows(xa,ya)
xstring(5.5,2.,"xarrows(xa,ya)")
xarc(0.,5.,4.,2.,0.,64*300.)
xstring(0.5,4,"xarc(0.,5.,4.,2.,0.,64*300.)")
xset("pattern",2);
xfarc(5.,5.,4.,2.,0.,64*360.)
xstring(0.,4.5,"WRITING-BY-XSTRING()",-22.5)
xnumb([5.5 7.2 8.9],[5.5 5.5 5.5],[33 14 15],1)
xset("font",2,2)
xset("pattern",34);
```

```
xfrect(5.4,4.5,3.2,0.8);
xset("pattern",2);
xset("font",4,5)
A=["   1"  "        2" "  3";"  4" "         5" " 6";..
   "68" "  17.2" "  9"];
rect=xstring1(7,8,A)
xrect(rect(1)-1,rect(2),rect(3),rect(4))
xset("font",4,2);
xset("font",4,5)
xstring(6.,8.,A)
xset("font",2,2)
isoview(0,12,0,12)
xarc(-2.,12.,5.,5.,0.,64*360.)
xstring(-1.5,9.25,"isoview + xarc")
xstring(7.,3.9,"xfarc then xclea")
xset("pattern",34);
xstring(3.8,0.5,"xfrect(3.1,1,3,1)",0)
```

The results of the above example are displayed in Figure 3.8.

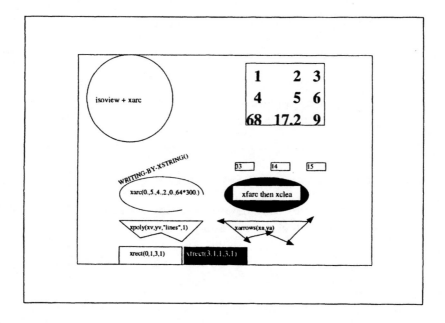

Figure 3.8: *Geometric graphics and comments*

3.3.5   Some Graphics Functions for Automatic Control

Scilab has a particularly rich collection of functions for automatic control (see Chapter 6). Similarly, Scilab contains a set of specialized graphics utilities relevant to automatic control. The suite of specialized automatic control graphics commands is detailed in the following list:

- **bode**: plots the magnitude and phase of the frequency response of a linear system. Both magnitude and phase are plotted on a log frequency scale. The magnitude is plotted in decibels and the phase in degrees. The linear system can be continuous or discrete and can be represented by a transfer function or in state space form. The **bode** command can plot the response of multiple linear systems simultaneously.

- **gainplot**: same as **bode** but plots only the magnitude of the frequency response.

- **nyquist**: plot of the imaginary part versus the real part of the frequency response of a linear system. The command **nyquist** takes the same arguments and functions as **bode**.

- **m_circle**: M-circle plot for use with **nyquist**.

- **chart**: plots the Nichols chart and is for use with Black's diagram (next command).

- **black**: plots Black's diagram for a linear system. The options are the same as for **bode** and **nyquist**.

- **evans**: plot the Evans root locus for a linear system.

- **plzr**: pole-zero plot of a linear system.

- **zgrid**: plots the lines of constant damping and the lines of constant frequency in the unit circle of the **z**-plane.

The following example illustrates these special graphics functions:

```
s=poly(0,'s')
h=syslin('c',(s^2+2*0.9*10*s+100)/..          ← create a linear system
                                                    using syslin
            (s^2+2*0.3*10.1*s+102.01))
h1=h*syslin('c',(s^2+2*0.1*15.1*s+228.01)/..
            (s^2+2*0.9*15*s+225))
```

```
xsetech([0.,0.,0.5,0.5],[-1,1,-1,1])
gainplot([h1;h],0.01,100)                                    ← Bode

xsetech([0.5,0.,0.5,0.5],[-1,1,-1,1])
nyquist([h1;h])                                              ← Nyquist

xsetech([0.,0.5,0.5,0.5],[-1,1,-1,1])
black([h1;h],0.01,100,['h1';'h'])                           ← Black
chart([-8 -6 -4],[80 120],list(1,0))                        ← Nichols chart

xsetech([0.5,0.5,0.5,0.5],[-1,1,-1,1])
H=syslin('c',352*poly(-5,'s')/poly([0,0,2000,200,25,1],..
        's','c'))
evans(H,100)                                                 ← Evans
```

Figure 3.9 illustrates the results of the above example.

3.3.6 Interactive Graphics Utilities

There are several particularly useful interactive graphics utilities in Scilab. They are described in the following list:

- **edit_curv**: an interactive graphics curve editor. This function allows the interactive insertion and displacement of points on a curve.

- **gr_menu**: interactive graphics editor. This function is a simple Xfig-like graphics editor. The function is useful for the interactive placement of text and other graphical objects.

- **locate**: used to interactively obtain the coordinates of one or more points selected with the mouse in a graphics window.

Figure 3.10 shows the use of **gr_menu** for commenting the plot and adding simple plane curves (here an ellipse).

3.4 3-D Plotting

Scilab provides a variety of functions for displaying 3-D graphics. This section begins by describing the primary 3-D plotting tool, **plot3d**. This is followed by a presentation of a variety of specialized commands for graphical representations in 3-D. The final section describes how to mix 2-D and 3-D graphics.

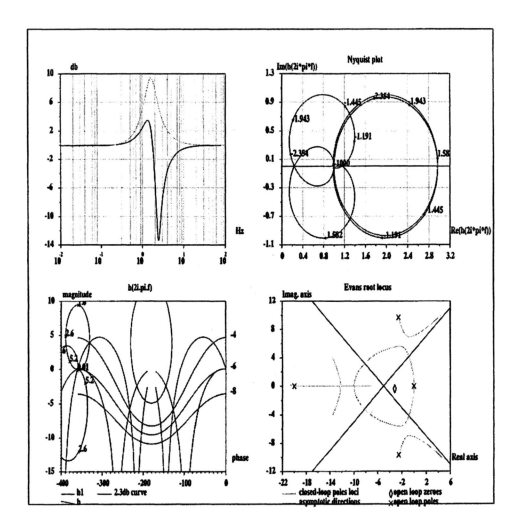

Figure 3.9: *Some specialized plots for automatic control*

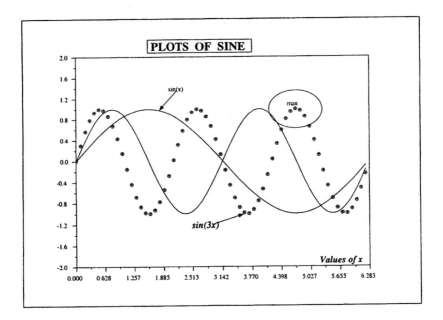

Figure 3.10: *Presentation of plots*

3.4.1   3-D Plotting

In this section four 3-D graphics commands are described. The primary 3-D plotting function is **plot3d**. The remaining three functions are variations of this function. Consequently, a complete description is presented only for **plot3d**. The four 3-D graphics commands are given in the following list:

- **plot3d**: the main 3-D plotting function, which can be used to plot a surface defined by point values or facets.

- **plot3d1**: 3-D plotting with color or grayscale values depending on the **z**-level of the surface.

- **fplot3d**: 3-D plotting of a surface described by a Scilab function **z=f(x,y)**.

- **fplot3d1**: 3-D plotting of a surface described by a function with color or grayscale levels.

The simplest syntax for **plot3d** is given by

```
plot3d(x,y,z)
```

where **x** and **y** are vectors giving the (x, y) locations of the z-values contained in the matrix **z**. The number of elements in **x** and **y** must be equal to the number of rows and columns of **z**, respectively.

The exact same syntax for **plot3d** is used when the surface is represented by facets. In this case **x**, **y**, and **z** are all $V \times N$ matrices, where N is the number of facets and V is the number of vertices per facet. Thus, the triple **[x(i,n),y(i,n),z(i,n)]** gives the location of the **i**th vertex of the **n**th facet. Plotting 3-D surfaces with facets is particularly useful when **z** is passed as the first element of a list where the second element is a vector of colors:

list(zf,colors)

where **colors** is a vector of integers that are indices into the colormap (see **xset()**). It should be noted that **genfac3d** (see below) is a useful function for generating the x, y, and z coordinates for facets given the vectors **x** and **y** and the matrix **z**.

The generic syntax for **plot3d** is

plot3d(xf,yf,list(zf,colors) [,theta,alpha,leg [,flag,ebox]])

where the optional arguments are:

- **theta,alpha**: scalars specifying the spherical coordinates of the observation point. These values are easily adjusted with the use of the **3D Rot.** button in the Scilab graphics window. The desired angle is indicated at the top of the window, and you can use it for a subsequent plot.

- **leg**: this parameter is used to label the x, y, and z axes of the 3-D plot. The format of **leg** is a string of the form **'x-label@y-label@z-label'**.

- **flag**: a vector of 3 real integers, **mode**, **type**, **box**, where **mode** determines how the hidden surfaces of the 3-D plot are displayed (**mode<0** only the shadow of the surface is drawn, **mode=0** the hidden parts are displayed, **mode>0** the hidden parts of the surface are not displayed), **type** determines the framing of the plot (**type=0** uses the framing set by a previous call and **type=1** sets the frame based on the values of the argument **ebox=[xmin,xmax,ymin,ymax,zmin,zmax]**), and **box** specifies how axes are displayed (**box=0** specifies no axes, no labels, and no box; **box=2** specifies only axes; **box=3** specifies a box and labels; and **box=4** specifies a box, labels, and axes).

3.4.2   Specialized 3-D Plots and Tools

This section describes several additional 3-D plot functions and utilities. The list of plot functions is

- **param3d**: plots parametric curves in 3-D. The function **param3d1** is the same except that it allows each curve to be drawn using a specified color.

- **contour**: 2-D plot of level curves for a 3-D function given by a matrix. The plot is composed of lines of constant data values. These contours can be drawn on the surface of 3-D plot or simply in a 2-D frame. In the case of the presentation in a 2-D frame, the function **contour2d** permits the use of an ensemble of arguments similar to those described for **plot2d** and also allows the use of the **2-D Zoom** button on the graphics window.

- **grayplot**: 2-D color or grayscale plot of a 3-D surface.

- **fcontour**: same as **contour** but where the surface is specified by a function. The function **fcontour2d** also exists.

- **hist3d**: 3-D plot of a 2-D histogram (i.e., a histogram of 2-D data).

The 3-D plot utilities are:

- **secto3d**: conversion of a surface described by cut sections to **plot3d** compatible data (see the on-line help for more details).

- **genfac3d**: converts a surface specified by a vector of x coordinates, a vector of y coordinates, and a matrix of $z(x, y)$ values to a set of three $4 \times N$ matrices giving the coordinates of the N associated rectangular facets.

- **eval3d**: evaluates a function on a regular grid. This function is used by **fplot3d**.

- **eval3dp**: computes facets of a parametrized surface.

3.4.3   Mixing 2-D and 3-D Graphics

3-D plot functions fix default graphics boundaries in $\mathbb{R}^3$. In the event that it is desired to work outside of these boundaries it is possible to create 3-D graphics using 2-D graphics functions. This provides greater flexibility.

However, it is necessary to project a desired 3-D surface onto the 2-D plane from the appropriate point of observation. This can be accomplished using the function **geom3d**, which gives the 2-D projection of a 3-D point after a 3-D plot has been made. The following example illustrates the use of **geom3d**:

```
r=(%pi):-0.01:0;x=r.*cos(10*r);y=r.*sin(10*r);
deff("[z]=surf(x,y)","z=sin(x)*cos(y)");
t=%pi*(-10:10)/10;
fplot3d(t,t,surf,35,45,"X@Y@Z",[-3,2,3]);
z=sin(x).*cos(y);
[x1,y1]=geom3d(x,y,z);
xpoly(x1,y1,"lines");
[x1,y1]=geom3d([0,0],[0,0],[5,0]);
xsegs(x1,y1);
xset("font",4,3);
xstring(x1(1),y1(1),' The point (0,0,0)');
```

In the example, **geom3d** is applied only after the use of **fplot3d**. This is required, since **geom3d** needs to know what the current angles of the observation point are to function. Figure 3.11 displays the result of this example.

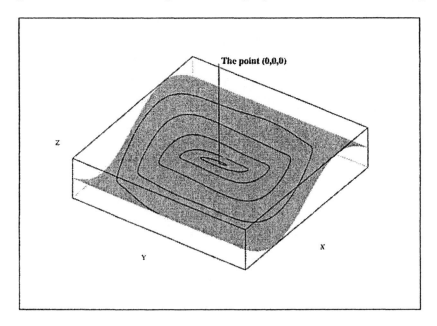

Figure 3.11: *2-D and 3-D plot*

3.5 Examples

This section presents two examples. The first example illustrates how to define subwindows within a graphics window. The second example illustrates a set of classic 2-D and 3-D plots.

3.5.1 Subwindows

Placing multiple plots in separate areas of the same graphics window is a common task for many applications. For example, the function **bode** does this to display the magnitude and phase of a linear system. In Scilab it is the function **xsetech**, which permits the subdivision of a graphics window. The following example illustrates how **xsetech** is used:

```
t=(0:.05:1)';st=sin(2*%pi*t);
xsetech([0,0,1,0.5]);
plot2d2("onn",t,st);
xsetech([0,0.5,1,0.5]);
plot2d3("onn",t,st);
xsetech([0,0,1,1]);
```

The results of this example are shown in Figure 3.12.

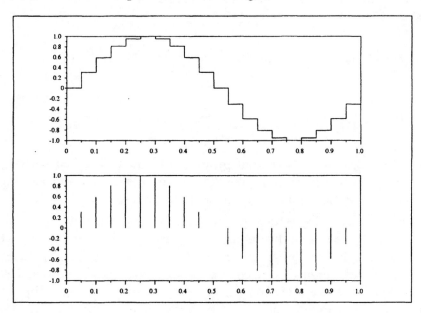

Figure 3.12: *Use of* **xsetech**

3.5.2   A Set of Figures

Here, a small sampling of the types of graphics outputs that can be created in Scilab is illustrated. The following code is what is used to create the output displayed in Figure 3.13. Note that the figure has been inserted in this document with the help of the command **Blatexprs** (see below):

```
plot3d1();
xtitle('plot3d1 : z=sin(x)*cos(y)')

contour();
xtitle('contour')

champ();
xtitle('champ')

t=%pi*(-10:10)/10;
deff('[z]=surf(x,y)','z=sin(x)*cos(y)');
rect=[-%pi,%pi,-%pi,%pi,-5,1];
z=feval(t,t,surf);
contour(t,t,z,10,35,45,'X@Y@Z',[1,1,0],rect,-5);
plot3d(t,t,z,35,45,'X@Y@Z',[2,1,3],rect);
xtitle('plot3d and contour')
```

where **feval(t,t,surf)** returns the matrix (**surf(t(i),t(j))**).

3.6   Printing Graphics and Exporting to LaTeX

This section describes the use of programs for handling Scilab graphics, using previewers, and printing results. These programs are located in the **bin** subdirectory of Scilab's main installation directory.

Before proceeding to the description of print and export commands it is important to recall that the default graphics driver in Scilab is **Rec**, which records all graphics commands contained in a window. These commands are erased when **xbasc()** is executed or when the **Clear** command is executed from the **File** menu on the graphics window. On the other hand, the **xclear()** command erases the plot, but the record of commands is preserved. Thus, it is possible when using the **xclear()** command to obtain printed or exported graphics output that does not correspond to what is displayed on the screen.

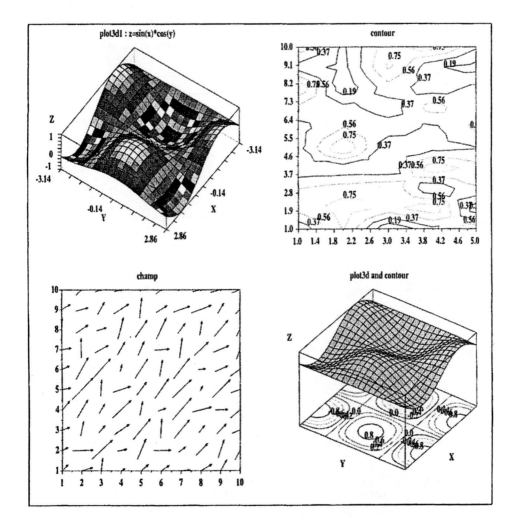

Figure 3.13: *Group of figures*

3.6.1   Window to Printer

The simplest way to get graphics output to a printer is to select the command **Print** from the **File** menu on the graphics window. This brings up a dialog box that allows the selection of the printer. In the Unix environment, the list of printers can be defined by the environment variable **PRINTERS** or directly at the beginning of the shell script **SCIDIR/bin/scilab**. Under Windows 95/NT the choice of the printer is driven by the system.

3.6.2   Creating a Postscript File

At the beginning of this chapter it was seen that a Postscript file of a graphics window could be obtained using

```
driver('Pos')
xinit('foo.ps')
plot3d1();
xend()
```

or

```
driver('Rec')
plot3d1()
xbasimp(0,'foo1.ps')
```

However, the Postscript files **foo.ps** and **foo1.ps.0** generated by Scilab cannot be directly sent to a Postscript printer (or to a previewer such as Ghostview or GSview), since they need a preamble. These files contain only the plotting statements in the Postscipt language, and they are what you get when you open the menu **File** at the top of a graphics window and choose **Export** and then **Postscript No Preamble** from the **Format Selection** option.

The appropriate preambles are obtained by using a set of programs that are provided with Scilab in the **SCIDIR/bin** directory. For example, the program **Blpr** is used to print a set of Scilab graphics on a single sheet of paper and is used as follows:

```
Blpr string-title file1.ps file2.ps ... > result.ps
```

You must give a string for the title of your plot (this string may be the empty string " "). The file **result.ps** can then be printed with the command

```
lpr result.ps
```

or **ghostview** (**GSview** under Windows 95/NT). A Postscript interpreter and viewer can be used to display the result on-screen.

The creation of a new file **result.ps** can be avoided by using a pipe. For example, to send the graphics directly to the printer:

```
Blpr string-title file1.ps file2.ps ... | lpr
```

The best result (best-sized figures) is obtained when printing two normal graphics per page.

The easiest way to obtain results that are identical to what is seen in the graphics window is with the menu buttons on the window. This includes obtaining output that captures the use of the **2-D Zoom** and **3D Rot.** buttons. However, resizing of the graphics window has no effect on the size of the Postscript output.

3.6.3   Including a Postscript File in LATEX

The programs **Blatexpr**, **Blatexpr2**, and **Blatexprs** are provided to help in inserting Scilab graphics into LATEX.

Taking the previous mentioned file, **foo.ps**, and typing the command

```
Blatexpr 1.0 1.0 foo.ps
```

creates two files, **foo.epsf** and **foo.tex**, replacing the initial Postscript file. To include the figure in a LATEX document the following code should be inserted:

```
\input foo.tex
\dessin{The caption of your picture}{The-label}
```

The syntax must be exactly as given here. Note that the second line with command **\dessin** is absolutely necessary, and of course, the absolute path of **foo.tex** must be given if you are working in a different directory (see below). The file **foo.tex** is only the definition of the command **\dessin**, which takes 2 parameters: the caption of the plot and the label. The command **\dessin** can be used with one or two empty arguments **""** to avoid the caption or the label. The Postscipt files are inserted in LATEX with the help of the **\special** command and with a syntax that works with the **dvips** program. If it is desired to modify some parameters in the file **foo.tex**, it is only about ten lines of LATEX code, and it can be easily edited to change the size of the graph or its position on the page. The program **Blatexprs** does the same thing as **Blatexpr** except that it is used to insert a set of Postscript graphics in a single LATEX figure.

The following example illustrates the use of **Blatexprs** to put four figures on a single page. The Postscript driver **Pos** is chosen followed by the successive initialization of the four Postscript files **fig1.ps**, ... , **fig4.ps**. The process is terminated by returning to the **Rec** driver:

```
driver('Pos')

t=%pi*(-10:10)/10;
xinit('fig1.ps')
xset("font",4,2);
plot3d1(t,t,sin(t)'*cos(t),35,45,'X@Y@Z',[2,2,4]);
xset("font",4,4);
xtitle('plot3d1 : z=sin(x)*cos(y)');
xend()
xinit('fig2.ps')
xset("font",4,2);
contour(1:5,1:10,rand(5,10),5);
xset("font",4,4);
xtitle('contour');
xend()
xinit('fig3.ps')
xset("font",4,2);
champ(1:10,1:10,rand(10,10),rand(10,10));
xset("font",4,4);
xtitle('champ');
xend()
xinit('fig4.ps')
xset("font",4,2);
t=%pi*(-10:10)/10;
deff('[z]=surf(x,y)','z=sin(x)*cos(y)');
rect=[-%pi,%pi,-%pi,%pi,-5,1];
z=feval(t,t,surf);                      ← feval(t,t,surf) returns
                                          matrix (surf(t(i),t(j)))
xset("font",4,2);
contour(t,t,z,10,35,45,'X@Y@Z',[1,1,0],rect,-5);
plot3d(t,t,z,35,45,'X@Y@Z',[2,1,3],rect);
xset("font",4,4);
xtitle('plot3d and contour');
xend()

driver('Rec')
```

After the creation of the files, the figures are prepared for insertion into LaTeX using the command

```
Blatexprs multi fig1.ps fig2.ps fig3.ps fig4.ps
```

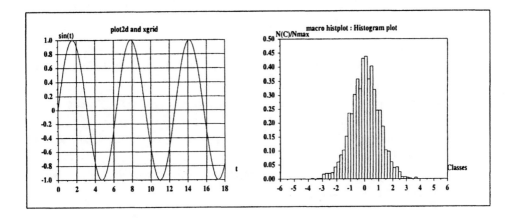

Figure 3.14: *Blatexpr2 example*

which yields the two files **multi.tex** and **multi.eps**, which are included into the LaTeX source file with

```
\input multi.tex
\dessin{The caption of your picture}{The-label}
```

The program **Blatexpr2** is used when two graphics are to be positioned side by side:

```
Blatexpr2 Fileres file1.ps file2.ps
```

An example of the result of **Blatexpr2** is shown in Figure 3.14.

It is sometimes convenient to have a main LaTeX document in a directory and to store all the figures in a subdirectory. The proper way to insert a picture file in the main document when the picture is stored in the subdirectory **figures** is the following:

```
\def\Figdir{figures/} % My figures are in the
              % "figures" subdirectory.
\input{\Figdir fig.tex}
\dessin{The caption of you picture}{The label}
```

The declaration

```
{\def\Figdir{figures/}}
```

is used twice, first to find the file **fig.tex** (when the **latex** command is used), and second to produce a correct pathname for the **special** LaTeX command found in **fig.tex** (when **dvips** is used).

3.6.4   Scilab, Xfig, and Postscript

In the Unix environment, a very useful tool for modifying and annotating graphics created in Scilab is Xfig, which is a GNU public license vector drawing program that can be obtained at

`ftp://ftp.x.org/contrib/applications/drawing_tools/xfig/`

From Scilab the command `xs2fig(active-window-number,file-name)` can be used to create a file in Xfig format. This command needs the use of the driver `Rec`.

As an example, the following graphics commands are executed to the graphics window `ScilabGraphic0`:

```
t=-%pi:0.3:%pi;
plot3d1(t,t,sin(t)'*cos(t),35,45,'X@Y@Z',[2,2,4]);
xs2fig(0,'demo.fig');
```

This produces a file `demo.fig`, which contains the Xfig file for reproducing the graphics in window `0`.

After loading the above file into Xfig, modifications and annotations can be made. This can be subsequently exported to a Postscript file by Xfig. Figure 3.15 is an example of a graphics created by Scilab and annotated in Xfig.

3.6.5   Creating Encapsulated Postscript Files

The program `Blatexpr` creates 2 files: a `.tex` file to be inserted in the LaTeX file and a `.epsf` file. We note that the `.epsf` file is not encapsulated Postscript, since it contains no bounding box. To obtain an encapsulated Postscript file the program `BEpsf` must be used. This program takes a Postscript graphic and adds a bounding box.

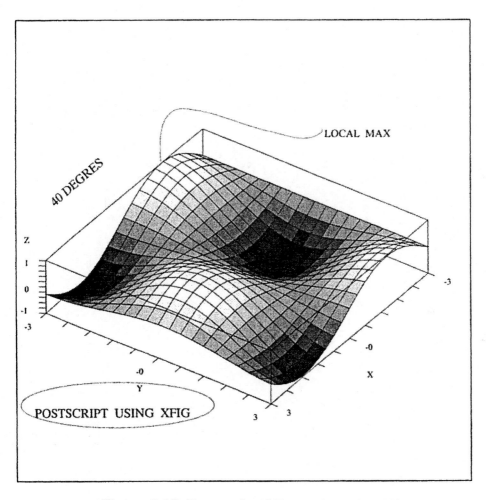

Figure 3.15: *Encapsulated Postscript using Xfig*

Chapter 4

A Tour of Some Basic Functions

This chapter describes a selected set of general-purpose functions that can be used as building blocks by the Scilab user. They also serve as beginning points to many related functions that can be found using the on-line help.

4.1 Linear Algebra

The basic linear algebra functions such as Householder's transform, singular value decomposition, Lyapunov equations, Choleski factorization, are implemented in Scilab as built-in primitives based on reliable numerical libraries. In addition, a number of linear algebra functions are constructed to perform specialized computations for, for example, control or signal processing applications.

Most of the linear algebra algorithms aim at transforming a given matrix (or set of matrices) into *canonical* form. This amounts to selecting appropriate bases in $\mathbb{R}^n$ or $\mathbb{C}^n$ such that the matrix representation after transformation yields zero entries at particular locations. Diagonal, block diagonal, and upper or lower triangular matrices are common examples of such canonical forms. Canonical forms often allow algebraic properties to be immediately recognized. For example, the eigenvalues of a diagonal matrix are immediately read along its main diagonal.

The set of column vectors $X = [x_1, x_2, \ldots, x_n]$ forms a basis of $\mathbb{R}^n$ if and only if X is a nonsingular matrix. Also if the bases $X = [x_1, x_2, \ldots, x_n]$ and $Y = [y_1, y_2, \ldots, y_m]$ of $\mathbb{R}^n$ and $\mathbb{R}^m$ respectively are selected, the $m \times n$ matrix A is transformed into $A_c = Y^{-1}AX$, i.e., the matrix representation of A with respect to the pair of bases X and Y. Choosing X as a basis in the domain space R^n amounts to a right multiplication by the nonsingular matrix X, and choosing Y as basis in the codomain space R^m amounts to a left multiplication by the nonsingular matrix Y^{-1}. For numerical stability it is generally best to select, when possible, orthonormal bases, that is, matrices X and Y such that $XX^T = I$ and $YY^T = I$. In the complex case $XX^H = I$ and $YY^H = I$, where X^H stands for the complex conjugate of X (both transpose and conjugate transpose operations are obtained in Scilab by the notation **x'** or **x.'**). Orthogonal transformations maintain numerical stability, since they leave invariant the L_2 (Euclidean) norm of vectors in $\mathbb{R}^n$ and $\mathbb{C}^n$. That is to say that they do not increase the L_2 norm of computation errors and can be inverted by a simple transposition operation, which also introduces no numerical error.

All the basic linear algebra functions in Scilab can be transparently used with complex matrices. Specific algorithms using complex arithmetic are automatically used for complex computations when needed.

4.1.1 QR Factorization

The QR factorization of an $m \times n$ matrix A carries out a left multiplication on A by a square $m \times m$ orthogonal matrix Q^T in such a way that the resulting matrix $R = Q^T A$ (the canonical form) is upper triangular or upper trapezoidal. The matrix Q^T is obtained by a finite sequence of Householder transforms, which at the kth step sets to zero the entries $k + 1$ to m of the kth column of A. In Scilab the QR factorization of the A matrix is obtained by **[Q,R]=qr(A)**.

A typical application of the QR factorization is for solving least squares problems of the type $\min \|Ax - b\|_2$. Since Q is orthonormal, this is equivalent to $\min \|Q^T(Ax - b)\|_2$, and the resulting triangular system is easily solved. Note that for nonsquare matrices, Scilab left and right matrix division operations (**x=A/B** or **x=A\B**) use the QR factorization for finding a least squares solution **x**.

If in addition to the row operations performed by the matrix Q^T, appropriate column permutations of A are performed, the QR algorithm yields a basis of the image of A. In this case, the calling sequence is

```
[Q,R,rk,E]=qr(A,tol)
```

where **rk** is the number of entries along the diagonal of **R=Q'*A*E** that are larger than **tol**. In this way, selecting a small **tol** (for example, **sqrt(%eps) * norm(A,'fro')**) can be used to determine the rank of A. Here is a simple example:

```
-->A=rand(4,2)*rand(2,3);
```
                                          ← a generic 4×3 rank 2 matrix

```
-->A(2,:)=A(1,:);                          ← A  has two identical columns

-->[Q,R]=qr(A);R                           ← calling qr: R is upper triangular
 R  =
! - 0.8186056  - 0.8186056  - 0.8294177 !
!   0.           0.           0.0872455 !
!   0.           0.         - 0.1004085 !
!   0.           0.           0.        !

-->[Q,R,rk,E]=qr(A,1.d-10);                ← calling qr with column pivoting

-->norm(Q'*A*E-R)                          ← E is a permutation matrix
 ans  =
     2.826E-16

-->R                                       ← R has now two zero rows
 R  =
! - 0.8400163  - 0.8082771  - 0.8082771 !
!   0.         - 0.1296271  - 0.1296271 !
!   0.           0.           0.        !
!   0.           0.           0.        !

-->rk                                      ← the numerical rank is clearly determined
 rk  =
     2.
```

We note that for A an $m \times n$ matrix, when column pivoting is required, the first rk diagonal entries of Q'*A*E are nonzero, whereas the last m-rk rows are zeros. This means that the rk first columns of Q form an (orthonormal) basis for the image of A. In addition, we can test whether a given vector b belongs to the image of A (i.e., whether the linear equation A*x=b has a solution). For that, one would form the vector y=Q'*b and check that its m-rk last components are zero (within numerical roundoff error).

More generally, subspace inclusion Im(B) ⊂ Im(A), i.e., the existence of a matrix z such that A*z=B, should be tested by calculating the matrix product Q'*B and checking that the m-rk last rows of this matrix are zero. The matrix z can be obtained from the command z=E'*(R1\B1), where R1=R(1:rk,:) and B1=(B'*Q(:,1:rk))'. If rk is equal to the rank of A, the linear system R1*E'*z=B1 is well-posed, and the numerical rank of A can be determined if the left rk × rk triangular part of the R matrix is well conditioned.

4.1.2 Singular Value Decomposition

The singular value decomposition (SVD) of matrix A performs orthogonal row and column operations on A in such a way that the resulting matrix is diagonal and the diagonal entries (singular values) are ordered in decreasing value and coincide with the square root of the eigenvalues of $A^T A$. In particular, the largest singular value σ_1 is the L_2 norm of A.

The basic calling sequence for singular value decomposition is [U,S,V]=svd(A). The decomposition is obtained by an iterative process based on orthogonal row and column operations, which are accumulated in U and V, respectively. The numerical rank of A is defined as the number of singular values larger than a given small tolerance parameter. To obtain the numerical rank, one should use the long syntax [U,S,V,rk]=svd(A,tol). The default value of tol is $\epsilon_M \max(m, n) \|A\|_2$. The singular value decomposition of A performs both row and column "compressions" of A and calculates orthogonal bases for the kernel and image of A.

As with the QR decomposition, to test whether a given vector b belongs to the image of A, one should examine the vector y=U'*b. Both the QR decomposition and the SVD give reliable rank estimates, but the SVD calculations are much more computationally intensive. Thus, QR decomposition is always preferable in problems where the singular values are not explicitly needed. The SVD is used in problems that involve L_2 norms. For example, the best L_2 approximation of a given rank to a matrix A is obtained by the simple truncation of the singular values of A (see the sva function).

The SVD-based cond function gives the conditioning of x, a square real or complex nonsingular matrix. The function cond(X) gives the ratio of the largest to the smallest singular value, i.e., norm(X)*norm(inv(X)). Note that the rcond function returns a rough estimate of the inverse of the conditioning, which can be used as a fast and reliable test for matrix inversion. The following illustrates a simple example of the use of SVD:

```
-->A=rand(4,2)*rand(2,3);              ←  a generic  4 ×  3 matrix of rank 2

--> [U,S,V,rk]=svd(A);                                    ←  calling svd

-->U'*A*V
 ans   =
!    1.7340173        0.            0. !
!    0.                .3369339      0. !
!    0.               0.            0. !
!    0.               0.            0. !
```

```
-->rk
 rk  =
     2.

-->[sort(sqrt(spec(A'*A))),svd(A)]
 ans  =
!   1.7340173      1.7340173 !
!    .3369339       .3369339 !
!   1.394E-08     0.         !

-->b=A*rand(3,1);                              ← a vector in Im(A)

-->U'*b                          ← last components are zero
 ans  =
! - 0.9216596 !
! - 0.1047227 !
!   0.        !
!   0.        !

-->A=rand(3,3);s=svd(A);                   ← a nonsingular matrix

-->[cond(A),max(s)/min(s),norm(A)*norm(inv(A))]   ← equivalent
 ans  =                                      evaluations of condition
!   8.259676      8.259676      8.259676 !
```

4.1.3   Schur Form and Eigenvalues

The Schur form of a square $n \times n$ matrix A is obtained by selecting an orthogonal basis U such that A is transformed into a quasi-triangular form. Since A and $T = U^T A U$ are similar, the eigenvalues of A are readily obtained from T. Isolated diagonal entries correspond to real eigenvalues, and 2×2 blocks along the diagonal correspond to pairs of complex conjugate eigenvalues. The algorithm used to implement the **schur** function (known as the QR algorithm [29]) generates a sequence of matrices A_n beginning with $A_0 = A$. These matrices are all orthogonally similar to A and in such a way that A_n tends to T. Note that the **spec** function calculates the eigenvalues of a matrix using the same algorithm, but the matrix is first balanced using the **balanc** function.

The order in which the eigenvalues appear along the diagonal of T is, in general, random. In some applications it is required that eigenvalues follow a specified ordering (e.g., the eigenvalues with smallest magnitude appear first). This can be obtained in a second step, by a finite number of orthogonal transformations. It is also possible to select a particular order

for the eigenvalues by passing a flag or a function as the second parameter
to the **schur** function. For example, eigenvalues with magnitudes less than
one can be selected by passing the **schur** function the flag **'d'** (corresponding to stable eigenvalues for discrete systems), and eigenvalues with negative real parts can be selected by passing the flag **'c'** (corresponding to
stable eigenvalues for continuous systems). User-defined functions can also
be used to select custom sets of eigenvalues (for more information see the
on-line help).

The following session shows a simple example of using the **schur**
function:

```
-->A=diag([0,0.5,3]);X=rand(A);A=X\A*X          ← random matrix
 A  =                                            with known eigenvalues
!     .4243287   - 2.5733812   - 3.7338407 !
!     .3248999     4.6122591     6.3414766 !
! -   .2318276   - 1.1529746   - 1.5365878 !


--> [U,k]=schur(A,'d');                          ← selecting eigenvalues with
                                                          magnitude < 1

-->U'*A*U          ← Triangular form: k = 2 eigenvalues with magnitude < 1
 ans  =
!    .5      .3176508   - 1.8060291 !
!   0.      0.            8.5713643 !
!   0.      0.            3.        !


-->Usmall=U(:,1:k);


-->spec(Usmall'*A*Usmall)                        ← selecting eigenvalues with
 ans  =                                                   magnitude < 1
!    .5        !
! - 6.754E-16 !
```

4.1.4   Block Diagonalization and Eigenvectors

We have seen that the ordered Schur form permits the computation of the
eigenspace associated with a given set of eigenvalues. The eigenvectors associated with eigenvalue λ can be obtained by computing the kernel of
$A - \lambda I$ (by QR or SVD factorization). If the A matrix decomposes into n
distinct eigenvectors and U is the nonsingular matrix whose columns are
these eigenvectors, then the transformed matrix $U^{-1}AU$ is diagonal. In the
case of repeated eigenvalues it is numerically difficult to choose a basis for
which A is put into canonical form. The reason for this is that there is no
numerically good canonical form that reveals the fine structure (number

and size of Jordan blocks) of the matrix and that can be obtained through orthogonal transformations.

The **bdiag** function tries to diagonalize *A* with a well-conditioned *U* matrix. Note that the structure of the blocks returned by **bdiag** is different according to whether real or complex arithmetic is used. The following is an example:

```
-->A=diag([0,0.5,3]);X=rand(A);A=X\A*X;          ← random matrix
                                                    with known eigenvalues
--> [Ab,U,bs]=bdiag(A);                           ← diagonalizing A

-->Ab
 Ab   =
!    .5      0.      0. !
!    0.      0.      0. !
!    0.      0.      3. !

-->bs                                             ← 3 simple eigenvalues
 bs   =
!    1. !
!    1. !
!    1. !

-->A=sysdiag(diag([0,0.5,3]),[0,1,0;0,0,1;0,0,0]);

-->X=rand(A);A=X\A*X;                             ← adding a 3 × 3 Jordan block

--> [Ab,U,bs]=bdiag(A,100);                       ← block-diagonalizing A

-->bs                                             ← eigenvalue 0 put into a 4 × 4 block
 bs   =
!    1. !
!    1. !
!    4. !
```

The **bdiag** function is used by elementary matrix functions **sqrtm**, **cosm**, **logm**, and others. For instance, if *A* is a square matrix, **B=sqrtm(A)** returns a square matrix *B* such that $B^2 = A$. When *A* is symmetric, the calculation is done by a call to **schur**, which diagonalizes *A*. Then the square roots of the diagonal entries are evaluated, and the resulting *B* is computed by returning to the canonical basis. When *A* is not symmetric, the **bdiag** function is used instead of **schur** to try to diagonalize *A*. If **bdiag** is unable to diagonalize *A*, an error message is displayed. For the computation of the matrix exponential, which has many practical applications, a specific built-in function is used called **expm**. This function, like **sqrtm**, tries to diagonalize A as best as it can, using the Padé approximation on the diagonal blocks. The result can be numerically sensitive for nonsymmetric matrices.

4.1.5   Fine Structure

It is difficult to evaluate the Jordan form of a matrix numerically. The problem is ill-posed, since small variations in the data can drastically modify the Jordan form of the matrix. However, it is possible, for a singular matrix, to examine the "fine structure" (i.e., the number and size of the Jordan blocks) of the zero eigenvalues.

The **projspec** function evaluates P, the eigenprojection of the zero eigenvalues of a singular matrix; D, the associated eigennilpotent; and S, the Drazin inverse matrix. The algorithm performs a sequence of full-rank factorizations of singular matrices computed recursively from A, and at each step, the conditioning of the matrix obtained is tested as a stopping criterion. The matrices returned by **projspec** determine the singularity characteristics of the rational matrix $(sI - A)^{-1}$ around $s = 0$. It accomplishes this by expanding $(sI - A)^{-1} = s^{-\nu}D^{\nu-1} + \cdots + s^{-2}D + s^{-1}P + S + sS^2 + \cdots$ around $s = 0$. This series defines the matrices P, D, and S (where ν is the index of A). The **projspec** function also returns the ranks of A^k for $k = 1, 2, \ldots$, from which the Jordan structure at zero can be easily inferred.

In some applications it is desired to compute the eigenprojection associated with a prescribed set of eigenvalues. Functions **pbig** and **psmall** can be used for that purpose. These algorithms call an ordered Schur decomposition in triangular form followed by an application of Sylvester's equation in block diagonal form.

In the following simple example the A matrix has two real eigenvalues, $s_0 = 0$ (index two) and $s_1 = 1$ (simple). We compute the two eigenprojections P_0 and P_1, and we compare $(sI - A)^{-1}$ calculated by the function **coff** with its partial fraction expansion given by

$$(sI - A)^{-1} = \frac{P_0}{s} + \frac{AP_0}{s^2} + \frac{P_1}{s-1}.$$

We also compare e^{At} calculated by the **expm** function with the explicit formula obtained from taking the inverse Laplace transform of the partial fraction expansion. Note that the calculations are possible here because the eigenvalues of A are known explicitly. In general, it can be difficult to perform such calculations due to numerical roundoff errors.

```
-->A=[0,0,1,0;0,0,0,1;0,0,0,0;0,0,0,1];
```

```
    X=rand(A);A=X\A*X;                              ← index 2 matrix

-->[S0,P0,D0,index]=projspec(A);        ←  eigencharacteristics at zero

-->[Q,M]=psmall(A,0.5,'d');             ←  another equivalent calculation
                                               for P0 : P0=Q0*M0
-->inv(A-P0-D0)+P0;              ←  another equivalent calculation for S0

-->[S1,P1,D1,i1]=projspec(A-eye);       ←  eigencharacteristics at s=1

-->[N,d]=coff(A);                       ←  the resolvent: (sI − A)⁻¹ = N/d

-->s=%s; W=P0/s+D0/s^2+P1/(s-1);  ←  partial fractions decomposition
                                               of (sI − A)⁻¹
-->clean(W-N/d)
  ans  =
!    0      0      0      0   !
!    -      -      -      -   !
!    1      1      1      1   !
!                            !
!    0      0      0      0   !
!    -      -      -      -   !
!    1      1      1      1   !
!                            !
!    0      0      0      0   !
!    -      -      -      -   !
!    1      1      1      1   !
!                            !
!    0      0      0      0   !
!    -      -      -      -   !
!    1      1      1      1   !
```

$$(sI - A)^{-1} = N/d$$

$$W = P0/s + D0/s^2 + P1/(s-1)$$

```
-->t=3;P0+t*D0+exp(t)*P1               ←  explicit formula for expm(A*t)
  ans  =
!    8.3052281      17.173611      15.973481      6.4191762 !
!    4.8150677      6.8968122      6.9503053       .7569095 !
!    5.5171016      13.017869      13.095205      4.8786215 !
! - 13.205987    - 22.62834    - 23.321873    - 5.2117085 !

-->expm(A*t)
  ans  =
!    8.3052281      17.173611      15.973481      6.4191762 !
!    4.8150677      6.8968122      6.9503053       .7569095 !
!    5.5171016      13.017869      13.095205      4.8786215 !
! - 13.205987    - 22.62834    - 23.321873    - 5.2117085 !
```

4.1.6 Subspaces

A subspace of $\mathbb{R}^n$ or $\mathbb{C}^n$ is usually represented by a real or complex matrix. The subspace is expressed as the set of all linear combinations that can be generated from the vectors forming the columns of this matrix. Scilab provides specific functions that can be used to determine various subspaces. These functions are based on two basic functions: **rowcomp** and **colcomp** . These functions perform a row, or respectively column, compression of a given matrix, which subsequently characterizes the range, or respectively the kernel, of a matrix. This is analogous to Section 4.1.1 on how QR factorization with column pivoting performs row compression and characterizes the image subspace of the span of the columns of a matrix.

Subspace addition, i.e., the subspace generated by the column concatenation $[A, B]$, is obtained by the **spanplus** function. This function returns an orthogonal matrix X such that

$$X \begin{bmatrix} A & B \end{bmatrix} = \begin{bmatrix} A_1 & B_1 \\ 0 & B_2 \\ 0 & 0 \end{bmatrix},$$

where both A_1 and B_2 have full row rank. Denoting by n_1 (respectively n_2) the number of rows of A_1 (respectively B_2), we see that the first n_1 columns of X^T span $\text{Im}(A)$ and the first $n_1 + n_2$ columns of X^T span $\text{Im}(A)+\text{Im}(B)$. The matrix X is obtained by two calls to **rowcomp**.

The subspace $A^{-1}(\text{Im}(B))$ is given by the function **im_inv** . This function returns two orthogonal matrices Y and X such that

$$[Y AX, YB] = \begin{bmatrix} A_{11} & A_{12} & B_1 \\ 0 & A_{22} & 0 \end{bmatrix}, \tag{4.1}$$

where B_1 has full row rank and A_{22} has full column rank (Y and X are obtained by calling **rowcomp** and **colcomp**, respectively). If n denotes the number of columns of A_{11}, then the first n columns of X span $A^{-1}(\text{Im}(B))$.

Subspace intersection is calculated using the well-known formula

$$\text{Im}(A) \cap \text{Im}(B) = A(A^{-1}(\text{Im}(B))).$$

This can be accomplished by an additional call to **rowcomp** to compress the rows of A_{11} in (4.1), obtaining updated matrices, where A_{11} now has full row rank. If A_{11} has k rows, then an orthogonal basis for $\text{Im}(A) \cap \text{Im}(B)$ is spanned by the first k columns of Y^T. A given vector b is in $\text{Im}(A) \cap \text{Im}(B)$ iff $w = Yb$ is partitioned accordingly: w having at most k nonzero entries in its first rows and zero last entries. The **spaninter** function, which returns Y and k, performs these calculations.

Finally, given two matrices having the same number of rows, it is often convenient to find a basis such that A, B, $\text{Im}(A) \cap \text{Im}(B)$, and $\text{Im}(A) + \text{Im}(B)$ are all simultaneously obtained. A typical example is the Kalman form in linear control theory. Unfortunately, this calculation cannot be done using only orthogonal transformations. The **spantwo** function returns a nonsingular matrix Y (which in general is not orthogonal) such that

$$Y \begin{bmatrix} A & B \end{bmatrix} = \begin{bmatrix} A_1 & 0 \\ A_2 & B_2 \\ 0 & B_3 \\ 0 & 0 \end{bmatrix},$$

where

$$\begin{bmatrix} A_1 \\ A_2 \end{bmatrix}, \begin{bmatrix} B_2 \\ B_3 \end{bmatrix}, \begin{bmatrix} A_2 & B_2 \end{bmatrix}, \text{and} \begin{bmatrix} A_1 & 0 \\ A_2 & B_2 \\ 0 & B_3 \end{bmatrix}$$

are all full row rank matrices. From this form, one can build bases for A, B, $\text{Im}(A) \cap \text{Im}(B)$, and $\text{Im}(A) + \text{Im}(B)$ by selecting the appropriate columns of Y^{-1} and testing whether a given vector belongs to one of these subspaces.

In using **spantwo** several particular cases can occur. For instance, the lack of a first row in the canonical form $Y \begin{bmatrix} A & B \end{bmatrix}$ (where Y is returned by **spantwo**) is equivalent to $\text{Im}(A) \subset \text{Im}(B)$. If $\text{Im}(A) \oplus \text{Im}(B) = \mathbb{R}^n$, the second and fourth rows of $Y \begin{bmatrix} A & B \end{bmatrix}$ disappear, and we can calculate P, the projection on $\text{Im}(A)$ parallel to $\text{Im}(B)$ by `P=X(:,k+1:$)*Y(k+1:$,:)` with $X = Y^{-1}$ and $k = \text{rank}(A)$. Alternatively, P can be obtained by the **proj** function.

All these linear algebra functions return the nonsingular matrices that combine the row and column operations necessary to put the matrices into canonical form and return the dimensions of the blocks that appear in the canonical form. The following is an example using the **spantwo** function. Two 2-dimensional subspaces in $\mathbb{R}^4$ are constructed and represented by the matrices **A** and **B** such that dim(Im(**A**)+Im(**B**))=3 and dim(**A**∩**B**)=1:

```
-->A=[3,2;-3,3;-6,-4;3,2]; B=[0,0;-6,-5;0,-1;0,0];          ← two
                                    subspaces defined by matrices
--> [Y,dimA,dimB,dimAB]=spantwo(A,B);          ← call to spantwo

--> [dimA,dimB,dimAB]                          ← dimension information
 ans  =
!   2.    2.    3. !
```

```
-->Y*[A,B]                                    ← the canonical form
  ans  =
!    4.2426407      2.8284271      0.             0.          !
!    5.1502352      3.2037372    - 0.4595058    - 0.7434961   !
!    0.             0.             0.           - 4.3525018   !
!    0.             0.             0.             0.          !
```

4.2   Polynomial and Rational Function Manipulation

Polynomials can be built in Scilab using the **poly** function (see Section 2.4.1) and can be defined by their coefficients or by their roots. The **roots** function calculates the roots of a polynomial using the algorithm described in [37]. Matrix polynomials are built-in Scilab objects that can be defined using the same syntax as with constant matrices and can be combined, as with constant matrices, with operators such as multiplication and addition.

4.2.1   General Purpose Functions

As with constant matrices, the determinant and inverse of a polynomial matrix are obtained using **det** and **inv**. The determinant of a polynomial matrix is calculated using an FFT-interpolation approach. The method computes the determinant of a sequence of constant matrices, which are obtained by evaluating the polynomial matrix at the Fourier points. Another approach for computing the determinant of a polynomial matrix is the Leverrier algorithm, which is valid for any matrix but which yields poor numerical results. This algorithm is implemented with the function **detr**.

By default, the inversion of a polynomial matrix is made by the method of cofactors, evaluating the determinants using the preceding method. The function **[N,D]=coffg(P)** is used to get separately the numerator matrix, $N(s)$, and the common denominator polynomial, $D(s)$, of the polynomial matrix $P(s)$. Alternatively, the polynomial matrix inverse can be computed by a "state-space" approach common in control theory with Silverman's algorithm. This is accomplished by constructing a linear system where the A, B, and C matrices are empty and the D matrix is the polynomial matrix to be inverted.

The following example illustrates some of the above methods for computing determinants and inverses of polynomial matrices:

```
-->s=%s; P=[1,s+1;(s+2)^3,(s-1)^2]            ← a polynomial matrix
```

```
-->det(P)                                     ← its determinant
```

```
 ans  =
                      2     3     4
 - 7 - 22s - 17s - 7s - s
```

```
-->[N,D]=coffg(P);N          ← cofactors matrix and determinant
 N  =
!                 2                    !
!    1 - 2s + s           - 1 - s  !
!                                     !
!                 2     3             !
! - 8 - 12s - 6s - s       1       !
```

```
-->inv(P)                                        ← inverse of P
 ans  =
!                      2                                              !
!          1 - 2s + s                      - 1 - s                    !
!   ----------------------      ----------------------               !
!                 2     3     4                2     3     4 !
! - 7 - 22s - 17s - 7s - s    - 7 - 22s - 17s - 7s - s  !
!                                                                     !
!                 2     3                                             !
!     - 8 - 12s - 6s - s                      1                       !
!   ----------------------      ----------------------               !
!                 2     3     4                2     3     4 !
! - 7 - 22s - 17s - 7s - s    - 7 - 22s - 17s - 7s - s  !
```

```
-->Sys=syslin('c',[],[],[],P);          ← P as a component of a linear
                                                            system
-->Sysinv=inv(Sys);                      ← the standard inverse system of the
                                                    original linear system
-->[spec(Sysinv('A')),roots(det(P))]                    ← eigenvalues of
 ans  =              inverse dynamics coincide with roots of determinant
! - 1.2632052 + 1.5198265i  - 1.2632052 - 1.5198265i !
! - 1.2632052 - 1.5198265i  - 1.2632052 + 1.5198265i !
! - 4.0287028               - 4.0287028              !
! -  .4448868               -  .4448868              !
```

```
-->ss2tf(Sysinv)
 ans  =
!                 2                                       2 !
!     - 1 + 2s - s              1 + s + 3.450E-10s  !
!   ----------------------      ---------------------- !
!                 2     3     4                2     3     4 !
!   7 + 22s + 17s + 7s + s     7 + 22s + 17s + 7s + s  !
!                                                          !
!                 2     3                                  !
```

```
!        8 + 12s + 6s + s            - 1 - 2.021E-10s        !
!        ----------------------      ---------------------   !
!                    2    3    4                 2    3    4  !
!        7 + 22s + 17s + 7s + s      7 + 22s + 17s + 7s + s  !
```

A rational function is defined as the quotient of two polynomials: **r=p/q**. The numerator and denominator polynomials can be extracted from a rational function by using **r('num')** and **r('den')**. Rational matrix manipulations such as concatenation or inversion can also be performed just as with constant and polynomial matrices using the usual operators and functions. This useful property for rational matrices is due to the fact that operators and functions in Scilab can be overloaded (see Section 2.5).

Calculations with rational matrices can give rise to many numerical difficulties and should be used with caution. It is often necessary to set small coefficients of high-degree terms (corresponding to roots at infinity) to zero. This can be done with the function **clean** (this function can receive any type of argument including rational matrices and sparse constant matrices).

Pole zero simplifications for rational matrices are done by the function **simp**. This built-in function is already used by most functions that make computations on polynomials and rational functions. The function **simp** has a conservative behavior. In particular, for low-degree polynomials only "exact" (i.e., to machine precision) simplifications are done. The function **trfmod** can be used to interactively force simplifications of common factors.

It is often useful to manipulate a polynomial or a rational function as a set of prime factors of first and second degree. The **factors** function is used for this. It returns a list containing the real factors of a polynomial.

As for constant and polynomial matrices, determinants and inverses of rational matrices are obtained with **det** and **inv**. Here calculations are made by evaluating the least common multiple (LCM) of each column's (or row's) denominators, transforming the rational matrix into the form $N(s)/D(s)$, where $N(s)$, is a polynomial matrix and $D(s)$ a diagonal polynomial matrix. Then polynomial matrix calculations are performed. Note that the evaluation of the determinant of a polynomial matrix is crucial to many polynomial and rational calculations. An important point is that when a polynomial matrix determinant is evaluated, "small" coefficients are set to zero. The meaning of small can be defined by the user by the specification of either absolute or relative thresholds using the the utility function **determ**.

4.2.2 Matrix Pencils

A number of functions deal with matrix pencil algorithms. A matrix pencil is a linear polynomial matrix $sE - A$. If this matrix is square and nonsingular, the pencil is called *regular*. For a regular pencil, the determinant and inverse can be computed by the **det** and **inv** functions, respectively.

For regular pencils, the function **gschur** yields the generalized Schur form. As with the **schur** function, a specific order of the eigenvalues can be prescribed. The function **gspec** returns the eigenvalues of a regular pencil (i.e., the roots of the polynomial $\det(sE - A)$). The function **gspec** is a generalization of the function **spec**.

If the regular pencil $sE - A$ of size $n \times n$ has a determinant of order k, it can be transformed, using nonorthogonal change of bases, into the canonical form

$$Q \left[sE - A \right] Z = \left[\begin{array}{cc} sI - A_r & 0 \\ 0 & sN - I \end{array} \right], \tag{4.2}$$

where N is nilpotent (i.e., has only zero eigenvalues) and A_r is a $k \times k$ matrix. The eigenvalues of A_r coincide with the eigenvalues of the pencil $sE - A$. It is the function **pencan** that performs this transformation. The bases Q and Z are obtained by examining the behavior of the rational matrix $(sE - A)^{-1}$ at $s = \infty$. This is done by the function **penlaur**, which itself calls the function **projspec**. During calculations, several matrices are inverted. Since these could be ill conditioned, the outputs of these functions should be checked and should be used with care, particularly if the underlying pencil is close to singular.

The structure of a general pencil is obtained by the function **kroneck**, which returns the orthogonal (triangular) Kronecker form of the pencil. This is a built-in function based on [8]. The function **kroneck** returns two orthogonal bases Q and Z such that

$$Q \left[sE - AZ \right] = \left[\begin{array}{cccc} sE_\epsilon - A_\epsilon & * & * & * \\ 0 & sE_\infty - A_\infty & * & * \\ 0 & 0 & sE_f - A_f & * \\ 0 & 0 & 0 & sE_\eta - A_\eta \end{array} \right].$$

The two central blocks are square and correspond to the decomposition (4.2). The ϵ and η blocks are due to the existence of right and left kernels for $sE - A$.

In the following session, a 5×6 pencil is defined that has the polynomial $s^2 + 1$ as an elementary divisor. The **randpencil** utility function can be used to construct matrix pencils having an arbitrary structure. First we find its structure by calling the **kroneck** function. Then we extract **Freg**, the regular part of the pencil, and we show three different ways of calculating the finite eigenvalues of **F**. The first is by using **gspec** for the (E, A) pair of matrices associated with **Freg**, the second by computing the determinant of **Freg**, and the third by performing a finite/infinite block diagonalization of **Freg** followed by the evaluation of a determinant. Finally, we show two different ways of calculating the inverse of **Freg** using **inv** and using the **glever** function. This last function calculates the polynomial part of the inverse matrix and separately its strictly proper part as the ratio of a polynomial matrix divided by the characteristic polynomial.

```
-->F                                              ←  F is a singular pencil
 F  =

!- 1 + s     4 - 4s     5 - 8s    - 4 + s   - 1 + 4s    - 6 + 3s!
!                                                                !
!- 1 + 2s    4 + s      8 + 2s    - 1 - s   - 4 - s     - 6 - 3s!
!                                                                !
!  s         3 + 2s     6 + 4s    - 2s      - 3 + s     - 3 - 6s!
!                                                                !
!- 1 + s     1 - s    - 1 - 2s    - 1 + s   - 1 + s        0     !
!                                                                !
!- 2 + s     2 - s      4 - 2s    - 2 + s   - 2 - 2s    - 3 + 3s!
```

```
-->[Q,Z,Qd,Zd,eps,eta]=kroneck(F);            ←  Kronecker form of F
```

```
-->Fcano=Q*F*Z;Freg=Fcano(Qd(1)+1:$,Zd(1)+1:$);  ←  regular part
```

```
-->det(Freg)                                   ←  determinant
 ans  =
                                    2
     93.530744 + 93.530744s
```

```
-->[E,A]=pen2ea(Freg); [al,be]=gspec(A,E);           ←  eigenvalues
```

```
-->[al(3:$)./be(3:$) roots(det(Freg))]        ← eigenvalues compared
 ans  =                                          to roots of determinant
! - 1.130D-15 + i    i !
! - 1.267D-15 - i   -i !
```

```
-->[Q,M,i]=pencan(Freg);                       ←  block diagonal form
```

```
-->Fregcan=clean(M*Freg*Q)
 Fregcan  =
```

```
!- 0.528626 + s      1.088625          0                0          !
!                                                                  !
!- 1.175285          0.528626 + s      0                0          !
!                                                                  !
!      0              0           - 1 - 0.400436s    0.167377s     !
!                                                                  !
!      0              0           - 0.958007s     - 1 + 0.400436s!
```

```
-->det(Fregcan(1:i,1:i))                    ← determinant of finite part
  ans  =
            2
      1 + s
```

```
-->Fi=inv(Freg);
```

```
-->[Bfs,Bis,chis]=glever(Freg);             ← two ways of calculating
                                                          the inverse
```

```
-->Fi(1,4)
  ans  =
                                        2              3
   - 0.1402258 + 0.8169308s - 0.1545431s + 0.5937487s
   -------------------------------------------------
                            2
                      1 + s
```

```
-->W=Bfs(1,4)/chis-Bis;W(1,4)               ← check the (1,4) entry
  ans  =
                                        2              3
   - 0.1402258 + 0.8169308s - 0.1545431s + 0.5937487s
   -------------------------------------------------
                            2
                      1 + s
```

4.3 Sparse Matrices

Sparse matrices (see Section 2.2.2) are created by the **sparse** function (see Section 2.4.1). The simplest syntax is **S=sparse(ij,values)**, where **ij** is a two-column integer matrix giving the indices of the nonzero entries, and **values** is the corresponding vector of the nonzero entries. Conversely, the **spget** function returns the pair **ij** and **values** associated with a sparse matrix. Matrix manipulations of sparse matrices are performed using the usual syntax. Note, however, that left and right divisions are valid only with square matrices.

To solve a sparse linear system the functions **lufact** and **lusolve** are used. The algorithms used in these two functions are from the sparse numerical package developed by K. Kundert and A. Sangiovanni-Vincentelli [39]. The **lufact** function performs a trapezoidal LU factorization with pivoting. The sparse factors are obtained with the function **luget**, and a basis for the left or right kernel can be easily found.

There are many applications of sparse matrices in Metanet, a graph handling toolbox (see Chapter 11). In this section a bandwith reduction algorithm that operates on sparse matrices is illustrated. The example concerns the triangularization of the 2-D disk with two holes, as shown in Figure 4.1.

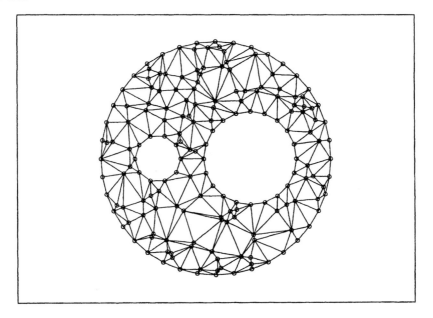

Figure 4.1: *2-D mesh*

The figure corresponds to the second example given in the help file for the **mesh2d** function. Associated to this structure is a sparse matrix with 0-1 entries defined by $A(i, j) = 1$ if $i = j$ or if the pair (i, j) is a side of a triangle of the finite-element discretization shown in the figure. Thus, the sparse matrix **A** is constructed from a set of points with coordinates (x_i, y_i) and a vector giving the indices of points that belong to the boundaries: **[nut,A]=mesh2d(x,y,boundary)**. The structure of the sparse matrix can be plotted as follows:

```
[nutr,A]=mesh2d(x,y,boundary);        ←construction of a sparse matrix
```

```
[i,j]=find(A~=0);                                    ←finding nonzero entries

N=size(A,1);xsetech([0,0,1,1],[1,0,N+1,N])

xrects([N-i+1;j;ones(i);ones(i)],ones(i));       ←plotting nonzero
                                                                    entries
xrect(1,N,N,N);xtitle('A sparse matrix structure');            ←
                                                              Figure 4.2
```

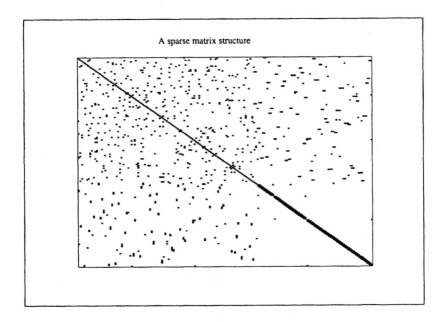

Figure 4.2: *Nonzero entries in A*

The function **bandwr** is used to reduce the bandwith of sparse matrices such as **A**. It returns a permutation vector such that the matrix obtained by applying this permutation on the rows and columns of **A** has a reduced bandwidth. That is, it transforms the original matrix into a banded matrix of reduced bandwidth. This is illustrated by the following commands and the result shown in Figure 4.3:

```
[perm1,mrepi,profile]=bandwr(triu(A));   ←  bandwr uses the upper
                                                       triangular part of A
A=A(mrepi,mrepi);                          ←  applying the permutation to A

[i,j]=find(A~=0);                          ←  finding nonzero entries
```

```
xsetech([0,0,1,1],[1,0,N+1,N])
```

```
xrects([N-i+1;j;ones(i);ones(i)],ones(i));        ← plotting nonzero
                                                      entries
xrect(1,N,N,N);xtitle('Reducing the bandwidth by rows '+..
    'and columns permutation');                    ← Figure 4.3
```

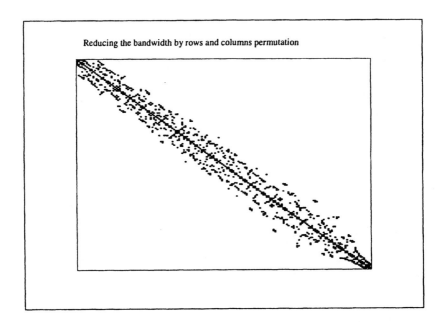

Figure 4.3: *Reduced banwidth matrix*

4.4 Random Numbers

The function **rand** is Scilab's standard random number generator. It provides random numbers from uniform and normal distributions. A new function, **grand**, which interfaces the Randlib library [16], has been added to the set of Scilab primitives.

The calling syntax of **grand** is as follows:

```
Y=grand(m,n,'dist-name' [,arg1,...,argn])
Y=grand(x,'dist-name' [,arg1,...,argn])
```

where **'dist-name'** is the name of the distribution and **[,arg1,...,argn]** are arguments specific to the chosen distribution.

The **grand** function provides random number generators for a large set of probability distributions according to the following value of the argument **'dist-name'**:

'bet' The beta distribution with parameters A and B.

'bin' The binomial distribution whose number of trials is N and whose probability of an event in each trial is P.

'chi' The χ^2 distribution with DF degrees of freedom.

'def' The uniform distribution of real numbers in $[0, 1]$.

'exp' The exponential distribution.

'f' The gamma distribution with fn degrees of freedom in the numerator and fd degrees of freedom in the denominator.

'gam' The gamma distribution.

'lgi' An integer uniform distribution in $(1, 2147483562)$.

'mn' The multivariate normal distribution.

'markov' The distribution of a Markov chain as specified by its transition matrix.

'mul' The multinomial distribution.

'nbn' The negative binomial distribution.

'nch' The noncentral χ^2 distribution.

'nf' The noncentral F distribution.

'nor' The normal distribution.

'poi' The Poisson distribution.

'prm' Random permutations of a vector.

'uin' The integer uniform distribution in $[a, b]$.

The random number generator implemented in Randlib is patterned after the work of P. L'Ecuyer and S. Cote [42] (the code was translated into Fortran from the their Pascal code). The generator contains 32 virtual random number generators. Each generator can provide 1,048,576 blocks of numbers, and each block is of length 1,073,741,824. Any generator can be set to the beginning or end of the current block or to its starting value. The current generator can be changed by **grand('setcgn',x)**, where **x** is an integer smaller than 32 giving the number of the generator, and the ordinal number of the current generator can be obtained from **grand('getcgn')**.

The user of the default generator can set the initial values of the two integer seeds with **grand('setall',s1,s2)** or **grand('setsd',s1,s2)**. If the user does not set the seeds, the random number generation will use the default values, 1234567890 and 123456789. The values of the current seeds can be obtained by a call to **grand('getsd')**. The call **grand('initgn',I)** can be used to reinitialize the state of the current generator. The next session shows how to use **grand** to generate 10 blocks of random numbers of size 1000 (from a uniform integer distribution in (1, 2147483562)) and then to reproduce exactly the same sequence:

```
-->NB=10                                    ← number of blocks

-->NR=1000                                  ← number of random numbers

-->grand('setall',12345,54321);             ← setting the seed for all
                                                the generators
-->grand('setcgn',10);                      ← using random number generator 10

-->l=list();

-->for i=1:2
-->   grand('initgn',-1);                   ← reset the current generator
-->   answer=ones(NB,NR);
-->   for iblock = 1:NB
-->       answer(iblock,1:NR)= grand(1,NR,'lgi');
-->       grand('initgn',1);                ← switch to next block
-->   end
-->   l(i)=answer;
-->end

-->norm(l(1)-l(2))                          ← checking the result
  ans   =
     0.
```

4.5 Cumulative Distribution Functions and Their Inverses

A set of cumulative distribution functions, **cdfbet** (beta), **cdfbin** (binomial), **cdfchi** (chi-square), **cdfchn** (noncentral chi-square), **cdff** (F), **cdffnc** (noncentral F), **cdfgam** (gamma), **cdfnbn** (negative binomial), **cdfnor** (normal), **cdfpoi** (Poisson), and **cdft** (Student's T) and their inverses are implemented in Scilab [17]. Each cumulative distribution function (CDF) has two calling syntaxes. The first computes the probability associated to the independent variable of the CDF, and the second computes the independent variable associated to a desired probability. For example, for the beta distribution the function implementing the CDF is called **cdfbet**, and its calling syntax for the two cases is as follows:

[P,Q]=cdfbet("PQ",X,1-X,A,B)   ← returns P, the integral from 0 to X of the beta distribution ($Q = 1 - P$)

[X,Y]=cdfbet("XY",A,B,P,Q)   ← returns X such that the integral from 0 to X of the beta distribution equals P

The cumulative distribution functions are mainly used for statistical tests. We provide here a short example that compares the cumulative distribution of a beta $(1.5, 4)$ distribution with the empirical distribution obtained with a random sample of size $N = 100$. The result is shown in Figure 4.4.

```
N=100;A=1.5;B=4;

Rdev=grand(N,1,'bet',A,B);                    ← 100 samples from beta(1.5, 4)

RdevS=gsort(Rdev,'g','i');              ← sorting the result for
                           an empirically computed cumulative distribution
X=(0:0.05:1)';

[P]=cdfbet("PQ",X,1-X,A*ones(X),B*ones(X));        ← computing
                                          the theoretical result
plot2d1("onn",X,P,2)      ← plotting empirical cumulative distribution

plot2d2("onn",RdevS,(1:N)'/N,1,"000");              ← overlaying
                                          the theoretical result
xtitle(['    Empirical and theoretical cumulative';
       '    distributions for a beta random variable']);
```

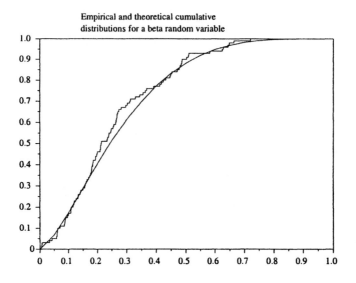

Figure 4.4: *Cumulative distribution for beta distribution*

Chapter 5

Advanced Programming

5.1 Functions and Primitives

Scilab is an interpreted language that is based on a set of powerful built-in functions called primitives (of which there are around three hundred) that are written in C or Fortran. Since the Scilab language is interpreted, it is easy to design, develop, and experiment with data and algorithms with Scilab. Furthermore, Scilab is very useful for prototyping, since it provides many tools that aid in the rapid development of programming tools. Application building in Scilab is accomplished by creating libraries of functions written in the Scilab language. The functions contained in a library are built up from other functions and Scilab primitives.

As an example, the Scilab function **balreal** is constructed in the Scilab programming language. This function computes a balanced realization of a linear system and does this by calling other Scilab functions from the linear algebra library such as **svd** and **chol**. The result is a high-level linear algebra function that is nevertheless very compact.

Scilab functions call basic primitives, but the contrary is also possible. A Scilab primitive can call a Scilab function that is given as an argument. For example, the general ordinary differential equation (ODE) solver is a Scilab primitive to which the user specifies a differential equation to be solved. The differential equation is actually computed by a high-level Scilab function.

The Scilab set of basic primitives is extensible. Each new version of Scilab introduces new primitives, and Scilab users can also add custom-designed primitives. The advantage of interfacing a new primitive is that the function implemented as a primitive (i.e., in C or Fortran) can often operate more efficiently than in the Scilab programming language. This is

119

especially true when the function requires a great deal of looping.

Primitives can be interfaced to Scilab permanently. However, it is also possible to add new primitives on the fly (i.e., during a running Scilab session) with dynamic linking (Section 5.6, page 145), which is available for most platforms Scilab runs on.

This chapter describes various possibilities for interfacing C and Fortran programs (i.e., new primitive functions) to Scilab. Let us begin with a simple example. Consider the following C code:

```
int ext1c(n, a, b, c)
     int *n;
     double *a, *b, *c;
{
  int k;
  for (k = 0; k < *n; ++k)
     c[k] = a[k] + b[k];
  return(0);
}
```

which performs the array operation `c=a+b`. We want to be able to call the C function `ext1c` from the Scilab interpreted environment.

- The first step is to add the C code to Scilab. This can be accomplished in two ways. The more permanent way is by building the C code into the executable for Scilab. For this it is necessary to have a source version of Scilab as well as the Scilab directory

  SCIDIR/routines/default

  (the default directory for storing user-created primitives). Alternatively, the `ext1c` code can be dynamically loaded into Scilab as needed. The Scilab command to dynamically import compiled code is called `link` (page 145). (The function `addinter` is also used for interface loading and is discussed later in this chapter.)

- The second step is to write a wrapper that will encapsulate the C code. A wrapper function must do the following: convert Scilab arguments into C objects, check arguments for consistency, call the primitive function, and convert the C outputs into Scilab objects. Wrapper programs can be implemented in one of the following ways:

  - With a call to the Scilab function `call` (Section 5.2).
  - With the help of a hand-written C or Fortran wrapper. In Scilab such wrappers are called interface programs (Section 5.3).

> – With an automatically generated wrapper (Intersci interface generation, Section 5.5, page 135).

In the next sections wrapper writing is discussed, and all the examples are given using dynamic linking. The last section (page 147) shows how permanent static linking of new primitives can be implemented.

5.2 The Call Function

The Scilab function **call** is the easiest way of creating a wrapper for a new primitive (note that **call** was formerly named **fort**). Recalling the C program **ext1c**, this function can be interfaced to Scilab using the following commands:

```
unix('make ext1c.o')                          ← compile C code

link('ext1c.o','ext1c','C')          ← build and load ext1c into Scilab

a=[1,2,3];b=[4,5,6];n=3;

c=call('ext1c',n,1,'i',a,2,'d',b,3,'d',..          ← call ext1c
    'out',[1,3],4,'d')
```

The argument list of **call** is quite long but is easy to decode. The name of the function that we want to call is **'ext1c'** (**link** loaded the object code for **ext1c** and maintains an entry point table for **call**). The arguments **n, a,** and **b** are three Scilab objects to be converted to input arguments to **ext1c**. The Scilab syntax of the sequence **a,2,'d'** means that **a** is to be converted as the second argument of the C routine **ext1c** and that the C routine is expecting an array of double-precision numbers. For each input argument to the C program the syntax in the argument list of **call** is the triple consisting of the Scilab object name, the object's position in the C program's calling list, and the object's C program type.

The variable **'out'** is a string that separates input and output parameters. Here, the output consists of one parameter described by the three arguments **[1,3],4,'d'**. These arguments indicate that the output is the fourth argument in the calling list to **ext1c**, that it, is a double-precision array that must be stored in a Scilab 1×3 matrix.

The wrapper provided by the naked call to the function **call** is the minimum necessary to interface the C code with Scilab. To provide a more sophisticated interface, for example to provide consistency checks on the passed variables, it is useful to construct a Scilab function that contains the call to `call`:

```
function [c]=ext1c(a,b)
  n=size(a,'*');
  if typeof(a)<>'constant' then
     error('ext1c: first argument has wrong type');return
  end
  if typeof(b)<>'constant' then
     error('ext1c: second argument has wrong type');return
  end
  if n<>size(b,'*) then
     error('incompatible arguments');return
  end
  c=call('ext1c',n,1,'i',a,2,'d',b,3,'d',..
        'out',size(a),4,'d');
```

The above example illustrates how the call to the C program `ext1c` can be made through a Scilab programmed function, `c=ext1c(a,b)`, which contains the call to `call`. The advantage here is that the value of `n` was obtained automatically from the Scilab `size` command, and the consistency of the input arguments was checked before calling the C program.

A Fortran version of the code for `ext1c` can be implemented as follows:

```
      subroutine ext1f(n,a,b,c)
      double precision a(*),b(*),c(*)
      do 1 k=1,n
         c(k)=a(k)+b(k)
1     continue
      end
```

This program can be loaded and called with almost the same sequence of Scilab commands described above. The only difference is the call to the function `link`:

```
link('ext1f.o','ext1f');
```

in which the third argument `'c'` is not used.

Fortran arguments are always passed by reference, and this convention was adopted in the `call` function. This means that a C function can be linked to Scilab without any modification only if its arguments are pointers. The data types accepted by the function `call` are `'i'`, `'r'`, `'d'`, and `'c'`, which correspond to pointers for integer, float, double, and char data types. Thus, the Scilab arguments that can be transmitted to a Fortran or

C function or subroutine are only real matrices or strings. Moreover, the dimensions of the output variables must be known when the `call` function is used. To bypass these limitations, it is necessary to use hand-written wrappers or to use the program Intersci (see Section 5.5).

A complete description of the syntax for `call` can be found in the on-line help. We mention here that there is a shorter calling sequence to the function `call`. This shorter syntax takes the form `c=call('ext14cI',a,b)`, which is different from what was described above, since the keyword `'out'` is not used. This syntax can be used only when a C or Fortran wrapper (here named `ext14cI`) has been written, as is illustrated below:

```
#include "../../routines/stack-c.h"              ← header file

int ext14cI(fname)        ← the wrapper function for a short call to the C
     char* fname;                              function ext1c
{
   int m1,n1,l1,m2,n2,l2,m3,n3,l3,n,l4;
   int minlhs=1, minrhs=3, maxlhs=1, maxrhs=3;
   Nbvars = 0;                         ← requested initialization
   CheckRhs(minrhs,maxrhs) ;           ← three arguments should be
                                    transmitted: 'ext14cI', a, and b
   CheckLhs(minlhs,maxlhs) ;           ← one return value: c
   GetRhsVar( 1, "c", &m1, &n1, &l1);  ← first argument is Scilab
                     string 'ext1cI' since it is the first argument of call
   GetRhsVar( 2, "d", &m2, &n2, &l2);          ← matrix a
   GetRhsVar( 3, "d", &m3, &n3, &l3);          ← matrix b
   if (m3*n3 != m2*n2)
      {
        sciprint("%s: Incompatible dimensions\r\n",fname);
        Error(999); return(0);
      }
   CreateVar(4, "d", &m2, &n2, &l4);         ← storage for output
   n = m3*n3;
   ext1c(&n,stk(l2),stk(l3),stk(l4));          ← call ext1c
   LhsVar(1) = 4;                    ← we return the 4th variable
   PutLhsVar();
   return(0);                        ← we do not use the return value
}
```

The full description of such wrappers is given in the next section, since they are used for building interface programs.

5.3   Building Interface Programs

Another way of using C or Fortran code in Scilab is by building an interface program. An interface program consists of two parts. First, a set of C or Fortran wrappers must be created that encapsulate the C or Fortran functions to be used in Scilab. Second, a C or Fortran function that manages a table of entry names (called the interface function) must be created. Note that a single interface program can be used to interface up to 99 functions. Interface programs are dynamically linked to Scilab by the **addinter** command.

We return now to the **ext1c** program and the short syntax call to **call** discussed above. Replacing the command **c=call('ext14cI',a,b)** with **c=ext1c(a,b)** makes for a more transparent interface between Scilab and the C program it calls. The following code shows how to accomplish this. First, let us consider the following C wrapper code:

```
#include "../../routines/stack-c.h"          ← header file

int intext1c(fname)                          ← the wrapper function for ext1c
      char *fname;
{
  int m1,n1,l1,m2,n2,l2,m3,n3,l3,n;
  int minlhs=1, minrhs=2, maxlhs=1, maxrhs=2;
  Nbvars = 0;
  CheckRhs(minrhs,maxrhs) ;                   ← two arguments should be
                                                transmitted: a and b
  CheckLhs(minlhs,maxlhs) ;                   ← one return value: c
  GetRhsVar( 1, "d", &m1, &n1, &l1);          ← matrix a
  GetRhsVar( 2, "d", &m1, &n2, &l2);          ← matrix b
  if (m1*n1 != m2*n2)
    {
      sciprint("%s: Incompatible dimensions\r\n",fname);
      Error(999); return(0);
    }
  CreateVar(3, "d", &m1, &n1, &l3);           ← storage for output
  n = m3*n3;
  ext1c(&n,stk(l1),stk(l2),stk(l3));          ← call ext1c
  LhsVar(1) = 3;                              ← we return the 3rd variable
  PutLhsVar();
  return(0);                                  ← we do not use the return value
}

static TabF Tab[]={          ← table with all the functions that comprise
   {intext1c, "ext1c"}          the interface. Here, just one function
        };                    intext1c and its Scilab name ext1c
```

```
int C2F(intext)()                        ←    the interface main entry point
{
  Rhs = Max(0, Rhs);
  (*(Tab[Fin-1].f))(Tab[Fin-1].name);
  return 0;
}
```

If the previous code were contained in a file named `intext1c.c` and the `ext1c` code in a file named `ext1c.c`, the Scilab command

`addinter(['intext1c.o','ext1c.o'],'intext',['ext1c'])`

would load a shared library composed of `intext1c.o` (the wrappers and tables) and `ext1c.o` (the program to be interfaced). The call to `addinter` also adds the name `ext1c` to the Scilab list of primitives (the information that is stored is that `ext1c` is the first function of the interface controlled by the function `intext`). Thus, `addinter` and `link` are similar in that they both perform dynamic linking and update entry-point tables. The difference is that `link` updates the tables used by `call` and the functions that use functional arguments, whereas `addinter` updates the table of Scilab primitives. The previous code has created a new Scilab primitive called `ext1c`. Now the simple and transparent Scilab command `c=ext1c(a,b)` can be performed.

Note that the last part of the interface program must contain a table, `TabF`, which itself contains a pair `{intext1c, "ext1c"}`. This pair gives the name of the interface program and the name of the associated Scilab function. If several functions are to be interfaced, then a pair of names must be given for each function. Finally, the entry point `intext` is used by the dynamic link command `addinter`. For new interfaces this function can simply be copied as is onto the end of the interface program file. The name, however, should be appropriately changed.

Examples of interface programs are given in the directory `SCIDIR/examples/addinter-examples`.
The files `template.c` and `template.f` are skeletons of interface programs. The simplest way to learn how to build an interface program is to customize these skeleton files and to look at the examples provided in this directory. The following describes the functions used to build interfaces. The functions are Fortran subroutines when the interface is written in Fortran and are C macros (defined in `stack-c.h`) when the interface is coded in C. The main functions are as follows (we cite here the C version):

- `CheckRhs(minrhs, maxrhs)`. Function `CheckRhs` is used to check that the Scilab function is called with a number of arguments in the range `[minrhs,maxrhs]`.

- **CheckLhs(minlhs, maxlhs)**. Function **CheckLhs** is used to check that the expected return values are in the range **[minlhs,maxlhs]** (typically **minlhs=1**, since Scilab functions can be called with fewer left-hand-side arguments than specified by the Scilab function).

- **GetRhsVar(k,ct,&mk,&nk,&lk)**. Here **k** (integer) and **ct** (string) are inputs and **mk**, **nk**, and **lk** (integers) are outputs of **GetRhsVar**. This function defines the type (**ct**) of the **k**th input variable in the calling sequence to the Scilab function. The pair **mk,nk** gives the dimensions (number of rows and columns) of the **k**th input if it is a matrix. If it is, a chain **mk*nk** is its length. The input **lk** is the address of the data associated to the **k**th input in Scilab's internal stack. In fact, **lk** points to a copy of the **k**th input variable, and thus the data it points to can safely be changed within the interface program. The type of the **k**th input, **ct**, should be set to **'d'**, **'r'**, **'i'**, or **'c'**, which stand respectively for double, float (real), integer, or character. Note that here, type is not the Scilab type of the **k**th argument (which can be a real matrix or a Scilab string) but the type to which its entries must be converted for use inside the interface function. The interface should call the function **GetRhsVar** for each of the rhs variables of the Scilab function with **k=1, k=2, ... , k=Rhs**. Note that if the Scilab argument does not match the requested type, Scilab calls an error function and returns from the interface function.

- **CreateVar(k,ct,&mk,&nk,&lk)**. Here **k,ct,mk,nk** are inputs of **CreateVar**, and **lk** is an output of **CreateVar**. The parameters are as above. The **k**th input variable is stored in Scilab's internal stack at address **lk**. When **CreateVar** is called, **k** must be greater than **Rhs**, i.e., **k=Rhs+1, k=Rhs+2, ...** , and the values used for **k** must be contiguous and increasing. If the argument cannot be created due to insufficient memory, a Scilab error function is called and the interface function returns.

- **CreateVarFromPtr(k,ct,&mk,&nk,&lk)**. Here **k,ct,mk,nk,lk** are all inputs to **CreateVarFromPtr**, and **lk** is the address of an object created by a call to a C or Fortran function. **CreateVarFromPtr** is the function used when a C object was created inside the interfaced function and the Scilab object corresponding to this object is to be created using a pointer to it (see the function **intfce2c** in the file **SCIDIR/examples/addinter-examples/Testc.c**). The function **FreePtr** must be used to free the pointer.

- **sciprint(message)**. This function prints the string **message** in the Scilab main window.

- **Error(k)**. This function is used to print an error message and to clean the Scilab stack after a detected error. This function call must always be followed by a **return** statement.

- **NumOpt(), IsOpt(k,name)**. The call **NumOpt()** returns the number of optional arguments (arguments given with the syntax **var=val** at the end of the argument list), and **IsOpt** checks whether the **k**th argument in the calling list is an optional argument returning its name if true.

- **GetFuncPtr(...)**. This function is used to deal with a functional argument (see Section 5.4.2).

Note that the functions described in the next section, Section 5.4, can also be used.

Fortran functions have the same syntax as their C counterparts. Fortran functions return logical values such as **.false.** when an error is encountered. If an error is encountered, an exit from the interface must then be performed. The C functions performing error handling are C macros (as already mentioned). For example, **GetRhsVar** is defined in the file
SCIDIR/routines/stack-c.h
as

```
#define GetRhsVar(n,ct,mx,nx,lx) \
 if (! C2F(getrhsvar) ((c_local=n,&c_local), \
    ct,mx,nx,lx,1L)) return 0;
```

Complete examples that illustrate how to use the above described functions are given in the Scilab directory
SCIDIR/examples/addinter-examples.
For each example, a Fortran and a C version are given (see Table 5.1).

Once the variables have been processed by **GetRhsVar** or created by **CreateVar**, the returned address, **lk**, can be used to read or change their values. Access to the data depends on the type of the variable. When a Fortran interface character is written, integer, real, and double type variables are stored in the **cstk**, **istk**, **sstk**, and **stk** Scilab internal arrays, respectively, and first values are at position **lk** in the corresponding array. For writing a C interface, macros are provided in the header file
SCIDIR/routines/stack-c.h.
These macros access variable data; for example, **sbtk(lk)** will give a pointer to an array of double-precision floating-point numbers.

```
GetRhsVar(1,"d",&m,&n,&lk);                     ← first argument is a Scilab matrix,
                                                stk(lk) points to the associated data
*stk(lk) = 10.0;                                ←  changing the first value
x = *stk(lk+m*n-1);                             ←  reading the last value
```

After the call to the C or Fortran implemented function, the variables
to be returned to Scilab must be delivered by the wrapper. We call these *lhs*
left-hand-side variables. Since the call from Scilab does not require all the
lhs variable to be returned, the interface should define the subset of variables that will be selected as the lhs variables. This is accomplished using
the global variable **LhsVar**. As an example:

```
  LhsVar(1) = 5;   LhsVar(2) = 3;
  LhsVar(3) = 1;   LhsVar(4) = 2;
  PutLhsVar();
```

means that the Scilab function can return up to 4 objects, and the returned
objects are selected in the variables numbered [5,3,1,2] (if one argument
is requested, [x1]=f(...), the fifth argument is returned, if two arguments
are requested, [x1,x2]=f(...), the fifth and third argument are returned,
...). It is worth pointing out that an increasing sequence $(\mathtt{Lhsvar(i)})_{i=1,n}$
will produce faster code.

Note that returned objects can be placed into input arguments. As an
example consider the following variation in the **intext1c** function:

```
int intext1c(fname)
     char *fname;
{
  int m1,n1,l1,m2,n2,l2;
  int minlhs=1, minrhs=1, maxlhs=1, maxrhs=1;
  Nbvars = 0;
  CheckRhs(minrhs,maxrhs) ;
  CheckLhs(minlhs,maxlhs) ;
  GetRhsVar(1, "d", &m1, &n1, &l1);             ←  matrix a
  n = m3*n3;
  ext1c(&n,stk(l1),stk(l1),stk(l1));            ←  call ext1c
  LhsVar(1) = 1;            ←  the first argument is used for input and output
  PutLhsVar();
  return(0);
}
....
```

Here **ext1c** has a single input value, and a single output value that is placed
in the same variable that syntactically is the first argument passed to **ext1c**.
The Scilab command **c=ext1c(a)** will return **c**, which contains **2*a**, but the

Scilab variable **a** will remain unchanged. This illustrates an important feature of the Scilab interpreter, that is, the arguments are transmitted by value. The command **GetRhsVar(1,"d",&m1,&n1,&l1)** inside the interface program returns the address of a copy of the **a** object, so we can safely use this pointer to change the data it points to and use this data as a return value.

file	highlighted feature
ext1*	Scilab matrix arguments (converted to integer, float, or double arrays)
ext2*	**NumOpt,IsOpt,numopt,isopt** optional argument passing with the syntax **f(...,var=val,...)**
ext3*	**CreateVarFromPtr,createvarfromptr FreePtr,freeptr** copying and freeing dynamically allocated objects created by the interfaced subroutines
ext4*	Scilab matrix arguments and an interfaced subroutine that is already inside Scilab (**dgemm**)
ext5*	**GetFuncPtr**: how to deal with argument functions
ext6*	string arguments and **GetMatrixptr, cmatptr** to read Scilab variables

Table 5.1: **addinter** *examples*

5.4 Accessing "Global" Variables Within a Wrapper

Wrapper functions described in the previous sections can also access internal Scilab variables selected by their names. This is accomplished with a set of useful functions that enable stack handling.

5.4.1 Stack Handling Functions

The following functions are used for stack handling:

- **ReadMatrix(aname,&m,&n,w)** (**creadmat** in Fortran). This function reads a matrix in Scilab's internal stack. The variable **aname** is a character string, which must be the name of the Scilab matrix (otherwise **ReadMatrix** calls the Scilab **error** function and returns). Outputs are integer pointers **m, n**, which gives the size of matrix, and the array **w**, which is filled with the entries of the matrix ordered *columnwise* (**w** must be large enough to contain the **m*n** entries).

- **ReadString(aname,&n,w)** (**creadchain** in Fortran). This function reads a string in Scilab's internal stack. The parameter **n** is used to input the maximum number of characters that can be stored in **w**, and on output it gives the number of characters stored in **w**, which is an array of characters.

- **WriteMatrix(aname,&m,&n,w)** (**cwritemat** in Fortran). This function creates a Scilab object with name **aname**, which is a matrix of size [**m,n**] filled (*columnwise*) with the C array of double-precision **w**.

- **WriteString(aname,&m,w)** (**cwritechain** in Fortran). This function creates a Scilab string with name **aname**, which is filled with **m** characters from **w**.

- **GetMatrixptr(aname,&m,&n,&l)** (**cmatptr** in Fortran). This function returns the dimensions **m**, **n** and the address **l** of the Scilab variable **aname**, which must be a real matrix. With this function it is possible to directly access matrix data in Scilab (read or write). It is imperative that when using this function care be taken not to write outside the boundaries of the Scilab variable.

Fortran functions have the same syntax as their C counterparts. Fortran functions return logical values such as **.false.** when an error is encountered. If an error is encountered, an exit from the interface must then be performed. For example, a typical use of **creadmat** should be

```
if(.not.creadmat(aname,m,n,a)) return
```

The C counterpart is **ReadMatrix(aname,&m,&n,a)**, where **ReadMatrix** is defined by

```
#define ReadMatrix(ct,mx,nx,w) \
 if (! C2F(creadmat)(ct,mx,nx,w,strlen(ct))) \
        return 0;
```

All the C "functions" are in fact C macros, which are defined in the header file **stack-c.h**.

Complete examples that show how to use the previously mentioned functions are given in the directory **SCIDIR/examples/link-examples**. For each example, a Fortran and a C version are given (see Table 5.2). In these examples the previous functions are used in conjunction with the **call** command, but in fact these functions can be used in all wrappers.

file	highlighted feature
`ext1c.c ext1f.f` `ext2c.c ext2f.f`	Simple example, double and integer arguments
`ext3c.c ext3f.f`	String argument
`ext4c.c ext4f.f`	`ReadSring, creadchain`: Reading a string in the Scilab stack
`ext5c.c ext5f.f` `ext6c.c ext6f.f`	`ReadMatrix, creadmat`: Reading a matrix in the Scilab stack
`ext7c.c ext7f.f`	`WriteMatrix, WriteString, cwritemat, cwritechain`: Writing a matrix or a string in the Scilab stack
`ext8c.c ext8f.f`	`GetMatrixPtr, cmatptr` : Getting a pointer to a Scilab matrix from which data are read and changed
`ext13c.c ext13f.f`	`call`: Two linked functions that share common data
`ext14c.c ext14f.f`	`call`: Short argument list form and C or Fortran wrapper

Table 5.2: `call` *examples*

The previous functions can be very useful in programming functional arguments (i.e., functions to be passed as arguments to other Scilab primitives) for which the calling syntax is explicitly specified by a Scilab primitive. For example, these functions can be used to pass additional arguments to the function. Illustrative examples for the **ode** function are provided in the directory **SCIDIR/examples/link-examples** in the files **ext9*** to **ex12***. The example **ext12c.c** is illustrated here:

```
#include "../../routines/stack-c.h"

int ext12c(neq, t, y, ydot)           ← imposed argument list for ode
                                       ← functional argument
     int *neq;
     double *t, *y, *ydot;

  static int m, n, lp;
  GetMatrixptr("param", &m, &n, &lp);     ← using GetMatrixptr
                           to pass the extra argument param to ext12c
  ydot[0] = - (*stk(lp)) * y[0] + (*stk(lp+1)) * y[1] * y[2];
  ydot[2] = (*stk(lp + 2)) * y[1] * y[1];
```

```
ydot[1] = -ydot[0] - ydot[2];
return 0;
```

5.4.2   Functional Arguments

The function **GetFuncPtr** is used to handle functional arguments in the
argument list of a Scilab primitive we want to add. Functional arguments
can be hard-coded functions (in C or Fortran) and called in Scilab by name
or can be Scilab functions. For example, we want to add the Scilab primitive **z=FuncEx(x,y,f)**. This function has three arguments: two vector arguments **x** and **y**, and one functional argument **f**. The function **FuncEx** returns the matrix **z** such that **z(i,j)=f(x(i),y(j))**. Here **f** is what we call
a functional argument. We expect the following different calls to **FunEx** to
work:

deff('[z]=f(x,y)','z=x+y');   ← a Scilab function

res=FuncEx(1:3,4:6,f);   ← first possibility: Scilab function as argument

res1=FuncEx(1:3,4:6,'fp1');   ← second possibility: string **'fp1'**
represents a predefined hard-coded function

link('/tmp/ex5cm.o','fp3','C')

res1=FuncEx(1:3,4:6,'fp3');   ← third possibility: string **'fp3'**
is a dynamically linked function

We will show below how to achieve the preceding task, i.e., implementing the new primitive **FunEx** with a functional argument.

First, we need to write the following C function:

```
#include "ex5.h"

int CFuncEx(x,nx,y,ny,z,f)
    double *x,*y,*z;
    int nx,ny;
    funcex f;
{
  double res;
  int i,j;
  for (i = 0 ; i < nx ; i++)
    for (j = 0 ; j < ny ; j++)
      (*f)(x[i],y[j],&z[i+nx*j]);
```

```
}

int fp1(x,y,z) double x,y,*z; { *z= x+y; }
int fp2(x,y,z) double x,y,*z; { *z= x*x+y*y; }
```

The above presents a C program that computes $z(i,j) = f(x(i),y(j))$ with **f** a pointer to a function and two potential predefined functions defined by **fp1** and **fp2**. Our aim is to write a wrapper for the **CFuncEx** function in order to create a new primitive, **FuncEx**, in Scilab. Furthermore, the objective is to do this in such a way so that the new primitive can accept a functional argument **f**. The interface program follows:

```
#include <string.h>

#include "../../routines/stack-c.h"          ← headers for interfaces
#include "../../routines/default/FTables0.h"
#include "../../routines/sun/link.h"

#include "ex5.h"                              ← headers to define funcex

FTAB FTab_FuncEx[] ={          ← table of predefined hard-coded functions
  {"fp1", (voidf) fp1},
  {"fp2", (voidf) fp2},
  {(char *) 0, (voidf) 0}};

#include <setjmp.h>
static  jmp_buf FuncExenv;                              ← error recovery

static int sci_f, lhs_f, rhs_f;            ← functional argument data

static int sciFuncEx (ARGS_FuncEx);               ← shadow function;
               this function is called when third argument is a Scilab function
static funcex ArgFuncEx;                    ← argument for CFuncEx

int cintFuncEx(fname)
    char *fname;
{
  int returned_from_longjump ;
  int m_x,n_x,l_x,m_y,n_y,l_y,m_res,n_res,l_res;
  static int minlhs=1, minrhs=3, maxlhs=1, maxrhs=3;
  Nbvars = 0;
  CheckRhs(minrhs,maxrhs) ;
  CheckLhs(minlhs,maxlhs) ;
  GetRhsVar(1, "d", &m_x, &n_x, &l_x);          ← first argument
  GetRhsVar(2, "d", &m_y, &n_y, &l_y);          ← second argument
```

```
ArgFuncEx  = (funcex) GetFuncPtr("FuncEx",3,FTab_FuncEx,
                     (voidf) sciFuncEx,&sci_f,&lhs_f,&rhs_f);
if (ArgFuncEx == (funcex) 0) return 0;           ← third argument,
        get a pointer to the function that will be used in CFunEx (see below)
m_res = m_x*n_x;   n_res = m_y*n_y;
CreateVar(4, "d", &m_res, &n_res, &l_res); ←  for storing result

if ((returned_from_longjump = setjmp(FuncExenv)) != 0)
   {
      Error(999);return 0;        ← if an error is detected while executing
                                    ArgFunEx, we quit the interface
   }

CFuncEx(stk(l_x),m_res,stk(l_y),n_res,stk(l_res),ArgFuncEx);

LhsVar(1) = 4;   PutLhsVar();           ←  return the fourth argument
return 0;
}

static TabF Tab[]={{cintFuncEx, "FuncEx"}};

int C2F(cfunc)()
{
  Rhs = Max(0, Rhs);
  (*(Tab[Fin-1].f))(Tab[Fin-1].name);
  return 0;
}
```

Here the return value of **GetFuncPtr** is stored in the variable **ArgFunEx**. The returned value depends on the Scilab type of the third argument of the function **FuncEx**. If this third argument is a string, then a function whose name matches the string is sought in the predefined table **FTab_FuncEx** or in the dynamic table maintained by **link**. If such a function is found, its value is stored in **ArgFunEx**. If the third argument is a Scilab function, we cannot directly send this macro to **CFunEx**, and again a wrapper is needed. In this last case, **ArgFunEx** will be bound to **sciFuncEx**:

```
int sciFuncEx(x,y,res)
      double x,y,*res;
{
  int scilab_i,scilab_j, un=1;
  CreateVar(5,"d",&un,&un,&scilab_i);  ←  5 is the first free position
  stk(scilab_i)[0] = x;
  CreateVar(6,"d",&un,&un,&scilab_j);
  stk(scilab_j)[0] = y;
```

```
PExecSciFunction(5,              ← calling the Scilab interpreter to execute
            the function described by sci_f and with first argument at position 5.
                                 If an error occurs, jump to FuncExenv
    &sci_f,&lhs_f,&rhs_f,"ArgFex",FuncExenv);
  *res = *stk(scilab_i);         ←  the result is in the first input location
  return 0;
}
```

Here, **sciFuncEx** is a C wrapper that converts its arguments to Scilab objects and stores them at a free positions in the Scilab stack, calls the Scilab interpreter (**PExecSciFunction**) for execution of the Scilab function described by **sci_f**, **sci_lhs**, and **sci_rhs**, and converts back the Scilab result to a C object (here, the result is a double).

5.5 Intersci

Intersci is a program that was designed to automatically generate an interface (more precisely, a Fortran interface) from a simple description file (file denoted by a **.desc** suffix). Fortran interfaces generated by Intersci use a different set of functions from those described previously, but the basic philosophy is the same. The descriptor file describes the argument list of the wrapped C or Fortran function and describes the mapping between the wrapped function's arguments and the new Scilab primitive's arguments.

5.5.1 A First Intersci Example

We begin with our C example **ext1c** stored in the file
SCIDIR/examples/intersci-examples:

```
#include "../../routines/machine.h"     ←  header file for C2F macro

int C2F(ext1c)(n, a, b, c)
    int *n; double *a, *b, *c;
{
  int k;
  for (k = 0; k < *n; ++k)  c[k] = a[k] + b[k];
  return(0);
}
```

Note that not every C function can be interfaced directly with Intersci. Indeed, all the arguments of the interfaced C function must be pointers to fit the Fortran convention. Moreover, as in **ext1c** above, we need to encapsulate the function name with a **C2F** C macro. This comes from the fact that

Intersci generates Fortran interfaces and **C2F** is used to change the names
of functions (by adding or not, depending on the operating system, an un-
derscore at the end and/or at the beginning of function names) for Fortran
and C names compatibility.

   To build a Fortran interface for the above program we must write a
descriptor file **ext1fi.desc** that contains all the information necessary to
build a wrapper for the **ext1c** function. This will create a new Scilab primitive for which we assign the same name as the name of the wrapped function
ext1c. The prefix **ext1fi** of the description file **ext1fi.desc** is used as the
name of the interface program.

```
ext1c    a b              ← Scilab function name and its input arguments: a,b
a        vector  m                            ← a is a vector of size m
b        vector  m                      ← b must have the same size as a
c        vector  m            ← the result and description of its Scilab type

ext1c    m a b c  ← description of the arguments of the wrapped function
m        integer          ← types of C or Fortran arguments: always pointers
a        double        ← double means array of double or pointer to double
b        double
c        double

out      sequence        c                          ← ext1c output
************************
```

Executing the command **intersci ex1fi.desc** in the directory containing
the above file (note that the program **intersci** is located in the directory
SCIDIR/bin) creates two files: **ex1fi.f**, the interface file, and **ex1fi.sce**, a
Scilab script. The interface file **ex1fi.f** contains an include statement for a
Scilab header that must be changed to fit the local Scilab installation. When
ex1fi.f and **ex1c.c** are compiled, they can be loaded and used as follows:

```
ifile='ex1fi.o'                ← interface file generated by intersci

ufiles=['ex1c.o'];                   ← object files for ext1c definition

exec('ex1fi.sce');                  ← addinter call, using ex1fi.sce
                                               generated by intersci
c=ext1c([1,2,3],[4,5,6]);       ← using the new Scilab primitive
```

5.5.2 Intersci Descriptor File Syntax

The file *<interface name>*.**desc** is a sequence of descriptions of variable
pairs for the Scilab function and the corresponding C or Fortran subroutine

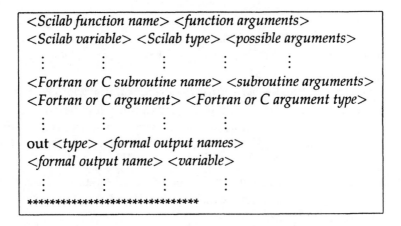

Table 5.3: *Description of a pair of a Scilab function and a programming language subroutine (various fields are separated by blanks or tabs)*

(see Table 5.3 for a summary of Intersci syntax). As noted before, an interface can contain up to 99 function descriptors.

Each description is made of three parts:

- a description of the Scilab function and its arguments;

- a description of the C or Fortran subroutine and its arguments;

- a description of the output of the Scilab function.

Description of the Scilab Function

The first line of the description is composed of the name of the Scilab function followed by the input arguments' names. Optional arguments can be added at the end of the list, and they are defined as follows:

- [c val]. Here c is an optional argument with default value **val** which can be a scalar, e.g., [c 10], an array, e.g., [c (4)/1,2,3,4/], or a character string, e.g., [c pipo] .

- {b xx}. Here b is an optional argument. If not found, the variable named xx is taken from the current Scilab environment.

For example:

```
f         a b [c 1] [d (3)/1,2,3/] {e Eflag}
```

means that the Scilab function **f** can be called with a minimum of 2 arguments and a maximum of 5; default values are specified for optional arguments. For example, the fourth argument, if not given, is set to a **3**-vector containing the values **[1,2,3]**.

The next lines of the descriptor file describe Scilab variables. The inputs and outputs of the Scilab function are described here. Furthermore, it is also possible to describe variables that are only used internally to the function, for example, those needed as working variables. These are identified by the **work** type keyword.

The description of a Scilab variable begins by its name followed by its type and finally by additional information depending on the variable type. Types of Scilab variables are:

any any type: only used for an input argument of a Scilab function.

bmatrix Boolean matrix: must be followed by its two dimensions. The Fortran types associated to a Boolean matrix can be **int** or an external data type.

column column vector: must be followed by its dimension.

imatrix complex matrix: must be followed by its two dimensions and a flag variable, which is used to test for the presence or absence of the imaginary part.

sparse sparse matrix: must be followed by its two dimensions, the name of the integer vector used for column compression coding, and a flag variable, which is used to test for the presence or absence of the imaginary part.

list list: must be followed by the name of the list, <*list name*>. This name must correspond to a file <*list name*>**.list** that describes the structure of the list (see Section 5.5.2).

tlist typed list: must be followed by the name of the list, <*list name*>. This name must correspond to a file <*list name*>**.list** that describes the structure of the list (see Section 5.5.2).

matrix matrix: must be followed by its two dimensions.

polynom polynomial: must be followed by its order and the name of the symbolic variable.

row row vector: must be followed by its dimension.

scalar scalar.

string character string: must be followed by its length.

stringmat string matrix: must be followed by its dimensions.

vector row or column vector: must be followed by its dimension.

work working array: must be followed by its dimension. It must not correspond to an input argument or to the output of the Scilab function.

A single blank line ends the description.

Dimensions of Nonscalar Variables   In defining nonscalar Scilab variables (vectors, matrices, polynomials, and character strings) dimensions must be given. There are a few ways to do this:

- The dimension can be given as an integer.

- The dimension can be the dimension of an input argument of a Scilab function. This dimension is then denoted by a formal name.

- The dimension can be defined as the output of the wrapper function. This means that the memory for the corresponding variable is allocated by the wrapper function. The corresponding Fortran or C variable must have an external type (see Section 5.5.2 below) or have a **Cstringv** type.

As an example:

```
ext12c   a
a        vector m
b        work m1      ← m1 is not valid; it cannot be bound to a numerical value
c        matrix m1 m1                              ← m1 is not valid
d        work m ← m is valid: bound to argument a, size when calling ext12c
e        matrix m2 n2     ← m2,n2 are valid, since e has an external type
ext12c   a b c d e m2 n2
e        cintf                           ← e is an external variable
....
```

The program Intersci is not able to treat the case where the dimension is specified as an algebraic expression of other dimensions. For this a Scilab variable corresponding to the algebraic expression must be defined.

Description of a Programming Language Subroutine

The first line of the description is composed of the name, of the Fortran or C subroutine followed by its arguments. The argument list is a suite of names, which can be Scilab variable names or formal dimension names. Three special names can be added to the argument list: **err**, **rhs**, and **lhs**. The **err** argument is used by the wrapper function to return an error status. If **err** is bound to a nonnull integer, the wrapper will enter a Scilab error function. The arguments **rhs** and **lhs** can be used inside the wrapper function to give the number of rhs and lhs arguments.

What follows describe Fortran or C variables: the arguments of the Fortran or C subroutine. The description of a Fortran or C variable is made up of its name and its type. Most variables correspond to Scilab variables (except for dimensions, see Section 5.5.2) and must have the same name as the corresponding Scilab variable. Types of Fortran or C variables are:

char character array.

double double-precision variable.

int integer variable.

real real variable.

Cstringv C matrix of strings that are dynamically allocated in conjunction with the **stringmat** Scilab type (see example: **SCIDIR/examples/intersci-examples/ex6***).

Other types also exist, called "external" types (see Section 5.5.2). Note also that for complex Scilab variables (**imatrix**) the corresponding C or Fortran variable is extended in the calling sequence into a sequence of three arguments: an integer flag, a pointer to the real part, and a pointer to the imaginary part. The C or Fortran type is used in that case for conversion of the real and imaginary parts. The same philosophy is used for sparse matrices (see example **SCIDIR/examples/intersci-examples/ex14** for details). A single blank line ends the description.

Description of the output of the Scilab function   The first line of this description must begin with the keyword **out** followed by the type of Scilab output. Types of output may be:

empty the Scilab function returns nothing.

list, tlist a Scilab list: must be followed by the names of Scilab variables that form the list.

sequence a Scilab sequence: must be followed by the names of Scilab variables elements from the sequence. This is the most common case.

This first line must be followed by other lines corresponding to output type conversion. This is the case where an output variable is also an input variable but with different Scilab type, for instance, when an input column vector becomes an output row vector. The line that describes this conversion begins with the name of the Scilab output variable followed by the name of the corresponding Scilab input variable.

A line beginning with an asterisk "*" ends the description of the Scilab function and programming language subroutine pair. This line is required even if it is the end of the file. Also, the file must end with a carriage return. Furthermore, additional blank lines must not be added at the end of the file.

Programming Language Variables with External Type

External types are used when the Fortran or C variable is a dynamically allocated object. The dimensions of the object must be an output of the subroutine. The dynamically allocated objects are copied to a Scilab variable and are eventually freed. In the hand-written interface section the same kind of arguments were treated with the **CreateVarFromPtr** function.

Existing external types:

cchar character string allocated by a C function to be copied into the corresponding Scilab variable.

ccharf the same as **cchar**, but the C character string is freed after the copy.

cdouble C double array allocated by a C function to be copied into the corresponding Scilab variable.

cdoublef the same as **cdouble**, but the C double array is freed after the copy.

cint C integer array allocated by a C function to be copied into the corresponding Scilab variable.

cintf the same as **cint**, but the C integer array is freed after the copy.

Consider the following simple example:

```
calc8
a       matrix m n
```

```
<comment on the variable element of the list>
<name of the variable element of list> <type> <possible arguments>
*******************************
```

Table 5.4: *Description of a variable element of a list*

```
ccalc8   a m n
a        cintf
m        integer
n        integer


out      a
**********************
```

Here the first argument of the wrapped function, **ccalc8**, is of type **cintf**, which means that **ccalc8** returns in its first argument (of type **int ****), a handle to an integer array dynamically allocated inside **ccalc8**. The wrapper will get this handle and the size of the array in **m** and **n**. It must create a Scilab matrix of size **m**×**n**, fill it with the integer array, and free the allocated memory associated to the array.

Looking at the generated wrapper, we can see that the fill and free operations are performed by a call to a C function. This C function has three arguments: the dimension of the variable, an input pointer, and an output pointer. Its name is the name of the external type, i.e., **cintf**. This means that external types can be defined by the user by creating tailored conversion and free functions. Intersci makes the call to the function. An example of such is provided in **SCIDIR/examples/intersci-examples/ex12***.

Using Lists as Input Scilab Variables

An input argument of the Scilab function can be a Scilab list. If *<list name>* is the name of this variable, a file called *<list name>*.**list** must describe the structure of the list. This file allows the association of a Scilab variable to each element of the list by defining its name and its Scilab type. The variable descriptions in the file are made according to the order shown in Table 5.4. Then an element of the list in the file *<interface name>*.**desc** is referred to as its name followed by the name of the corresponding list in parentheses. For instance, **la1(g)** denotes the variable named **la1**, which is an element of the list named **g**.

A Set of Examples

A set of examples illustrating the use of intersci is provided in the Scilab directory **SCIDIR/examples/intersci-examples** (see Table 5.5). For each example, a scilab script **ex*.sce** is given, and running this script (for example: **exec('ex1.sce')**) will perform all the necessary steps for running a complete demo. For example, the contents of **ex1.sce** are as follows:

```
host('../../bin/intersci ex1fi');     ←  call intersci for example 1

host('./sedprov ex1fi');        ← change SCIDIR in the Fortran generated
                                                                interface
ifile='/tmp/ex1fi.o';                      ←  object file for the interface

ufiles=['/tmp/ex1c.o'];             ←  object file for the wrapper function

host('make '+ifile+' '+ufiles);                      ←  compile sources

exec('ex1fi.sce');                   ←  call addinter with the Intersci
                                                            generated script
a=[1,2,3];b=[4,5,6];

c=ext1c(a,b);                              ←  call the new primitive

if norm(c-(a+b)) > %eps then pause,end          ←  test the result
```

Interfacing C Functions

C functions must be handled as procedures, i.e., their type must be **void** or the returned value must not be used. The arguments of a C function must be treated as Fortran arguments, i.e., they must be pointers. Finally, the name of a C function must be recognized by the Fortran interface. For that, the include file **machine.h** located in the directory **SCIDIR/routines** should be included in C functions, and the macro **C2F** should be used.

Writing Scilab Compatible Code

Messages and Error Messages   To write messages in the Scilab main window, the user must call the Fortran routine **out** or the C procedure **cout** with the character string of the desired message as input argument. To return an error flag of an interfaced routine the user must call the Fortran routine **erro** or the C procedure **cerro** with the character string of the desired message as input argument. This call will print the message in the

file	highlighted feature
ex1*	Scilab vectors
ex2*	Scilab vectors and type conversion to real
ex3*	Vectors and a Scilab internal function **daxpy**
ex4*	An argument of type list
ex5*	Strings and matrices
ex6*	Two examples with Scilab matrices of strings **Cstringv**
ex7*	Scilab string
ex8*	Dynamically allocated array of int inside the wrapped function (**cintf**)
ex9*	Dealing with optional arguments
ex10*	Simple example with lists
ex11*	Example with complex matrices (**imatrix**)
ex12*	Example with a user-defined external type
ex13*	Size conversion on output
ex14*	Examples with sparse matrices
ex15*	Examples with real matrices
ex16*	Examples with Boolean matrices

Table 5.5: *Intersci examples*

Scilab main window and subsequently provoke the exit of the associated Scilab function.

Input and Output

To open files in Fortran, it is recommended that the Scilab routine **clunit** be used. This routine is described below. If the interfaced routine directly uses the Fortran **open** instruction, logical units must be greater than 40 to avoid conficts with Scilab handled files:

```
call clunit(lunit, file, mode)
```

where:

- **file** is the name of the file.

- **mode** is a two-vector of integers defining the opening mode, where **mode(2)** defines the record length for a direct access file if positive, and **mode(1)** is an integer formed with three digits represented by **f**, **a**, and **s**, and where:

- **f** specifies whether the file is formatted (**0**) or not (**1**);
- **a** specifies whether the file has sequential (**0**) or direct access (**1**);
- **s** specifies whether the file status must be new (**0**), old (**1**), scratch (**2**), or unknown (**3**).

Files opened by a call to **clunit** must be closed by

```
call clunit( -lunit, file, mode)
```

In this case the **file** and **mode** arguments are not referenced.

5.6 Dynamic Linking

We describe some utilities concerning dynamic linking in this section. Dynamic linking is available on all platforms for which Scilab has been ported. However, some functionalities (for example, unlinking) are not always supported. (The return value of **link('show')** can be used to test the link status of your machine).

As previously noted, the command **link('path/pgm.o','pgm',flag)** calls the machine's linker to transform the first argument (a list of object files) into a shared library and then loads this shared library into a running Scilab session (note that on Windows 95 or NT the first argument must be a DLL name, not a list of object files). Then, names provided by the second argument of the **link** command are added to Scilab's internal function lookup table. More than one name can be specified by using a vector of character strings such as **['pgm1','pgm2']**. Each name in the vector denotes an entry point defined in the loaded shared library. Finally, **flag** must be set to **'C'** for a C-coded program and to **'F'** for a Fortran subroutine (**'F'** is the default flag and can be omitted).

If the link operation is successful, Scilab returns an integer, **n**, identifying the shared library. Subsequent calls to **link** with this integer can be performed to add new names to the Scilab table from the associated shared library. To remove all the entry points associated with a shared library, **ulink(n)** can be used.

The command **c_link('pgm')** returns the value of true if **pgm** is currently linked to Scilab and false if not. When an entry point is added with the link command, this entry point can be used by the **call** function or by any of the Scilab primitives that accept functional arguments in their argument list. The following is an example with the Fortran BLAS **daxpy** subroutine used in Scilab:

```
-->n=link(SCI+'/routines/calelm/daxpy.o','daxpy')
linking files
/usr/local/lib/scilab-2.3/routines/calelm/daxpy.o
to create a shared executable.
Linking daxpy (in fact daxpy_)
Link done
 n   =                    ← 0 is the Id of the first shared library loaded with link
    0.
```

```
-->c_link('daxpy')  ← test if daxpy is added in the dynamic entry points list
 ans   =
  T
```

```
-->link('show')                                     ← shows link status
Number of entry points 1
Shared libs: [0 ]: 1 libs
Entry point daxpy in shared lib 0
 ans   =
    1.
```

```
-->ulink(n)            ← unlink all entry points from shared library 0
```

```
-->c_link('daxpy')
 ans   =
  F
```

The **addinter** command also uses the link facility. Here the added entry point is the name of the interface (i.e., the radix name of the descriptor file when using Intersci and the interface name for hand-written interfaces). As an example:

```
-->addinter(['/tmp/ex1cI.o','/tmp/ex1c.o'],..
            'foobar','foubare');
linking files /tmp/ex1cI.o /tmp/ex1c.o
to create a shared executable
shared archive loaded
Linking foobar (in fact foobar_)
Interface 0 foobar
```

```
-->link('show')
Number of entry points 1
Shared libs: [0 ]: 1 libs
Entry point foobar in shared lib 0              ← the interface name
 ans   =                                  foobar is a new entry point
    1.
```

Note that this entry point may not be directly used. The third argument of

addinter gives the list of new primitives that are accessible for use.

5.7 Static Linking

5.7.1 Static Linking of an Interface

It is possible to add a set of new built-in functions to Scilab by creating a permanent link through a static interface program. A source version is required, and the files **fundef** and **callinterf.h**, located in the directory **SCIDIR/routines/default**, must be updated. The file **callinterf.h** contains the table of interface names (i.e., entry points), and the file **fundef** contains the names of the built-in functions contained in each interface as well as their positions in the interface table.

For hand-written interfaces, the new primitive names and the interface name are at the end of the interface file. For example, the tail of file **SCIDIR/examples/addinter-examples/ex3cI.c** is as follows:

```
....                                      ← end of ex3cI.c
static TabF Tab[]={
  {intfce1c, "funcc1"},                   ← primitive names funcc1
  {intfce2c, "funcc2"},                             ← funcc2
  {intfce3c, "funcc3"},
  {intfce4c, "funcc4"},
};

int C2F(testcentry)()                          ← interface name
{
  Rhs = Max(0, Rhs);
  (*(Tab[Fin-1].f))(Tab[Fin-1].name);          ← jump into the
  return 0;                                    selected wrapper
}
```

The interface name **testcentry** must be added to the file **callinterf.h** at the end of the interface list. The position in the interface array of this new entry will be the interface number. Using this interface number, the new primitives (which are listed in the **Tab** array above) must be added at the end of the file **fundef**:

```
....                                      ← end of fundef file
#define IN_testcentry 36             ← new interface: number 36
{ "funcc1",   IN_testcentry,1, 3},         ← {primitive-names,
                   interface-number,position-of-primitive, 3}
```

```
{ "funcc2",    IN_testcentry,2, 3},
{ "funcc3",    IN_testcentry,3, 3},
{ "funcc4",    IN_testcentry,4, 3}
```

When Intersci is used to generate an interface (by the shell command **intersci interface-name interface-number**), the **fundef** associated to the interface is also generated. Of course, when **fundef** has been updated with new entry names, the associated files must be added to Scilab (by changing the appropriate **Makefile**'s), and Scilab must be relinked (by typing **make all** or **make bin/scilex** in the Scilab root directory).

Finally, two unused empty interface routines called by default, **matusr.f** and **matus2.f** are predefined and may be replaced by the interface program. Their interface numbers are respectively **14** and **24**. They can be used as default interface programs. For instance, you can create your own **matusr.f** program interface and use it instead of the one in Scilab.

5.7.2   Functional Argument: Static Linking

The **SCIDIR/routines/default** directory contains a set of C and Fortran routines that can be customized by the user. Scilab primitives that accept functional arguments (for example **ode** function) and the **call** primitive can have a set of predefined functional arguments that are coded as C functions or Fortran subroutines.

All these predefined functional arguments are stored in the directory **SCIDIR/routines/default**. The **Ex-*.f** files are provided for all the Scilab primitives having functional arguments. In each of these files, subroutines are already provided. For example, in the file **Ex-ode.f** the subroutine **fex**, which describes a dynamic system to be solved by **ode**, can be given as argument to the **ode** function as **'fex'**. You can replace in this file the given **fex** subroutine by your own **fex** subroutine.

You can also add your own subroutines in the **Ex-*.f** files. After that, you must add the new corresponding entry point in the line corresponding to the Scilab primitive (for instance, the line containing **fydot_list** for function **ode**) in the file **Flist** in the same **SCIDIR/routines/default** directory.

Of course, if you need to insert new files in the default directory, adequate changes in the corresponding **Makefile.*** and in the main Scilab **Makefile.*** must be performed.

Part II

Tools

Chapter 6

Systems and Control Toolbox

This chapter describes the main functions available in Scilab for system analysis and control. The basic functions that require speed and reliability are made of built-in primitives coded in C or Fortran, and the specialized functions are coded in the Scilab language.

Many built-in control primitive functions are based on SLICOT, a general purpose control library realized by WGS [58]. SLICOT is a very powerful mathematical library for control-theoretical computations. In particular, it provides tools for most of the basic system analysis and synthesis tasks. The main emphasis in SLICOT is on numerical reliability and efficiency of implemented algorithms.

6.1 Linear Systems

A dynamical system can be seen as an input-output mapping, the input being an $\mathbb{R}^m$-valued function of time $u(t)$ and the output being an $\mathbb{R}^p$-valued function of time $y(t)$. For control applications, linear dynamical systems are particularly important, since most of the control algorithms are based on linear algebra tools. Linear systems are the basic objects of the Scilab control toolbox.

There are several ways to define and manipulate linear systems.

6.1.1 State-Space Representation

Systems in state-space representation make use of an internal state variable $x(t)$ (in $\mathbb{R}^n$), and the equations linking u and y are the following:

- for continuous-time linear systems (CLS):

$$\dot{x}(t) = Ax(t) + Bu(t),$$
$$y(t) = Cx(t) + Du(t),$$
$$x(0) = x_0;$$

- for discrete-time linear systems (DLS):

$$x(t+1) = Ax(t) + Bu(t),$$
$$y(t) = Cx(t) + Du(t),$$
$$x(0) = x_0.$$

We will assume in this chapter that (A, B, C, D) are constant matrices of appropriate dimensions. For general nonautonomous or nonlinear systems, nonlinear tools (see Sections 8.2 and 9) can be used for simulation.

Continuous and discrete systems are represented in Scilab as abstract objects using **tlist** (see Section 2.2.8) data types. These abstract data objects are referred to as **syslin** lists and may be built using the **syslin** function:

```
A=[-1 1 ; 1 -1] ;B=[-1 -1 ; -1 1];
C=[-1 0];D=[0,0];
sys=syslin('c',A,B,C,D);              ← a 2 input, 1 output linear system
```

A common way to describe a linear system is to represent it by its system matrix. Recall that the system matrix of the above linear system is the following matrix pencil:

$$\begin{bmatrix} -wI + A & B \\ C & D \end{bmatrix}, \tag{6.1}$$

where the symbol $w = s$ stands for the Laplace variable in the continuous-time case, and $w = z$ stands for the time-forward shift operator in the discrete-time case. All the structural properties of the linear system can be obtained from the Kronecker invariants of this pencil.

The system matrix pencil can be easily defined by, e.g.,

```
[-s*eye + A,B; C,D];                   ← system matrix
```

(the polynomial s can be defined by **s=poly(0,'s')**). Using **systmat(sys)**, where **sys** is a Scilab **syslin** list, is an equivalent way to define the system matrix of **sys**. Note that the system matrix of **sys** is a standard polynomial matrix, which can be manipulated as such in Scilab.

6.1.2   Transfer-Matrix Representation

Systems in transfer-matrix representation are functions of the Laplace transform or z-transform variables. The transfer matrix of a CLS is

$$H(s) := C(sI - A)^{-1}B + D, \tag{6.2}$$

where $H(s)$ is represented in Scilab by a rational matrix.

For example, $H(s)$ can be defined by its entries:

```
s=poly(0,'s');                              ←   the polynomial s
H=[1/s,1/(2+s)]                             ←   2 input, 1 output transfer matrix
```

or by using the function **ss2tf**, which transforms **syslin** lists into rational matrices:

```
-->sys=syslin('c',[-1,1;1,-1],[-1,-1;-1,1],[-1,0]);

-->H=ss2tf(sys)                             ←   state space to transfer matrix
  H   =
!   1           1      !
!   -         -----    !
!   s         2 + s    !
```

An arbitrary change of variables in transfer matrices can be obtained using the **horner** function. For instance, the bilinear transform defined by

$$s \leftarrow \frac{\alpha z + \delta}{\gamma z + \beta}$$

is easily achieved as follows:

```
-->sys=syslin('c',1/(s^2+1));

-->sys1=horner(sys,(3*z+1)/(2*z-2))         ←   replace s by
                                               (3z + 1)/(2z − 2) in sys

  sys1   =
                    2
      4 - 8z + 4z
      -----------
                    2
      5 - 2z + 13z
```

The transformed system is still one of continuous time. Usually, after the bilinear transform has been done, the system is to be interpreted as a discrete-time system. We need to explicitly make it a discrete-time system:

```
-->sys1('dt')                                    ← time field dt  is undefined
 ans  =
     []
```

```
-->sys1('dt')='d'                                ← making sys1  discrete time
```

6.2   System Definition

As said before, the basic function used for defining linear systems is
syslin. The calling sequence of **syslin** is either

```
S1=syslin(dom,A,B,C,D,x0)
```

or

```
S1=syslin(dom,H)
```

where **dom** is the character string **'c'** or **'d'**, which respectively stand for
a continuous-time and discrete-time system. Sampled systems can also be
defined using the same syntax but with **dom** a real number representing
the system sampling period. For instance, **syslin(0.2,A,B,C)** defines a
sampled system with sampling frequency $f_s = 1/0.2 = 5$ Hz in state-space
form, and **syslin(0.2,H)** defines a sampled system in transfer representation.

For state-space representations, **(A,B,C,D)** are constant matrices, and
x0 is the initial state at $t = 0$. For transfer-matrix representations, **H** is a
rational matrix ($Y = H(s)U$ gives the Laplace or z-transform of the system
assuming that **x0=0**).

The **syslin** function works by converting its arguments (e.g., the four
matrices **A,B,C,D**) into a typed list. This internal representation as a **tlist**
is not important from the user's point of view. What matters are the functions that operate on this new data type. For example, if **sys** is a state-space
syslin list, the user can easily access the **A** matrix by using the syntax
sys('A'), and the (i, j) entry of the **A** matrix by **sys('A')(i,j)**.

Similarly, for transfer representation, **H('num')** (resp. **H('den')**) returns the numerator (resp. denominator) polynomial matrix of **H**. The time
domain is returned in the string **H('dt')**.

Conversion from a representation to another is done by **ss2tf** and
tf2ss.

It is very useful to be able to manipulate systems as abstract data objects. This is done by the standard Scilab representation, which allows the
definition of new abstract data types that can be recognized by specific or
overloaded operators.

Linear systems with **m** inputs and **p** outputs, represented either in transfer form or state-space form, can be seen as **p**×**m** matrices. By operator overloading, the syntax used for ordinary constant matrices is valid for objects representing linear systems. For instance, if **i** and **j** are vectors of indices, then **s1(i,j)** denotes the subsystem obtained by selecting input indices **j** and output indices **i**. When **s1** is a transfer matrix, **s1(i,j)** is the submatrix of **s1** obtained by selecting rows **i** and columns **j**. When **s1** is a state-space system (**syslin** list), **s1(i,j)** is obtained from **s1** by selecting appropriate rows and columns in the **(B,C,D)** matrices **((B(:,j),C(i,:),D(i,j)))**.

6.2.1 Interconnected Systems

Interconnection is done with the following matrix operators: row concatenation **[;]**, column concatenation **[,]**, multiplication *****, addition **+**, and the feedback operation **/.** . The meaning of these operators is illustrated in Figure 6.1: For instance, **s2*s1** is interpreted as the series interconnection of **s1** and **s2**.

Of course, it is possible to mix constant matrices and linear systems, the former being interpreted as constant gains. Thus, this syntax is a natural extension of the usual Scilab matrix syntax.

For each of the possible interconnections of two systems **s1** and **s2** the command that makes the interconnection is shown on the right side of the corresponding block diagram in Figure 6.1. Note that feedback interconnection is performed by **s1/.s2**.

The representation of linear systems can be in state-space form or in transfer function form. These two representations can be interchanged using the functions **tf2ss** and **ss2tf**, which change the representations of systems from transfer function to state space and from state space to transfer function, respectively. An example of the creation, the change in representation, and the interconnection of linear systems is shown in the following Scilab session:

```
-->s=poly(0,'s');                              ← s polynomial

-->S1=1/(s-1); S2=1/(s-2);          ← two 1 × 1 rational functions

-->S1=syslin('c',S1); S2=syslin('c',S2);         ← two transfer
                                                      functions

-->Gls=tf2ss(S2);                          ← convert to state space
```

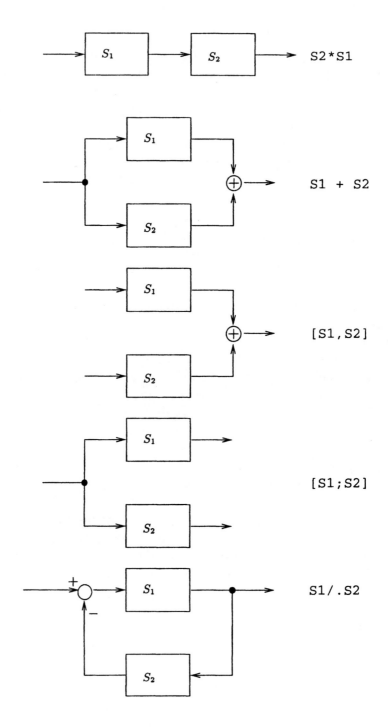

Figure 6.1: *Interconnection of linear systems*

```
-->hls=Gls*S1;                              ← series interconnection

-->ht=ss2tf(hls)                            ← display in transfer form
 ht   =
          1
     ---------
                2
     2 - 3s + s

-->S2*S1                                     ← transfer function series interconnection
 ans   =
          1
     ---------
                2
     2 - 3s + s

-->S1+S2                                     ← adding transfer functions
 ans   =
      - 3 + 2s
     ----------
                2
     2 - 3s + s
```

For simplicity, the above examples deal with transfer functions. Scilab can also handle rational matrices. In this case, the algorithms used for addition and multiplication are much more complex and involve LCM computation and simplification of rational functions. For polynomial matrices, calculations are done using built-in primitives, and in the rational function case the calculations are done with Scilab functions.

This brief session gives simple examples of manipulation of first-order linear systems:

```
-->[S1,S2]                                   ← 2 input, 1 output transfer function
 ans   =
!    1              1     !
!   -----         -----   !
! - 1 + s       - 2 + s   !

-->[S1;S2]                                    ← 1 input, 2 output transfer function
 ans   =
!    1     !
!   -----  !
! - 1 + s  !
!          !
```

```
!      1      !
!    -----    !
!  - 2 + s    !

-->S1/.S2                              ←   feedback interconnection
  ans  =
     - 2 + s
     ---------
                   2
     3  -  3s  +  s
```

It is also possible to invert linear systems both in transfer form and state-space form. In the multivariable case, this is a complex algorithm. For systems in transfer form, assume, for example, that we want to compute $X = (A(s)./B(s))(C(s)./D(s))^{-1}$, where $(A(s), B(s), C(s), D(s))$ are four square polynomial matrices. The basic steps of the algorithm are the following. First convert the denominator rational matrix

$$C(s)./D(s) = C_1(s)\text{diag}(D_1(s))^{-1},$$

where the diagonal entry $\text{diag}(D_1(s))(j,j)$ is the LCM of the polynomials in the jth column of matrix $D(s)$. Then

$$X = ((A(s)\text{diag}(D_1(s)))./B(s))C_1(s)^{-1}.$$

These calculations involve polynomial matrix inversion and multiplication of polynomial or rational matrices. The state-space inversion is based on Silverman's structure algorithm [53].

The following session illustrates some uses of inversion:

```
-->S1/(2*S2)                          ←   division of transfer functions
  ans  =
    - 2 + s
    -----
    - 2 + 2s

-->sys=ssrand(2,2,3);                 ←   random 2 × 2 system

-->sys1=inv(sys);                     ←   Silverman's algorithm

-->clean(ss2tf(sys*sys1))             ←   checking the result
              (function clean sets to zero entries that can be ignored)
   ans  =
```

```
!    1       0   !
!    -       -   !
!    1       1   !
!                !
!    0       1   !
!    -       -   !
!    1       1   !
```

6.2.2   Linear Fractional Transformation (LFT)

More complex interconnection operations can be realized with the flexible
lft function. Figure 6.2 shows the block diagrams that can be built with
lft for two-port systems. The partition of such systems is given by the
integer 2-vector that gives the size ([number of outputs, number of inputs])
of the $(2,2)$ block. The feedback K is wrapped around S_{22}, resulting in the
closed-loop system $u_1 \rightarrow y_1$:

$$S_{11} + S_{12}K(I - S_{22}K)^{-1}S_{21}. \tag{6.3}$$

The generalized **lft** operation involving two 2-port systems in the
bottom part of Figure 6.2 is often used to make a "preliminary feedback"
around S_{22} by choosing $\Sigma = \begin{bmatrix} K & I \\ I & 0 \end{bmatrix}$, where K has the same size as S_{22}^T.
Note that lft(S, Σ) is still a 2-port system partitioned as S.

6.2.3   Time Discretization

Sampled systems are also handled by Scilab. They are characterized by
the fact that in their internal representation the **'dt'** field is the sampling
period. A typical function that builds a sampled system is **dscr**. This function performs the time discretization of a linear system.

For a plant described by the transfer function $H(s)$ and preceded by a
ZOH (zero order hold), the sampled transfer function is $(1-z^{-1})\mathcal{Z}(H(s)/s)$,
where $\mathcal{Z}(F(s))$ is the z-transform of the time series obtained by sampling
the function whose Laplace transform is $F(s)$.

Let us illustrate a sampling operation on the system with transfer function $H(s) = 1/(1+s^2)$. The function $G(s) = H(s)/s$ is the Laplace transform
of the function $1 - \cos(t)$, and the z-transform of the sequence $1 - \cos(nT)$
is

$$\frac{z(1 - \cos(T)(z + 1))}{((z - 1)(z^2 - 2\cos(T)z + 1))}.$$

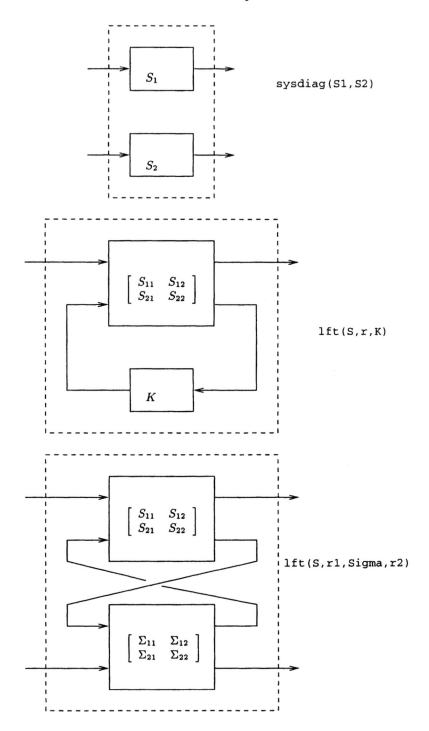

Figure 6.2: *Interconnection of linear systems*

Thus, the transfer matrix of the sampled system is

$$H(z) = \frac{1 - \cos(T)(z + 1)}{z^2 - 2\cos(T)z + 1}.$$

This result is easily obtained using the function **dscr**:

```
--> Sys=syslin('c', 1/(1+s^2))
 Sys  =
       1
     -----
          2
     1 + s

-->T=0.05;Sysd=ss2tf(dscr(tf2ss(Sys),T))
 Sysd  =
     .0012497 +   .0012497z
     --------------------
                         2
     1 - 1.9975005z + z

-->(1-cos(T))*(z+1)/(z^2-2*cos(T)*z+1)
 ans  =
     .0012497 +   .0012497z
     --------------------
                         2
     1 - 1.9975005z + z

-->Sysd('dt')                      ←  sampling period coded in Sysd
 ans  =
     .05
```

The function **dscr** uses a state-space approach to discretize the continuous-time system. The solution is obtained by solving the following linear system on $[0, T]$:

$$\dot{x}(t) = Ax(t) + Bu(t),$$
$$\dot{u}(t) = 0,$$
$$x(0) = x_T, \quad u(0) = u_T.$$

It is also possible to use the bilinear transform to discretize a linear system. In the following session, the frequency-response magnitude plot is given for the continuous system and for the sampled systems obtained by exact discretization and by using the Tustin transform. The sampling period is $T = 0.05$ sec. High-frequency distortions are clearly visible in Figure 6.3.

```
frq=logspace(-2,2.99,100);                    ← selected frequencies
[frq1,repf1]=repfreq(Sys,frq);                ← frequency response:
                                              continuous-time system
[frq2,repf2]=repfreq(Sysd,frq);               ← frequency response:
                                                exact discretization

Sysd1=horner(Sys,(2/T)*(z-1)/(z+1));          ← Tustin bilinear
                                                    transform
Sysd1('dt')=T;                                ← sampling period
[frq3,repf3]=repfreq(Sysd1,frq);              ← frequency response:
                                                Tustin transform
reps=20*log(abs([repf1;repf2;repf3]'))/log(10);
plot2d1('oln',frq',reps);                     ← Figure 6.3
xtitle('frequency response: effect of sampling','Hz','DB');
```

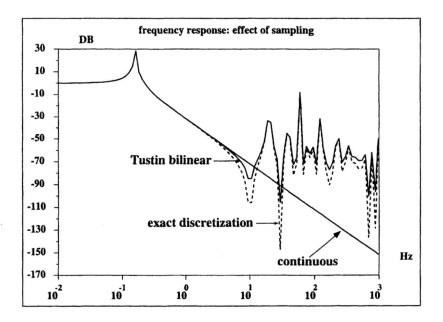

Figure 6.3: *Discretization and Tustin transform*

6.3   Improper Systems

In some applications it is necessary to handle improper systems, i.e., systems with an impulsive part. Although improper systems are not physically realizable, since Scilab aims at manipulating linear systems as ordinary matrices, it is necessary to be able to represent them in order to have a complete algebra (e.g., system inversion naturally leads to improper systems).

6.3.1 Scilab Representation

There are two ways of representing improper systems in state-space form: One is using descriptor systems, i.e., coding the improper transfer function $\Sigma(s)$ into $\{E, A, B, C\}$ such that

$$\Sigma(s) = C(sE - A)^{-1}B,$$

and the second is using state-space descriptions (A, B, C, D), allowing D to be a polynomial matrix, i.e.,

$$\Sigma(s) = C(sI - A)^{-1}B + D(s).$$

The latter is the representation that is used by default in Scilab. This representation allows a unique framework for both proper and improper systems. The reason for this choice is that the descriptor representation puts together finite and infinite modes, which can cause numerical difficulties (in some cases, it is difficult to distinguish a large mode from an infinite mode). To illustrate this point, consider the following (minimal) realization of the polynomial $p(s) = s$:

$$E = \begin{pmatrix} 0 & -1 \\ 0 & 0 \end{pmatrix}, \quad A = \begin{pmatrix} 1 & 0 \\ 0 & 1 \end{pmatrix}, \quad B = \begin{pmatrix} 0 \\ 1 \end{pmatrix}, \quad C = \begin{pmatrix} 1 & 0 \end{pmatrix}.$$

Suppose that the matrix E is perturbed as follows:

$$E = \begin{pmatrix} \epsilon & -1 \\ 0 & \epsilon \end{pmatrix}.$$

The perturbed transfer function is $s/(1 - s\epsilon^2)$, which for $\epsilon \neq 0$ is a proper system and, in many ways, very different from s.

In Scilab, the polynomial part of the system is coded explicitly into the D matrix, and the corresponding algebraic operations are in part strict state-space operations and in part polynomial, the polynomial part corresponding to the infinite modes. As a consequence, infinite modes can be handled much more easily, since they remain separate from the other modes of the system. Moreover, the realization algorithm for converting an improper transfer function becomes a trivial extension of the standard state-space realization algorithm.

6.3.2   Scilab Implementation

Since there are several possible representations of systems, functions are
needed to convert a representation into another one. Conversion flexibility
is obtained by the functions shown in Figure 6.4.

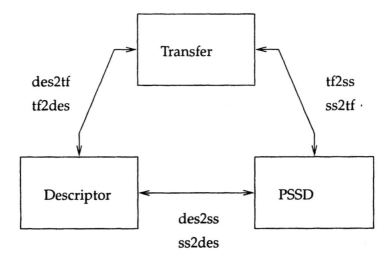

Figure 6.4: *Conversion functions*

Let us briefly describe these functions:

- **Ss=des2ss(des)**
  This function returns a state-space system possibly including a polynomial D matrix. The input is a regular descriptor system
  Des=(A,B,C,D,E) of arbitrary index. The shuffle algorithm [43] is applied to the regular pencil **s*E-A**, i.e., a polynomial matrix **Ws** is computed (using a finite sequence of SVDs and system multiplications)
  such that the pencil **s*E1-A1=Ws*(sE-A)** has a nonsingular **E1** matrix. The output **Ss** is the linear system

  Ss = C* [E1\A1,Id,Id,0] * Ws * B.

- **Des=ss2des(Ss)**
  A standard realization algorithm is performed to obtain a realization
  of the polynomial part **D(s)** of **Ss**: **D(s)= C1*(s N - I)* B1** with
  N nilpotent (SVD factorization of the Hankel matrix). The output is
  the descriptor system

  Des: ([I,0;0,A], [B1;B2] , [C1,C1] , 0 , [N,0;0,I]).

- **Des=tf2des(G)**

  The Euclidean division algorithm is applied to each entry of the transfer matrix **G**. This decomposes **G** into its strictly proper part and its polynomial part. Each part is converted into state-space form by **tf2ss** (the polynomial part being realized through a nilpotent matrix). The output **Des** is then constructed as above (**ss2des**).

- **G=des2tf(Des)**

  The conversion from descriptor form to transfer form is made by the algorithm described in [24]. This function may return three outputs **chi**, **Bfs**, **Bis**, which are respectively the characteristic polynomial **det(s*E-A)**, the "numerator" polynomial matrix, and the polynomial matrix "at infinity" such that **G = Bfs/chi + Bis**.

- **G=ss2tf(Ss)**

  The strictly proper part of the transfer matrix **G** is calculated by a characteristic polynomial method, and the polynomial part is added to it. This function can be invoked with three output arguments as above.

- **Ss=tf2ss(G)**

  Each column of the transfer matrix **G** is realized independently as a linear system, and these systems are stacked together by column concatenation. For each column, the LCM polynomial of the denominators and the associated numerator polynomial vector are calculated; this system is then realized in canonical controllable form, and the unobservable part of the overall system is removed.

The following session illustrates the use of some of these functions:

```
-->s=poly(0,'s');G=[1/(s-1),s;1,2/s^3];        ← an improper G

-->S1=tf2des(G);S2=tf2des(G,"withD");        ← improper transfer
                                to descriptor (with possibly a constant D matrix)
-->S1('D'),S2('D')
 ans   =
!   0.      0. !
!   0.      0. !
 ans   =
!   0.      0. !
!   1.      0. !

-->W1=des2ss(S1); W1('D')        ← standard Scilab representation
                                    with polynomial D matrix
 ans   =
```

```
!   0        s   !
!                !
!   1        0   !
```

6.4   System Operations

The basic properties of a linear system in state-space representation can be obtained by looking at its matrices in a particular basis. For instance, if X is the current basis of the state space, the system matrices are $(X^{-1}AX, X^{-1}B, CX, D)$. Most of Scilab's system and control functions calculate specific bases that put the system into canonical forms. **sys1=ss2ss(sys,X)** is the function that applies the above change of basis and returns a new **syslin** list.

This function can also be used to perform state feedback, output injection, or factorization of linear systems.

For example, let us consider state feedback: This operation amounts to transforming $G = (A, B, C, D)$ into $G_F = (A + BF, B, C + DF, D)$ where G_F can be factored into $G_F = G \times (A + BF, B, F, I)$. This can be illustrated as follows:

```
-->ss2tf(G)                      ← G  has 2 inputs, 1 output, and 2 states
  ans   =
!   1          1     !
!   -        -----   !
!   s        2 + s   !

-->F=diag([1,2]);[G_F,right,left]=ss2ss(G,eye(2,2),F);   ← state
                                                           feedback
-->clean(ss2tf(left*G*right-G_F))
  ans   =
!   0      0   !
!   -      -   !
!   1      1   !

-->W=left*G*right;size(W('A')),size(G_F('A'))            ← ss2ss
                                        returns a  minimal system
  ans   =
!    4.      4.  !
  ans   =
!    2.      2.  !
```

6.4.1 Pole-Zero Calculations

For a system **sys** in state-space representation, computation of the poles is done in the same way as the evaluation of the eigenvalues of the **A** matrix: **spec(sys('A'))**. For a system in transfer representation, the function **roots** can be used.

The calculation of the zeros is more complex. For SISO plants, the zeros are easily obtained as the roots of the numerator polynomial. For MIMO square nonsingular systems, the zeros can be computed as the roots of the determinant of the system matrix or by calculating the generalized eigenvalues of the system matrix pencil (using the **gspec** function). More generally, the **kroneck** function, based on the algorithm [8], can be used to extract the part corresponding to the finite eigenvalues of the system matrix. The function **trzeros** can also be used which is a built-in implementation of the algorithm [25]. Let us examine a simple example:

```
-->sys=[1/(s+1),1/(s+2);1/(s-2),1/(s+3)];        ← square regular
                                                      system
-->trzeros(tf2ss(sys))                            ←  zeros
 ans  =
  - 1.75

-->det(systmat(tf2ss(sys)))                       ←  zero polynomial
 ans  =
  - 7 - 4s

-->sys=[1/(s+1);1/(s-2)];sys1=(s-3)*sys*sys';     ←  a singular
                                                      system
-->trzeros(sys1)                                  ←  zeros
 ans  =
    3.

-->S=systmat(tf2ss(sys1));                        ←  system matrix is singular

--> [Q,Z,Qd,Zd,numbeps,numbeta]=kroneck(S);       ←  Kronecker form

-->S=Q*S*Z;

-->rowf=(Qd(1)+Qd(2)+1):(Qd(1)+Qd(2)+Qd(3));           ←  extracting
                                                      finite eigenvalues
-->colf=(Zd(1)+Zd(2)+1):(Zd(1)+Zd(2)+Zd(3));

-->roots(St1(rowf,colf))
 ans  =
    3.
```

6.4.2   Controllability and Pole Placement

The controllability properties of a linear system are given by the function
contr. The command [dim, X]=contr(A,B) puts the pair (A, B) into canonical controllable form $(X^{-1}AX, X^{-1}B)$, where the first dim columns of
X span the controllable subspace. Since controllability requires rank determination, it can be difficult to safely compute the controllable part of a pair
(A, B), and it is recommended that different approaches be compared.

In the following session, we build a noncontrollable pair, and we in-
vestigate its controllability. Recall that a random (A, B) pair is generically
controllable. To get a generic random noncontrollable system we must build
a noncontrollable system in canonical form and perform a random change
of basis. This is done using **ssrand**.

```
-->sys=ssrand(2,3,5,list('co',2));        ← generic noncontrollable
                                                             system

-->[A,B,C,D]=abcd(sys);                   ←  retrieving the matrices

-->[dim,X]=contr(A,B);                     ← controllable canonical form

-->dim                                              ←  great! dim value is 2
  dim  =
      2.

-->sys1=ss2ss(sys,X);

-->clean(sys1('A')(dim+1:$,1:dim))         ← checking canonical form
  ans  =
!   0.     0. !
!   0.     0. !
!   0.     0. !

-->clean(sys1('B')(dim+1:$,:))
  ans  =
!   0.     0.     0. !
!   0.     0.     0. !
!   0.     0.     0. !
```

The controllable subspace can also be obtained with the **kroneck** function (this subspace corresponds to the left kernel of the singular controllability pencil $[-sI + A, B]$). More details are given in the on-line help of
kroneck. We see in the following example that this approach gives compatible results to the previous example:

```
-->W=[-s*eye+A,B];               ← building the singular controllability pencil
```

```
-->[Q,Z,Qd,Zd,numbeps,numbeta]=kroneck(W);     ← Kronecker form

-->svd([X(:,1:dim),Q(1:Qd(1),:)'])     ← are the subspaces identical?
 ans  =
!    1.4 !
!    1.4 !
!    0.  !
!    0.  !                                       ← yes, they are!
```

A simple trick to check controllability is to compare the eigenvalues of
A+B*F for random **F**'s: common eigenvalues are the uncontrollable modes.
A useful test to check stabilizability is to build the the matrix

$$\begin{bmatrix} A & -BB^T \\ -I & -A^T \end{bmatrix}$$

and look at the conditioning of the matrix P where $\begin{bmatrix} P \\ Q \end{bmatrix}$ is the stable invariant subspace of H.

The controllable eigenvalues of a pair (A, B) can be placed using the
ppol function. This function assumes that the pair is controllable. More generally, the function **stabil** places the controllable eigenvalues and returns
a stabilizing full-state feedback gain if the pair (A, B) is stabilizable (stabilizability can be checked with the function **st_ility**). Let us illustate this:

```
-->sys=ssrand(2,3,5,list('st',2,4,4));     ← unstabilizable system

-->[A,B,C,D]=abcd(sys);

-->F=stabil(A,B,-3);                               ← stabilizing feedback

-->spec(A+B*F)                     ← check that two eigenvalues are set to −3
 ans  =
! - 5.           !                     ←     stable uncontrollable
! - 3.           !                     ←     controllable: set to −3
! - 3.           !                     ←     controllable: set to −3
!    .5          !                     ←   unstable uncontrollable
! - 3.1891026 !                        ←     stable uncontrollable
```

Another tool that can be used for calculating a stabilizing feedback is
lqr.

The LQR control problem is to find a feedback $u = Kx$ solution of the following problem:

$$\min_{u} \int_0^\infty \left(x^T Q x + u^T R u + 2 x^T S u \right) \, dt,$$
$$\dot{x} = Ax + Bu.$$

This problem can be rewritten as follows:

$$\min_{u} \quad \|z\|_2,$$
$$\dot{x} = Ax + Bu, \qquad\qquad\qquad (6.4)$$
$$z = Cx + Du,$$

where the matrices C and D are obtained from Q, R, and S by full-rank factorization:

$$\left[\begin{array}{c} C^T \\ D^T \end{array} \right] \left[\begin{array}{cc} C & D \end{array} \right] = \left[\begin{array}{cc} Q & S \\ S^T & R \end{array} \right].$$

The argument of the Scilab function **lqr** is the plant described in equation (6.4), and its return value is the optimal gain K. This gain stabilizes the pair (A, B) if the pair is stabilizable, and the auxiliary matrices Q, R, S (or equivalently C, D) are chosen appropriately (in particular, R must be invertible).

Let us construct an example of a full-state LQR stabilizing gain:

```
-->A=[1,2;3,4];B=[1;2];

-->Q=diag([4,4]);R=1;S=[0;0];

-->W=fullrf([Q,S;S',R]);C=W(:,1:2);D=W(:,3:$);          ← finding
                                                         C and D
-->Sys=syslin('c',A,B,C,D);              ← defining continuous-time system

-->K=lqr(Sys);                                         ← LQR gain

-->spec(A+B*K)                                  ← checking stability
  ans  =
!  -   .4047381 !
!  - 6.9882893 !
```

6.4.3   Observability and Observers

Observability properties of a pair (A, C) are easily checked by looking at the controllability properties of the pair (A^T, C^T). The function **unobs** returns the unobservable subspace, and the function **dt_ility** checks detectability. This function calculates an orthogonal change of basis U such that

$$\left[\frac{U^T A U}{CU}\right] = \begin{bmatrix} A_{11} & * & * \\ 0 & A_{22} & * \\ 0 & 0 & A_{33} \\ \hline 0 & 0 & C_3 \end{bmatrix}.$$

The function **lqe** is the counterpart of **lqr** and can be used for continuous time or discrete-time systems. Consider a linear system described by state-space matrices (A, B, C, D) with state variable x, input u, and output y. Given a signal $z = Hx$, a linear observer is a linear system $(u, y) \rightarrow \hat{z}$ such that the error $z(t) - \hat{z}(t)$ goes to zero as t tends to ∞. The function **observer** calculates an observer for a possibly unobservable system. A test is provided to check whether the signal $z = Hx$ can be tracked by the observer.

Now we design an observer for a nonlinear competition model.

The model is the following:

$$\begin{aligned} \dot{x}_1 &= p_r x_1(1 - x_1/p_k) - u p_a x_1 x_2, \\ \dot{x}_2 &= p_s x_2(1 - x_2/p_l) - u p_b x_1 x_2, \end{aligned}$$

where x_1, x_2 model two populations that grow on the same resource u. For a given resource level, say $u_e = 1$, an equilibrium point can be found by the function **fsolve** (see Section 8.4.3), and a linearized model can be obtained by the function **tangent**. We assume that x_1 is observed, and we build an observer for (x_1, x_2). The model is given by the function **compet**:

```
function xdot=compet(t,x,u)
xdot= [p_r*x(1)*(1-x(1)/p_k) - u*p_a*x(1)*x(2);
       p_s*x(2)*(1-x(2)/p_l) - u*p_b*x(1)*x(2)]
```

The constants chosen are **p_r=0.01, p_a=5d-5, p_s=5d-3, p_b=1d-4, p_k=1d3, p_l=500**.

```
u_e=1;

deff('[y]=f(x)','y=compet(0,x,u_e)');          ← function used
                                                   by fsolve
x_init=[20,200];                        ← must be chosen carefully!
x_e=fsolve(x_init,f);                     ← finding zero of f
[A,B]=tangent('compet',x_e,u_e);C=[1,0];    ← matrices of
                                            linearized model
x0=[1;1];Sys=syslin('c',A,B,C,0,x0);           ← linearized
```

	model with initial state **x0**
`Obs=observer(Sys);`	← observer for **Sys**
`xchap0=[1.3;1.3];Obs('X0')=xchap0;`	← setting observer initial state
`Sys2=syslin('c',A,B,eye(2,2),[0;0],x0);`	← viewing the states of **Sys**
`deff('ut=u(t)','ut=100*sin(t)');`	← input function
`T=0:0.1:15;x=csim(u,T,Sys2);`	← simulating a trajectory
`SObs=Obs*[1;Sys];`	← system $u \rightarrow \hat{x}$
`xhat=csim(u,T,SObs);`	← simulation of $\hat{x}$
`plot2d1("onn",T',[x(2,:);xhat(2,:)]')`	← Figure 6.5

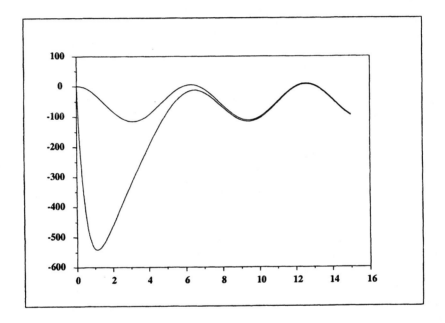

Figure 6.5: *Observer output tracks* x_2

6.5   Control Tools

Several functions are devoted to linear system analysis. For continuous-time systems, time responses can be obtained directly using the function **csim**, which calls an ODE solver. Another approach is to discretize the system (function **dscr**) and to simulate the discrete-time system.

Simulation of discrete-time systems is done using simple matrix recursion equations. Two built-in functions, **ltitr** and **rtitr**, are devoted to that purpose. For the simplest state-space equations, $x(t+1) = Ax(t) + Bu(t)$, the function **ltitr** performs the calculation of x, given u and possibly x_0. A utility function (**dsimul**) can be used for calculating an output $y = Cx + Du$. Consider now a discrete-time system with transfer matrix $H(z) = D^{-1}(z)N(z)$ where $D(z)$ is a square nonsingular polynomial matrix of size $r \times r$ and $N(z)$ is a polynomial matrix of size $r \times p$. Writing

$$D(z) = D_0 + D_1 z + \cdots + D_d z^d$$

and

$$N(z) = N_0 + N_1 z + \cdots + N_n z^n,$$

the output y of this system corresponding to input u can be computed by solving the following recursion equations:

$$D_d y(t+d) + D_{d-1} y(t+d-1) + \cdots + D_0 y(t)$$
$$= N_n u(t+n) + N_{n-1} u(t+n-1) + \cdots + N_0 u(t).$$

The function **rtitr** evaluates these equations for a given set of inputs($u(k)$, $u(k+1), \ldots, u(k+T)$) and returns ($y(k), y(k+1), \ldots, y(k+T+d-n)$). Let us look at a few simple examples:

```
-->N=1;D=z;u=1:5;                           ←   one-unit delay

-->y=rtitr(N,D,u)                           ←   y(t+1) = u(t)
 y  =
!   0.     1.     2.     3.     4.     5. !


-->N=z;D=1;u=1:5;                           ←   one-unit lead

-->y=rtitr(N,D,u)                           ←   y(t) = u(t+1)
 y  =
!   2.     3.     4.     5. !


-->N=[1;z];D=[z^2+1,z-1;z+1,z^2-2];

-->y=rtitr(N,D,u)
 y  =
!   0.     0.     0.     1.     0.     1.  !
!   0.     1.     2.     5.     7.    14.  !


-->Dy=rtitr(D,eye(2,2),y);

-->Nu=rtitr(N,eye(2,2),u);                  ←   same as Dy
```

The function **flts** performs the simulation of a linear system. Since Scilab state-space representation allows for a polynomial D matrix, we consider

$$
\begin{aligned}
x(t+1) &= Ax(t) + Bu(t), \\
y(t) &= Cx(t) + D_0 u(t) + D_1 u(t+1) + \ldots + D_p u(t+p), \qquad (6.5)\\
x(0) &= x_0.
\end{aligned}
$$

In transfer function form, Scilab uses the representation

$$
H_{ij}(z) = Num_{ij}(z)/Den_{ij}(z)
$$

where *Num* and *Den* are two polynomial matrices. The function **flts** transforms H into a form valid for **rtitr**. Each row of H is converted into $D_i^{-1}(s)N_i(s)$, where $D_i(s)$ is the LCM of the ith row of H. This can be done, for example, as follows:

```
-->H=1/z;u=1:5;
```
$\leftarrow$ one-unit delay

```
-->y=flts(u,H)
 y  =
!   0.    1.    2.    3.    4. !
```
$\leftarrow$ $y(t+1) = u(t)$

```
-->N=z;D=1;u=1:5;
```
$\leftarrow$ one-unit lead

```
-->y=flts(u,syslin('d',N,D))
 y  =
!   2.    3.    4.    5. !
```
$\leftarrow$ $y(t) = u(t+1)$

```
-->N=[1;z];D=[z^2+1,z-1;z+1,z^2-2];
```

```
-->y=flts(u,inv(D)*N)
 y  =
!   0.    0.    0.    1.    0. !
!   0.    1.    2.    5.    7. !
```

The function **csim** is used to calculate the time response of linear systems. The input can be a step function, an impulse, or a general function of time. The simulation of **sys** is performed by a block diagonalization of its **A** matrix followed by a call to the **ode** solver function. The impulse response of **sys** is calculated as the step response of **s*sys** assuming that **sys** is strictly proper.

Let us consider a simple example of step and impulse response calculations for the continuous-time second-order system with transfer function $H(s) = (1 - s)/(s^2 + 0.2s + 1)$. The response obtained is typical of a non-minimal phase system. The result is shown in Figure 6.6.

```
Sys=syslin('c',(1-s)/(s^2+0.2*s+1));            ← the linear system
T=0:0.05:50;                                    ← sampling times
y=[csim('step',T,Sys);csim('impulse',T,Sys)];          ← step
                                                and impulse responses
plot2d([T;T]',y');
plot2d([T;T]',[0*T;ones(T)]',[1,1],"000");            ← horizontal
                                                lines at zero and one
xstring(30,1.1,'Step response');          ← step response caption
xstring(30,0.1,'Impulse response');           ← impulse response
                                                          caption

xtitle('P=(1-s)/(s^2+0.2*s+1)','Time');
```

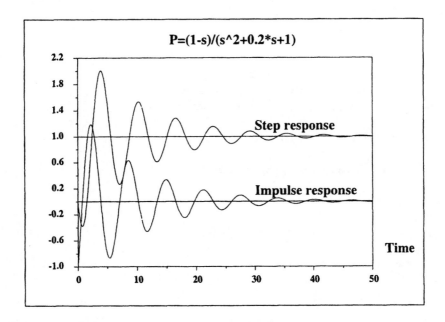

Figure 6.6: *Step and impulse responses for $H(s) = (1-s)/(s^2 + 0.2s + 1)$*

The time response to a square-wave input is now simulated for the same transfer function $H(s)$. First, we use the csim function with a functional input $u(t) = 1$ if $t/50 \in [2*k, 2*k+1[$ and 3 if not:

```
deff('[uu]=u(t)','uu=2+(-1).^(floor(t/50))');    ← square-wave
                                                          input
Sys=syslin('c',(1-s)/(s^2+0.2*s+1));
T=0:0.05:200;
y=csim(u,T,Sys);
```

The function csim uses an ODE (ordinary differential equation) solver to compute the time response. Although it is a reliable way to get the time response, in some cases, if the input function u is discontinuous, convergence cannot always be guaranteed, particularly for high-order systems.

Now we show how this problem can be circumvented using the function **odedc**, which can be used for simulating sampled dynamical systems. At sampling instants a discrete variable is updated, and between sampling instants an ODE with right-hand side depending on the value of the discrete variable is solved. A safer way to simulate the previous example follows:

```
-->ssSys=tf2ss(Sys);                        ← state-space representation

--> [A,B,C,D]=abcd(ssSys)


-->deff('xdu=phi(t,x,u,flag)',..
          'if flag=0 then xdu=A*x+B*u;
          else xdu=(1-u)+3;end');            ← output
                                   of phi is either xdot or updated u
-->x0=[0;0];u0=1;nx=2;nu=1;                 ←  two states, one input

-->xu=odedc([x0;u0],nu,[50,0],0,T,phi);     ←  sampling period
                                                equals 50 units
-->x=xu(1:nx,:);u=xu(nx+1:nx+nu,:);

-->plot2d1('onn',T',(C*x)',1,'061');        ←  Figure 6.7

-->plot2d2('onn',T',u',2,'000');   ←  displaying piecewise constant u

-->xtitle('Time response')
```

A classical problem is to simulate a system with discrete- and continuous-time components. For instance:

$$\begin{aligned}
\dot{x}_c(t) &= f_c(t, x_c(t), u(t)), \\
y(t) &= h_c(x_c(t), u(t)), \\
x_d(t_i^+) &= f_d(x_d(t_i), y(t_i)), \\
u(t) &= h_d(t, x_d(t_i^+)) \text{ for } t \in [t_i, t_{i+1}].
\end{aligned}$$

Here $x_c(t)$ is the continuous state of the controlled plant, $y(t)$ is the observation, and the piecewise control u is allowed to jump at discrete times (t_i) $(u(t) = u_i$ for $t \in [t_i, t_{i+1}])$. The updated u at jump times are functions of a discrete-time compensator (x_d).

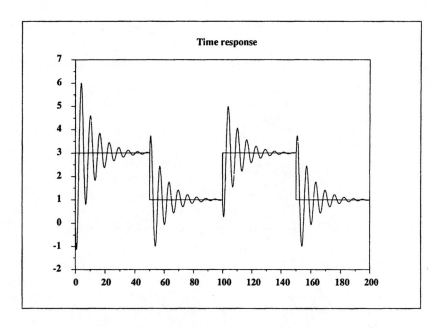

Figure 6.7: *Step and impulse responses for* $H(s) = (1 - s)/(s^2 + 0.2s + 1)$

Such systems can be simulated by using the function **odedc**, and an example follows:

```
Sys=syslin('c',(1-s)/(s^2+0.2*s+1));
ssSys=tf2ss(Sys);
[A,B,C,D]=abcd(ssSys);

deff('st=e(t)','st=sin(0.3*t)')
deff('xcd=f(t,xc,xd,flag)',..
        ['if flag==0 then'
        '   xcd=A*xc+B*(Cd*xd-0.1*e(t));'
        'else '
        '   xcd=Ad*xd+Bd*C*xc;'
        'end']);
Ad=-0.4;Bd=1;Cd=1;T=0:0.05:160;
stdel=[5,0];x0=[0;0];u0=1;
nx=2;
xcd=odedc([x0;u0],nu,stdel,0,T,f);
y=C*xcd(1:nx,:);u=Cd*xcd(nx+1:nx+nu,:)-0.1*e(T);
plot2d1('onn',T',y',1,'061');
plot2d2('onn',T',u',2,'000');
xtitle('Time response')
```

Note that the functional argument of **odedc** is a Scilab function, in this case **f(t,xc,xd,flag)**, which returns, depending on the value of the parameter **flag**, either $\dot{x}_c(t)$ or $x_d(t_i^+)$.

Figure 6.8 shows the output $y(t) = Cx_c(t)$ and the control $u(t)$ for the linear closed-loop system

$$\dot{x}_c(t) = Ax_c(t) + Bu(t),$$
$$y(t) = Cx_c(t),$$
$$x_d(t_i^+) = A_dx_d(t_i) + B_dy(t_i),$$
$$u(t) = C_dx_d(t_i^+) - \sin(0.3t) \text{ for } t \in [t_i, t_{i+1}[.$$

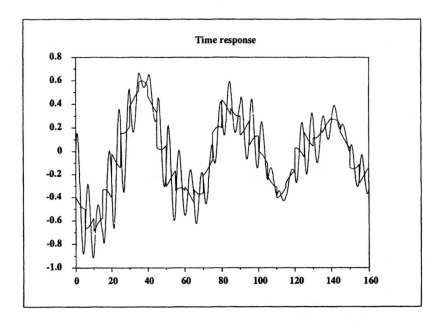

Figure 6.8: *A mixed continuous discrete-time system*

Digitally controlled systems usually have a compensator that is preceded by a zero-order hold (Zoh). The equations are

$$\dot{x}_c(t) = f_c(t, x_c(t), u_k) \text{ for } t \in [t_k, t_{k+1}],$$
$$y(t) = h_c(x_c(t), u_k),$$
$$\xi_{k+1} = f_d(\xi_k, y(t_k)),$$
$$u_k = h_d(\xi_k).$$

Note that such systems differ from the previous example only with regard to the equation $\xi_{k+1} = f_d(\xi_k, y(t_k))$, where $y(t_k)$ replaces $y(t_{k+1})$. These equations can also easily be put into a form for use with **odedc** as follows:

$$\dot{x}_c(t) = f_c(t, x_c(t), h_d(\xi_k)) \text{ for } t \in [t_k, t_{k+1}],$$
$$\xi_{k+1} = f_d(\xi_k, h_c(x_c(t_k)), h_d(\xi_k))).$$

Now, with $\Xi = (\xi, \mu)$, we get

$$\dot{x}_c(t) = F_c(t, x_c(t), \Xi_k) \text{ for } t \in [t_k, t_{k+1}],$$
$$\Xi_{k+1} = F_d(x_c(t_{k+1}), \Xi_k),$$
$$F_c(t, x, y) \equiv f_c(t, x, h_d(y_1)),$$
$$F_d(x, y) \equiv \left[\begin{array}{c} f_d(y_1, y_2) \\ h_c(x) \end{array} \right].$$

6.6 Classical Control

The classical control approach is based mainly on a set of graphical tools showing the time or frequency responses of the system. We will illustrate how to use these tools on a classical example of a simplified DC motor [27]. In this single-input single-output model, the angular velocity θ and the electric current i are the state variables. The input is the voltage V, and the output is the angular velocity. The classical equations

$$J\ddot{\theta} + b\dot{\theta} = Ki,$$
$$L\frac{di}{dt} + Ri = V - K\dot{\theta}$$

are easily transformed into a state-space representation:

$$\frac{d}{dt} \left[\begin{array}{c} \dot{\theta} \\ i \end{array} \right] = \left[\begin{array}{cc} -\dfrac{b}{J} & \dfrac{K}{J} \\ -\dfrac{K}{L} & -\dfrac{R}{L} \end{array} \right] \left[\begin{array}{c} \dot{\theta} \\ i \end{array} \right] + \left[\begin{array}{c} 0 \\ \dfrac{1}{L} \end{array} \right] V,$$

$$\dot{\theta} = \left[\begin{array}{cc} 1 & 0 \end{array} \right] \left[\begin{array}{c} \dot{\theta} \\ i \end{array} \right].$$

The DC motor system can be easily coded in state-space form as a **syslin** list:

```
J=0.01; b=0.1; K=0.01; R=1; L=0.5;
A=[-b/J    K/J;      -K/L    -R/L];
B=[0 ;     1/L];
C=[1    0];
Motor=syslin('c',A,B,C);
```

Let us examine the open-loop system step response. This is done by

```
T=0:0.01:3;                                      ←  time discretization range
plot(T,csim('step',T,Motor),..
    'Time(secs)','Magnitude','Open loop step response');
```

This produces the plot in Figure 6.9.

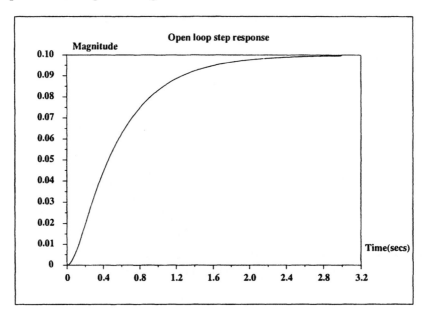

Figure 6.9: *DC motor: Open loop step response*

The function **csim** is used to calculate the time response of **Motor**.

Let us examine the effect of a PID control on the DC motor. The PID controller is defined here by its transfer function as an improper linear system. The closed-loop system is also obtained in transfer-function form, but we could do the same calculations in state-space form using the same syntax. The **csim** function automatically converts the single-input triple-output systems **[Closed1;Closed2;Closed3]** into a state-space form with a state in $\mathbb{R}^4$. Since we are dealing with low-order systems, no numerical difficulties will appear. This fact is shown by the following session:

```
unitGain=1;
Tf=ss2tf(Motor);                           ←  converts to transfer function
Ki=1; Kd=1; Kp=100;
Pid1=Kd*s+Kp+Ki/s;                                ←  PID controller definition
Closed1=(Tf*Pid1)/.unitGain;                         ←  closed-loop plant
```

```
Ki=200;Kd=1;Kp=100;Pid2=Kd*s+Kp+Ki/s;
Closed2=(Tf*Pid2)/.unitGain;
Ki=200;Kd=10;Kp=100;Pid3=Kd*s+Kp+Ki/s;
Closed3=(Tf*Pid3)/.unitGain;
T=0:0.01:1.5;
w=csim('step',T,[Closed1;Closed2;Closed3]);      ← step response
plot2d1("onn",T',w')                             ← making Figure 6.10
xtitle('Three PID control strategies','Time(secs)',..
       'Magnitude');
```

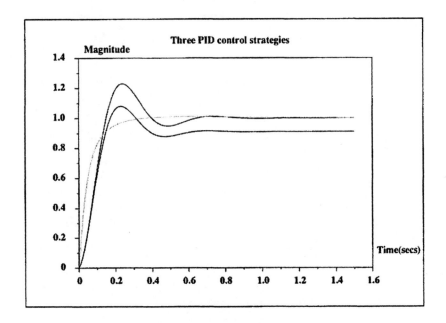

Figure 6.10: *DC motor: PID controls*

Note that the steady-state value of the output for **Closed1** is one, although the convergence to steady state is much slower than for **Closed2** and **Closed3**. The final value theorem stipulates that the step response, for large values of t, converges to the limit of **s*Closed(s)*(1/s)** as **s** goes to zero. This is easily checked in transfer or state-space form:

```
-->horner(Closed1,0)                    ← evaluates Closed1(s) at s=0
 ans =
    1.

-->[A,B,C,D]=abcd(Closed1);-C*inv(A)*B              ← in state-space
```

```
ans   =
    1.
```

Let us now design a root-locus-based controller. The root-locus plot is given by the **evans** function (see Figure 6.11):

```
-->evans(Tf);
```

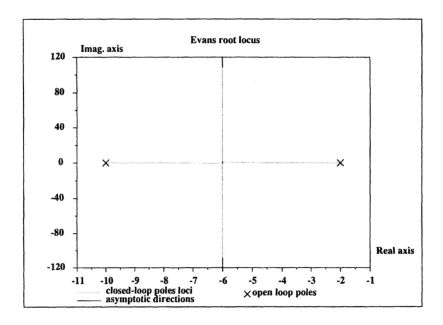

Figure 6.11: *DC motor: root locus*

The function **evans(Tf)** gives the locus in the complex plane of the roots of the function $1+kTf(s)$ as k goes from 0 to ∞. It is easy to determine the gain value **k** corresponding to a given point in the complex plane. The function returns the coordinates (x, y) of a mouse-selected point. The value of **k** corresponding to that point solves the equation $1 + kTf(s) = 0$ for $s = x + iy$ and is given by -1/horner(Tf,x+%i*y).

The selection of the gain **k** is often done in conjunction with a lead or lag compensator. This is a trial-and-error process in which the quality of a selected gain and that of the lead/lag compensator are evaluated by looking at the closed-loop response. This process is easily performed. For instance, the closed-loop response for the system with the controller $k(s-z_0)/(s-p_0)$ is

```
P=x+%i*y;                                          ← selected point
```

```
k=-1/real(horner(Tf,P));                          ← corresponding gain
w=csim('step',T,(Tf*k*(s-z0)/(s-p0))/.unitGain);
```

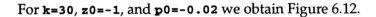

For **k=30**, **z0=-1**, and **p0=-0.02** we obtain Figure 6.12.

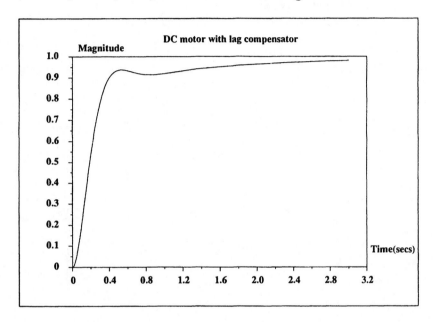

Figure 6.12: *DC motor: effect of lag compensator*

6.6.1 Frequency Response Plots

In practice, linear systems are often represented by their frequency response, i.e., $H(i\omega)$, $\omega \in \mathbb{R}^+$, for continuous-time systems and $H(\exp(i\omega))$, $\omega \in [0, 2\pi]$, for discrete-time systems.

The frequency response can be computed by the **repfreq** function. The vector of frequencies $f = (f_1, \ldots, f_n)$ is given in Hz. For instance, the command **repf=repfreq(SYS,f)** returns the frequency response of the system **SYS** in the complex vector **repf** (i.e., **repf(k)=Sys(i*2*%pi*f(k))** for a continuous-time system, and **repf(k)=Sys(exp(i*2*dt*%pi*f(k)))** for a discrete-time system, $dt = 1$, or sampled-time system). It may be difficult to select an appropriate f, and the function **calfrq** can be used just for that purpose. When the command **repfreq(SYS,fmin,fmax)** is issued, **calfrq** is automatically called. The graphical representation of the frequency response is obtained by the functions **bode**, **nyquist**, and **black**.

Let us visualize the Bode plots for the DC motor for the open-loop case and also for the closed-loop cases obtained using the two PID strategies described above:

```
leg=['DC motor: open loop';
     'DC motor: Ki=200;Kd=1;Kp=100;';
     'DC motor: Ki=200;Kd=10;Kp=100;'];
sys=[Motor;Closed2;Closed3];
bode(sys,1.e-2,1.e2,leg);                              ←  Figure 6.13
```

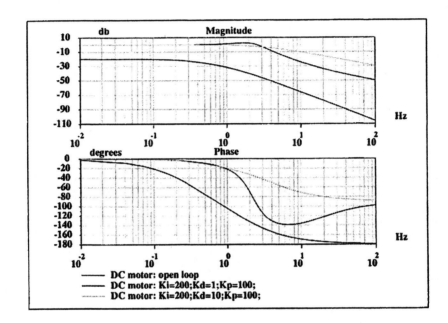

Figure 6.13: *DC motor: Bode plot*

Bode plots of sampled systems are drawn with a frequency range that does not exceed the Nyquist frequency $2\pi/T$, where T is the sampling period.

Let us now show a typical Nyquist plot for a randomly generated system:

```
Sys=ssrand(1,1,8);
nyquist(Sys);                                          ←  Figure 6.14
fmin=-0.23;fmax=0.23;frq=linspace(fmin,fmax,100);
frep=repfreq(Sys,frq);                     ←  zoomed frequency response
nyquist(frq,frep);                         ←  Nyquist plot from precomputed
                                                 data (Figure 6.15)
```

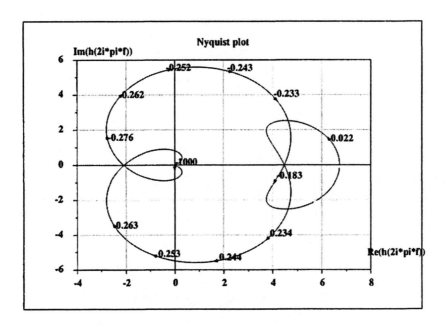

Figure 6.14: *Nyquist plot of the system* **Sys**

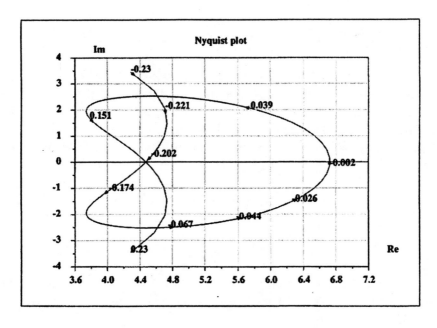

Figure 6.15: *Zoomed Nyquist plot*

The preceding session shows how Nyquist diagrams can be built from a given set of frequencies and response pairs. Typically, Nyquist diagrams are used to check the stability of a plant whose frequency response is known only through experimental measurements.

6.7   State-Space Control

On page 180 we saw how to build a compensator $K(s)$ in the SISO case, using PID and lead/lag compensators. With the general MIMO plant case, it is necessary to carry out the calculations in state-space form, for which powerful algorithms exist. This section is devoted to that purpose. The classical feedback control system is described in Figure 6.16.

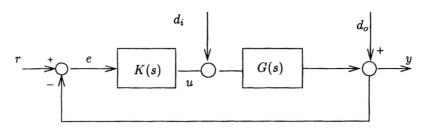

Figure 6.16: *Feedback control system*

In this diagram r denotes the command or reference variable, $K(s)$ is the to-be-designed compensator, and $G(s)$ is the nominal plant. Furthermore, d_o is the plant output disturbance, and d_i is the plant input disturbance. The linear system $L(s) = G(s)K(s)$ is called the open-loop transfer function. Figure 6.17 introduces three transfer functions of interest, called output sensitivity functions:

$$S(s) \equiv (I + L(s))^{-1},$$
$$R(s) \equiv K(s)(I + L(s))^{-1},$$
$$T(s) \equiv L(s)(I + L(s))^{-1} = I - S(s).$$

By definition, $S(s)$ is the transfer function from d_o to y in Figure 6.16 (assuming $r = 0$ and $d_i = 0$). Note that it is also the transfer function from $w = r$ to $z_1 = e$ in Figure 6.17.

Similarly, input sensitivity functions (transfer functions from d_i) can be defined. They are displayed in Figure 6.18, and defining $L_i = K(s) * G(s)$,

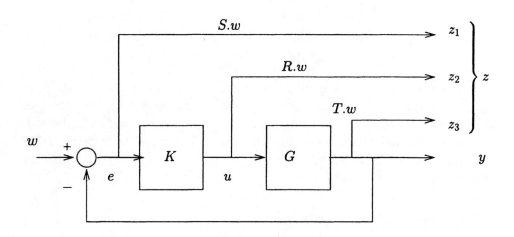

Figure 6.17: *Output sensitivity functions*

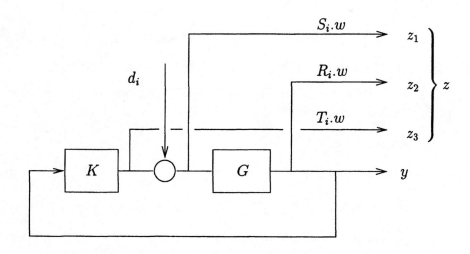

Figure 6.18: *Input sensitivity functions*

they may be written as follows:

$$S_i(s) \equiv (I + L_i(s))^{-1},$$
$$R_i(s) \equiv G(s)(I + L_i(s))^{-1},$$
$$T_i(s) \equiv L_i(s)(I + L_i(s))^{-1} = I - S_i(s).$$

A common way of designing a compensator $K(s)$ is to give specifications for the sensitivity functions. Typically, the frequency responses of $T(i\omega)$, $R(i\omega)$, and $S(i\omega)$ are to be shaped. For instance, their singular values should remain within specified bounds. This amounts to minimizing a given function of S, R, and T (e.g., the L_2 norm of S under a stability constraint) with respect to K.

6.7.1 Augmenting the Plant

Sensitivity functions can be obtained by lft operations (see equation (6.3) and Figure 6.2) on an "augmented plant." An easy calculation shows that

$$\begin{bmatrix} S \\ R \\ T \end{bmatrix} = \text{lft}(P_{SRT}, K), \qquad P_{SRT} \equiv \left[\begin{array}{c|c} I & -G \\ 0 & I \\ 0 & G \\ \hline I & -G \end{array} \right], \qquad (6.6)$$

$$\begin{bmatrix} S_i \\ R_i \\ T_i \end{bmatrix} = \text{lft}(P_{S_i R_i T_i}, K), \qquad P_{S_i R_i T_i} \equiv \left[\begin{array}{c|c} I & -I \\ G & -G \\ 0 & I \\ \hline G & -G \end{array} \right]. \qquad (6.7)$$

For instance, for the DC motor and the PID compensator introduced on page 179, $R(s)$ can be obtained as follows:

```
-->G=Motor;

--> [ny,nu]=size(G);

-->K=tf2ss(Pid1);                        ← state-space Pid compensator

-->R=lft([zeros(nu,ny),eye(nu,nu);
          eye(ny,ny)    ,    -G    ],[ny,nu],K);

-->ss2tf(R)
  ans   =
```

```
                    2          3     4
20.02 + 2014s + 1221.02s + 112s + s
-----------------------------------
                         2     3
     2 + 220.02s + 14s + s
```

In general, reckless application of formulas (6.6) or (6.7), using the Scilab concatenation operations for state-space systems, may result in non-minimal realizations for P_{SRT} or $P_{S_i R_i T_i}$, since the plant dynamics, G, may be duplicated in the formula for P_{SRT} or $P_{S_i R_i T_i}$. The function **augment** returns an "augmented plant," which is minimal and from which the sensitivity functions may be safely calculated through an **lft** operation. The function **augment** computes P_{SRT} and $P_{S_i R_i T_i}$ (in state-space form) by

$$P_{SRT} = \begin{bmatrix} I & 0 & 0 \\ 0 & I & 0 \\ -I & 0 & I \\ I & 0 & 0 \end{bmatrix} \times \begin{bmatrix} I & -G \\ 0 & I \\ I & 0 \end{bmatrix},$$

$$P_{S_i R_i T_i} = \begin{bmatrix} I & -I \\ 0 & 0 \\ 0 & I \\ 0 & 0 \end{bmatrix} + \begin{bmatrix} 0 \\ I \\ 0 \\ I \end{bmatrix} G \begin{bmatrix} I \\ -I \end{bmatrix}^T.$$

Let us illustrate the use of **augment** to compute the T sensitivity function for the previous example.

```
-->[P,r]=augment(G,'T');                    ←  r: size of G = size of P22

-->T=lft(P,r,K);                            ←  state-space T sensitivity function

-->clean(ss2tf(T))
 ans   =
                    2
        2 + 200s + 2s
        --------------------
                      2     3
     2 + 220.02s + 14s + s

-->unitGain=1;ss2tf(G*K/.unitGain)          ←  check
 ans   =
```

```
                  2
    2 + 200s + 2s
    ---------------------
                  2   3
    2 + 220.02s + 14s + s

-->clean(ss2tf(G*K*inv(eye+G*K)))              ←  another check, not
                                                  numerically safe

 ans  =
                       2
      1.9999965 + 200s + 2s
    ----------------------------
                        2   3
    1.9999962 + 220.02s + 14s + s
```

6.7.2 Standard Problem

Figure 6.19 shows how the feedback control system in Figure 6.16 is naturally embedded into a "standard problem" (assuming $d_i = 0$).

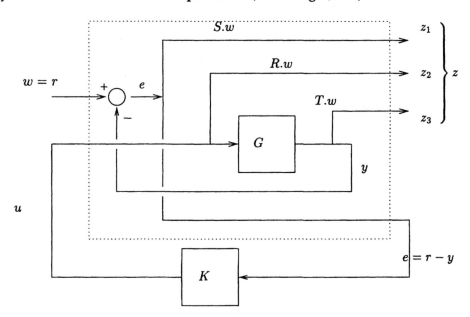

Figure 6.19: *Formulation as a standard problem*

This standard form is shown in Figure 6.20 and is very useful for representing, in a common framework, various control problems.

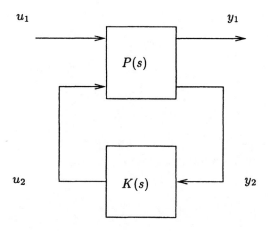

Figure 6.20: *Standard problem*

In this configuration u_1 is an external input (noise or external reference), u_2 is the control, y_1 is the controlled output, and y_2 is the measurement. The closed-loop transfer function $u_1 \to y_1$ is obtained by an `lft` operation.

For instance, in the previous section, output sensitivity functions have been computed from the formulation in Figure 6.19. The LQG control problem can also be formulated in the same framework.

6.7.3 LQG Design

The LQG equations are

$$
\min_{u} \mathbb{E} \left[\int_0^\infty x^T Q x + u^T R u + 2 x^T S u \; dt \right],
$$
$$
\dot{x} = Ax + Bu + \xi_1,
$$
$$
y = Cx + Du + \xi_2,
$$

where $\xi = (\xi_1, \xi_2)$ is white noise with covariance W:

$$
W = \left[\begin{array}{cc} W_{11} & W_{12} \\ W_{21} & W_{22} \end{array} \right].
$$

This problem can be rewritten in a standard form with P_{LQG} as follows:

$$\min_{u} \quad \|z\|_2, \tag{6.8}$$

$$\dot{x} = Ax + B_1 w + B_2 u, \tag{6.9}$$

$$z = C_1 x + D_{12} u, \tag{6.10}$$

$$y = C_2 x + D_{21} w + D_{22} u, \tag{6.11}$$

where $B_2 = B$, $D_{22} = D$, and w is white noise with unit covariance. Matrices C_1 and D_{12} can be obtained from Q, R, and S by a full-rank factorization:

$$\begin{bmatrix} C_1^T \\ D_{12}^T \end{bmatrix} \begin{bmatrix} C_1 & D_{12} \end{bmatrix} = V = \begin{bmatrix} Q & S \\ S^T & R \end{bmatrix}.$$

Similarly, B_1 and D_{21} can be obtained from W by

$$\begin{bmatrix} B_1 \\ D_{21} \end{bmatrix} \begin{bmatrix} B_1^T & D_{21}^T \end{bmatrix} = W = \begin{bmatrix} W_{11} & W_{12} \\ W_{21} & W_{22} \end{bmatrix}.$$

Equations (6.9), (6.10), and (6.11) define a (two-port) standard plant P_{LQG} that can be used to build an LQG controller K_{LQG}. Given the nominal plant G and the two weighting matrices W and V, the function `lqg2stan` returns the standard form P_{LQG}. The two weighting matrices W and V must be symmetric nonnegative definite. Moreover, W_{22} and $V_{22} = R$ must be strictly positive definite.

Let us consider an example of LQG design for the DC motor:

```
[r,nx]=syssize(Motor);              ←  r=[ny,nu]
V=diag([1,2,3]);                    ←  size nx+nu
W=diag([2,3,5]);                    ←  size nx+ny
[Plqg,r]=lqg2stan(Motor,V,W);       ←  defining the LQG problem
K=lqg(Plqg,r)                       ←  LQG controller
cl=lft(augment(Motor,'T'),r,K)      ←  closed-loop system
N=-cl('C')*inv(cl('A'))*cl('B')     ←  DC gain
y=csim('step',T,cl/N);              ←  closed loop with DC gain step response
```

This classroom example illustrates an LQG synthesis. This example is much too simple to be realistic: The choice of V and W to obtain a desired step response requires more detail, including the signal model to be tracked. In addition, we have been forced to add a constant multiplicative gain to the reference in order to obtain good asymptotic tracking of the step input.

We propose here to use an LQG approach for a model following problem. We are given an "ideal" model T_{id}, and we want to find a controller such that the closed-loop response matches T_{id}.

If we define the augmented plant as

$$P_{aug} \equiv \left[\begin{array}{c|c} -T_{id} & G \\ \hline I & -G \end{array} \right],$$

then the closed-loop system becomes $lft(P_{aug}, K) = GK(I + GK)^{-1} - T_{id}$, and we can try to minimize its L_2 norm. This is done by directly using P_{aug} in the function **lqg** as follows:

```
Tid=tf2ss(syslin('c',1/(1+1.2*s+s^2)));      ← the "ideal" model
Paug=augment(Motor,'T')-[Tid,0;0,0];              ← augmented plant
Paug('D')(1,2)=0.001;                    ← D₁₂ must be full column rank
r=[1,1];K=lqg(Paug ,r);            ← r=[ny,nu] computing the controller
Tzw=lft(augment(Motor,'T'),K);                    ← the closed loop
T=0:0.02:20;y=csim('step',T,[Tid;Tzw]);                    ← step
                                    response and ideal step response
plot2d([T',T'],y',[-1,1]);                        ← Figure 6.21
```

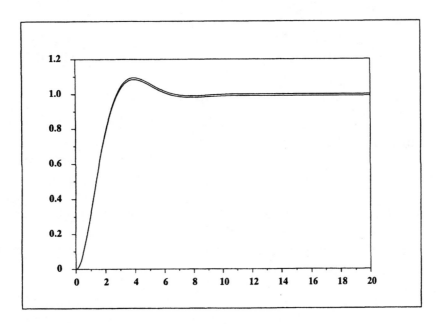

Figure 6.21: *Matching ideal time response*

We see in Figure 6.21 that we obtain the exact "ideal" response. In fact, the LQG controller obtained is close to the controller that exactly inverts the plant, i.e., the controller that solves the equation

$$lft(P_{aug}, K_{inv}) = GK_{inv}(I + GK_{inv})^{-1} - T_{id} = 0$$

(note that `lqg` requires $D_{12} \neq 0$). By pure chance, the K_{inv} controller is proper and stabilizing. It is easy to calculate K_{inv} and compare it with K:

```
-->Q=inv(Motor)*Tid;
```

```
-->Kinv=inv(eye-Q*Motor)*Q;
```                         $\leftarrow lft(P_{aug}, K_{inv}) = 0$

```
-->Tzw1=lft(augment(Motor,'T'),Kinv);
```

```
--> [h2norm(Tid-Tzw),h2norm(Tid-Tzw1)]
```                    $\leftarrow$ comparing
H_2 norms of LQG and invert the plant closed loops

```
 ans   =
!     .0085529      7.414E-09 !
```

In general, the previous approach requires scaling P_{aug} by an appropriate weighting function $w(s)$ as follows:

$$P_{aug} = \begin{bmatrix} -wT_{id} & wG \\ I & -G \end{bmatrix}.$$

This can be accomplished using the command **sysdiag(w,I)*Paug**.

6.7.4   Scilab Tools for Controller Design

An observer-based compensator design can also be obtained using the standard problem formulation. Figure 6.22 shows the classical observer-based compensator diagram. Here the observer $O(s)$ is a linear system that constructs an estimate $\hat{x}$ of the state of G, given u and y, i.e., a mapping $u, y \to \hat{x}$ ($r = 0$). Here F is a full-state feedback gain that stabilizes the pair (A, B) of the nominal plant G. Figure 6.22 also shows how this diagram can be redrawn as the usual feedback control system of Figure 6.16.

The following session illustrates two different methods for building an observer-based controller.

1. The first method follows the second diagram in Figure 6.22. We first build the controller K, which is given by the function **obscont**:

   ```
   --> [A,B,C,D]=abcd(Motor);
   ```         $\leftarrow$ motor state-space matrices

   ```
   -->nx=size(A,'c');
   ```         $\leftarrow$ number of states

   ```
   -->F=-ppol(A,B,[-1]*ones(1,nx));
   ```         $\leftarrow$ stabilizing full-state gain

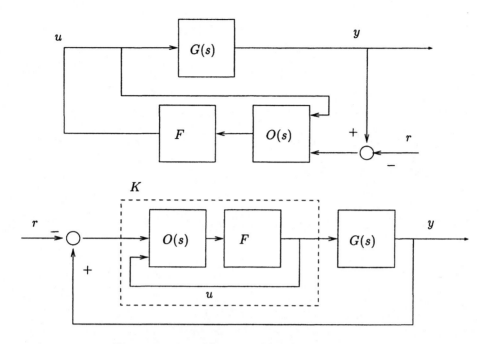

Figure 6.22: *Observer-based compensator*

```
-->J=-ppol(A',C',[-5]*ones(1,nx))';          ←  observer gain

-->K1=  obscont(Motor,F,J);          ←  observer-based controller

-->ss2tf(K1)
 ans   =


     247.5598 + 125.88s
     ------------------
                2
         25.98 + s

-->cll=-lft(augment(Motor,'T'),-K1);          ←  closed-loop
                                              system in Figure 6.22
```

2. The second method is almost the same; we do not use function **obscont** but recompute the result using the definition of K provided in the small subdiagram of Figure 6.22:

```
-->Obs=observer(Motor,J);          ←  building an observer
```

```
-->[ny,nu]=size(Motor);

-->FO=F*Obs;

-->FO1=FO(:,1:nu); FO2=FO(:,nu+1:nu+ny);

-->K2=inv(eye-FO1)*FO2;                    ← dashed box in Figure 6.22

-->ss2tf(K1-K2)
 ans  =
     0
     -
     1
```

This can also be achieved using the **lft** operation:

```
-->Switch=[zeros(nu,ny),eye(nu,nu);
            eye(ny,ny),zeros(ny,nu)];

-->K3=lft([eye(nu,nu);eye(nu,nu)]*FO*Switch,[nu,nu],..
            eye(nu,nu));

-->ss2tf(K1-K3)
 ans  =
     0
     -
     1
```

We can also build the closed-loop system in Figure 6.22 by literally following the diagram and using Scilab interconnection operations:

```
-->nr=ny;Iu=eye(nu,nu);Iy=eye(ny,ny);Ir=Iy;

-->G0=Switch;                                     ← (r,u) → (u,r)

-->G1=sysdiag([Motor;Iu;Motor],Ir)*G0;     ← (r,u) → (y,u,y,r)

-->G2=sysdiag(Iy,Iu,[Iy,-Ir])*G1;          ← (r,u) → (y,u,y−r)


-->G3=sysdiag(Iy,F*Obs)*G2;                   ← (r,u) → (y,F(x̂ − w))
                                              where w = Obs(u,r)′
-->cl2=-lft(G3,[nu,nu],Iu);  ← closing the loop with an lft operation
```

```
-->ss2tf(c12-c11)                          ← checking the result
 ans  =
    0
    -
    1
```

6.8 H_∞ Control

The H_∞ framework is convenient for translating many design problems into standard form (see Section 6.7.2). We will consider here the simple and classical problem of controlling a SISO double integrator

$$G(s) = \frac{1}{s^2}.$$

We are going to define the control problem in view of Figure 6.18, which defines input sensitivity functions. The desired closed-loop properties are expressed in terms of constraints put on the frequency response $L(s) = G(s)K(s)$. We want the Nyquist plot of the open loop $L(s)$ to be outside a constant-magnitude circle at around $m = 6$ dB to satisfy stability robustness criteria. We also want L to have significant roll-off at high frequencies, since the "true" plant's dynamics are not properly modeled there. Finally, the closed-loop system should reject the plant input disturbances d_i. These specifications can be converted into the following problem: Find a controller $K(s)$ that stabilizes $G(s)$ and such that

$$\left\| \begin{array}{c} w_R(s)R_i(s) \\ w_T(s)T_i(s) \end{array} \right\|_\infty \leq 1.$$

Here w_T and w_R are weighting functions that should be selected to satisfy the design specifications. These closed-loop functions are obtained by an lft operation from the augmented plant (see Section 6.7):

$$P_{R_iT_i} = \left[\begin{array}{c|c} w_RG & -w_RG \\ 0 & w_T \\ \hline G & -G \end{array} \right]. \tag{6.12}$$

We choose $w_T(s) = 1/Q(1 + (s/a)^3)$ with $Q = 10^{m/20}$, i.e., $m = 20\log_{10}(Q)$. The constant a is chosen by a simple search using the bisection method. We look at the feasibility of the associated H_∞ problem.

The standard H_∞ problem is solved by the Riccati equation approach, which requires regularity assumptions not necessarily satisfied by the augmented plant. This difficulty is overcome by left and right multiplications of the augmented plant by appropriate functions.

   A common requirement on the controller is that it provide an integral behavior, i.e., that it have a pole at zero. This is easily accomplished by designing a controller K_1 for $G_1 = G/s$ and using the controller $K = K_1/s$ for G. This is easily done in Scilab as shown here using **h_inf** function:

```
G=syslin('c',1/s^2);                              ← double integrator
G1=G/s;                                  ← introducing an additional integrator
Q=2; w_R=1;                                        ← design parameters
r=[1,1];                                     ← dimension of the plant
M=(1+s)^3;L=sysdiag(1,1,-M);R=sysdiag(1,1/M);              ← left and
                                                        right multipliers
romin=0.8;romax=1.2;niter=1;                      ← iteration constants
amin=0;amax=20;
while (amax-amin)/amax > 1.e-2;                    ← bisection search
  a=(amin+amax)/2; disp('a='+string(a));
  w_T=(1+(s/a)^3)/Q;
  P=sysdiag(w_R,-w_T,1)*augment(G1,'RT','i');            ← weighted
                                                    augmented plant
  P=L*P*R;                                ← regularized augmented plant
  [K1,ro]=h_inf(P,r,romin,romax,niter);          ← testing feasibility
  if ro==[] then amin=a; else amax=a; end
end
```

   At the end of this first step we have obtained a good weighting function $w_T(s)$. The $w_R(s)$ weighting function is kept constant ($w_R(s) = 1$). Now we can approach the optimal gain by the usual γ-iterations (the parameter $\rho = 1/\gamma^2$ is optimized). Since we know that the optimal closed-loop function is all pass, obtaining $\rho = 1$ as the optimal gain implies that the functions R_i and T_i are shaped as desired. This is the well-known H_∞ loop-shaping design illustrated below.

```
-->w_T=(1+(s/amin)^3)/Q;                          ← weighting function

-->P=sysdiag(w_R,-w_T,1)*augment(G1,'RT','i');

-->P=L*P*R;

-->romin=0.9;romax=1.1;[K1,mu]=h_inf(P,r,romin,romax,20);
                                                    ← γ-iterations

-->spec(h_cl(P,r,K1))            ← checking stability of closed-loop poles
  ans  =
! - 3.7287885 + 9.5514863i !
! - 3.7287885 - 9.5514863i !
! - 0.3533392 + 0.4892652i !
```

```
! - 0.3533392 - 0.4892652i !
```

`-->K=K1/s;`   ← forcing the integral behavior

`-->L=K*G;`   ← optimal open-loop function

`-->frq=0.10:0.01:20;nyquist(frq,repfreq(L,frq));`   ← Nyquist plot, Figure 6.23

`-->m_circle(20*log(Q)/log(10))`   ← 6 dB M-circle

`-->P=augment(G,'T');T=lft(tf2ss(P),tf2ss(K));`   ← closed loop **T** function

`-->t=0:0.01:10;[y,x]=csim('step',t,T);`   ← step response

`-->plot2d1('onn',t(:),[y(:),ones(y(:))])`   ← Figure 6.24

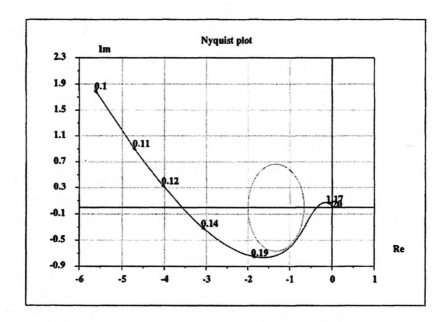

Figure 6.23: *Nyquist plot and M-circle*

6.9 Model Reduction

Model reduction is an important aspect of control design. In this section we
will give just the basic ideas, illustrated by a simple example.

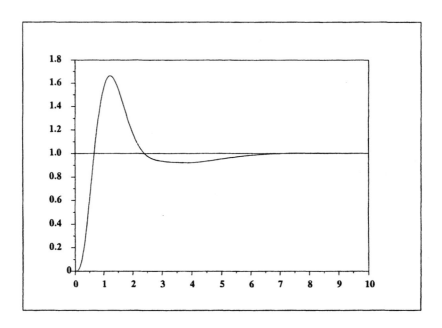

Figure 6.24: *Step response*

The objective of model reduction is to find a reduc-d order model that approximates a higher-order one. The reduced model is usually obtained by a projection that preserves a given set of characteristics. The simplest case is to preserve a set of eigenvalues of the A matrix of the higher-order model. Thus, the projection matrix is chosen to project onto the subspace spanned by the selected eigenvalues of A. As an example we consider a system made of the series interconnection of five second-order systems,

$$H(s) = \sum_{n=1}^{5} \frac{\omega_n^2}{s^2 + 2\zeta\omega_n s + \omega_n^2}.$$

The first reduced model selects the first three modes of the A matrix of H. The second reduced model selects the first three eigenvalues of the matrix $W = W_c W_o$, where W_c (resp. W_o) is the controllability (resp. observability) gramian matrix of H. These eigenvalues are also the singular values of the Hankel matrix. The following session illustrates two model reduction strategies:

```
-->om=logspace(-2,3,5);zeta=0.01*(1:5);          ← parameters

-->H=0;for i=1:5                                  ← construction of H(s)
```

```
-->H=H+tf2ss(syslin('c',om(i)^2/(s^2+2*om(i)*zeta(i)*s+..
                              om(i)^2)));
```

```
-->abs(spec(H('A')))                          ← poles of H
 ans  =
!   0.01      !
!   0.01      !
!   0.1778279 !
!   0.1778279 !
!   3.1622777 !
!   3.1622777 !
!   56.234133 !
!   56.234133 !
!   1000.     !
!   1000.     !
```

```
-->thres=10;[Q1,M1]=psmall(H('A'),thres,'d');     ← keeping
```
first three modes

```
-->Hred1=projsl(H,Q1,M1);                     ← reduced-order model
```

```
-->abs(spec(Hred1('A')))                      ← checking eigenvalues
 ans  =
!   0.01      !
!   0.01      !
!   0.1778279 !
!   0.1778279 !
!   3.1622777 !
!   3.1622777 !
```

```
-->Wc=ctr_gram(H);Wo=obs_gram(H);spec(Wc*Wo)     ← gramian
```
matrices and squared Hankel singular values
```
 ans  =
!   637.66016 !
!   612.65121 !
!   162.65571 !
!   150.14116 !
!   73.768179 !
!   65.414847 !
!   22.52142  !
!   36.070398 !
!   42.338041 !
!   27.492053 !
```

```
-->thres=50;[Q2,M2]=pbig(Wc*Wo,thres,'d');            ← selecting
```

largest Hankel singular values

```
-->Hred2=projsl(H,Q2,M2);
```
← another reduced order model

```
-->gainplot([H;Hred1;Hred2]);
```
← Figure 6.25

```
-->Wcr=ctr_gram(Hred2);Wor=obs_gram(Hred2);spec(Wcr*Wor)
```
←
checking Hankel singular values

```
 ans   =
 !    637.66016 !
 !    65.414847 !
 !    73.768179 !
 !    150.14116 !
 !    162.65571 !
 !    612.65121 !
```

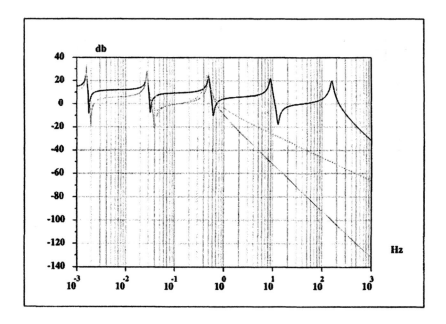

Figure 6.25: *Magnitude responses*

Note that the second reduction policy assumes that the system is stable for the gramian matrices to be well-defined. It is also possible to obtain a model in which controllability and observability gramian matrices are equal and diagonal. This is obtained using the function **balreal**.

6.10 Identification

Most of the time, real systems are known only via recorded input–output trajectories. An identification step is necessary to find a model compatible with these trajectories. Many algorithms have been developed to perform this identification step, and identification is still an active research area. In this section we will describe just a few Scilab identification tools.

The function **imrep2ss** builds a discrete-time linear model from an impulse response. The construction is in state space and is based on a standard algorithm of an SVD factorization of the impulse response's Hankel matrix. To illustrate the use of **imrep2ss**, we simulate the impulse response of a system and try to reconstruct a state-space realization:

```
-->A=[0,2;0.25,0.33];B=[1;2];C=[-2,-1];

-->Sysd=syslin('d',A,B,C);

-->N=20;u=[];u(N)=0;u(1)=1;impulse=flts(u',Sysd);

-->sli=imrep2ss(impulse(2:$));

-->[ss2tf(Sysd),ss2tf(sli)]
 ans  =
!    - 7.59 - 4z              - 7.5654538 - 4z            !
!    ---------------    ------------------------------    !
!                 2                                  2 !
! - 0.5 - 0.33z + z    - 0.4803621 - 0.2944459z + z  !
```

If a continuous-time model is required, the function **bilin** can be invoked to transform the identified discrete model into a continuous-time one. For instance, **bilin(s1, [T/2,1,-T/2,1])** performs the change of variable $s = (z + 2/T)/(-z + 2/T)$, where T is the sampling period. The function **arl2** returns a stable discrete-time system whose impulse response is the closest to the data in the L_2 norm. This function uses a transfer function approach. The order of the approximating system should be given. Let us examine the behavior of **arl2** for the previous example:

```
-->sla=arl2(impulse(2:$),%z,2);

-->za=flts(u',sla);norm(za-impulse)
 ans  =
    0.7554359

-->zi=flts(u',sli);norm(zi-impulse)
```

```
ans   =
    2.8491441

-->plot2d2("onn",[1:N]',[za;zi]',..
[1,2],"051"," ",[0,-10,N,0]);
```
                                                                    ← Figure 6.26

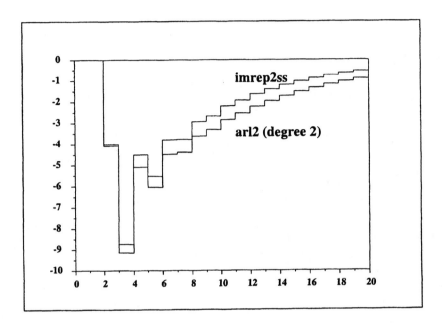

Figure 6.26: *Impulse response identification* **arl2** *and* **imrep2ss**

To identify a SISO model knowing its frequency response, the function **frep2tf** can be used. This function performs a least squares fit. If the transfer function is $n(s)/d(s)$, **frep2tf** adjusts the coefficients of n and d in order to minimize

$$\sum_k (\|n(\phi_k) - d(\phi_k)r_k\|_2^2),$$

where $\phi_k = 2i\pi f_k$ for the continuous-time domain or $\phi_k = \exp(2i\pi f_k \Delta t)$ for sampled systems:

```
Sys=ssrand(1,1,20);                          ←  random system
frq=logspace(-3,2,100);                      ←  vector of frequencies
[frq,rep]=repfreq(Sys,frq);                  ←  frequency response
Sys2=frep2tf(frq,rep,10);                    ←  identified model of order 10
[frq,rep2]=repfreq(Sys2,frq);bode(frq,[rep;rep2]);              ←
                                                comparing responses
```

Note that **frep2tf** does not return a stable system. If one is interested only in the magnitude response, the system can be converted into a stable and minimum phase system by taking the mirror image of the unstable poles and zeros with respect to the imaginary axis (or the unit circle in the discrete-time case). This conversion, which does not change the magnitude response, is achieved using the function **factors** as follows:

```
Sys3=factors(Sys2,'c');              ←  stabilization of system Sys2
[frq,rep3]=repfreq(Sys3,frq);
bode(frq,[rep;rep3]);                          ←  Figure 6.27
```

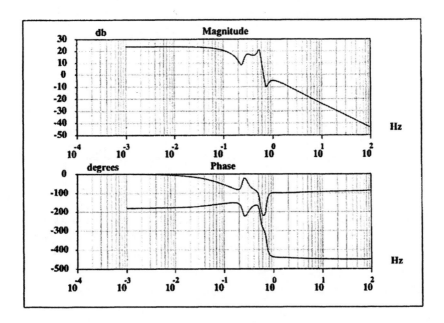

Figure 6.27: *Matching magnitude with stable system*

Many control applications seek a system that matches a given magnitude response. In order to apply the function **frep2tf**, a compatible phase must be identified. A first step toward this end can be accomplished using the function **cepstrum** [46].

A few arma process identification algorithms are also implemented in Scilab. We give a simple example.The problem is to identify model parameters given an input trajectory u and an output trajectory y, where y and u are related by a model of the following form:

$$A(z^{-1})y = B(z^{-1})u + D(z^{-1})e(t),$$

where A, B, and C are polynomials and e is discrete-time white noise.

```
--> a=[1,-2.851,2.717,-0.865];                          ← coefficients
                                              of the A, B, and C  polynomials
-->b=[0,1,1,1];d=[1,0.7,0.2];

-->ar=armac(a,b,d,1,1,1)                       ← a Scilab object of type arma
 ar  =
  A(z^-1)y=B(z^-1)u + D(z^-1) e(t)

  A(s)=| 1 |s^0 + |-2.851 |s^1 + | 2.717 |s^2 + |-0.865 |s^3
  B(s)=| 0 |s^0 + | 1 |s^1 + | 1 |s^2 + | 1 |s^3
  D(s)=| 1 |s^0 + | 0.7 |s^1 + | 0.2 |s^2
  e(t)=Sig*w(t); w(t) 1-dim white noise
  Sig=  | 1 |

-->n=300;                                      ← the given input size

-->u=-prbs_a(n,1,int([2.5,5,10,17.5,20,22,27,35]*100/12));
                                               ← the given input
-->rand('seed',0);

-->zd=narsimul(ar,u);                          ← simulation of the output

-->plot2d2("enn",1,zd',[1,1],"121","Simulated output");

-->plot2d2("enn",1,1000*u',[3,3],"100","Input [scaled]");

-->[arc,la,lb,sig,resid]=armax(3,3,zd,u,1,1);  ← a least squares
                        identification algorithm (white noise assumption)
-->arc                                   ← the identified arma process
 arc  =
  A(z^-1)y=B(z^-1)u + D(z^-1) e(t)

  A(s)=| 1 |s^0 + |-2.8677932 |s^1 + | 2.7500177 |s^2 +
       |-0.8814121 |s^3
  B(s)=| 0 |s^0 + | 1.20468 |s^1 + | 0.7736701 |s^2 +
       | 0.7874893 |s^3
  D(s)=| 1 |s^0
  e(t)=Sig*w(t); w(t) 1-dim white noise
  Sig=  | 1.2053576 |

-->rand('seed',0);zd1=narsimul(arc,u);             ← simulation of
                              identified model (same noise sequence)
-->plot2d2("enn",1,zd1',[2,4],"000","Identified output");
                                               ← Figure 6.28
-->[arc1,resid]=armax1(3,3,2,zd(1:n),u,1);         ← general
                        identification algorithm (colored noise assumption)
```

```
-->arc1                                    ← the identified arma process
 arc1  =
  A(z^-1)y=B(z^-1)u + D(z^-1) e(t)

  A(s)=| 1 |s^0 + |-2.8680452 |s^1 + | 2.7506248 |s^2 +
       |-0.8817700 |s^3
  B(s)=| 0 |s^0 + | 1.2032827 |s^1 + | 0.8060807 |s^2 +
       | 0.7129602 |s^3
  D(s)=| 1 |s^0 + | 0.2292066 |s^1 + | 0.0095127 |s^2
  e(t)=Sig*w(t); w(t) 1-dim white noise
  Sig=  | 1.1886908 |
```

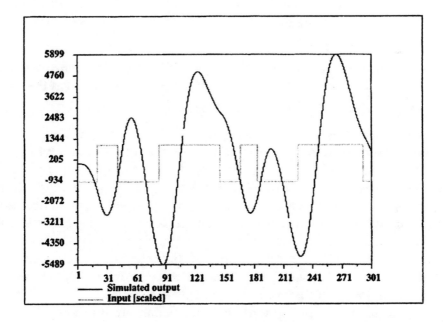

Figure 6.28: *Arma simulation*

6.11   Linear Matrix Inequalities

Scilab has an interface with the recently developed SP code (Lieven Vandenberghe and Stephen Boyd, [56]) for solving semidefinite optimization problems. Many problems in systems and control can be formulated as follows (see [15]):

$$\Sigma: \begin{cases} \text{minimize} & f(X_1,\dots,X_M) \\ \text{subject to} & \begin{cases} G_i(X_1,\dots,X_M) = 0, & i = 1,2,\dots,p, \\ H_j(X_1,\dots,X_M) \geq 0, & j = 1,2,\dots,q, \end{cases} \end{cases}$$

where

- $X_1,\dots,X_M$ are unknown real matrices,

- f is a real linear scalar function of the entries of the unknown matrices $X_1,\dots,X_M$ and is called the *objective function*,

- The G_i's are real matrices that are affine functions of the unknown matrices, $X_1,\dots,X_M$ and are called the linear matrix equality (LME) functions,

- The H_j's are real symmetric matrices that are affine functions of the unknown matrices $X_1,\dots,X_M$, and they are referred to as the "linear matrix inequality" (LMI) functions.

The interactive function `lmitool` is an interface to describe an LMI problem. This function is built around the function `lmisolver`, which computes the solution $X_1,\dots,X_M$ of problem Σ, given functions f, G_i, and H_j. To solve Σ, the user must provide a function that "evaluates" f, G_i, and H_j given values for $X_1,\dots,X_M$. The user must also supply an initial guess for the unknown matrices. The function `lmisolver` can be either invoked directly by providing the necessary information in a special format or used through the interactive interface function `lmitool` which automates most of the steps required before invoking `lmisolver`. In particular, `lmitool` generates a file (with the extension `.sci`) that contains the solver function and the evaluation function written in the Scilab programming language. The solver function is used to define the initial guess and to modify optimization parameters (if needed).

Several examples of LMIs are given as Scilab demos, which can be run by clicking on the demo button. A specific user guide is also available that describes several control examples.

Chapter 7

Signal Processing

The purpose of this chapter is to illustrate the use of the Scilab software package in a signal-processing context. This section discusses the core signal-processing tools, which will include discussions on signal representation, FIR and IIR filter design, and spectral estimation.

Scilab contains many signal processing tools other than those cited in the above list. Among these tools are those for stochastic realization, Kalman and Wiener filtering, and time frequency representation of signals (e.g., Wignerville). The user may use Scilab's online help for the signal-processing toolbox, as well as the code contained in signal processing functions as guidelines to the syntax and utilization of these tools.

7.1 Time and Frequency Representation of Signals

In this section a variety of ways in which signals can be represented in Scilab are described. The section begins by explaining how sampled signals can be resampled using interpolation and decimation procedures. This is followed by a discussion of the discrete Fourier transform and its computation using the FFT. The representation of signals in the Fourier transform domain allows for the analysis of magnitude, phase delay, and group delay. This section describes Scilab tools for analyzing these quantities. Polynomial transforms of sampled data, such as the z-transform, can be easily computed in Scilab. Finally, this section describes Scilab's object-oriented approach to signal representation and how signal processing tools can be transparently used on these objects without regard to their particular representation.

7.1.1   Resampling Signals

Signals often need to be resampled. For example, two signals that are obtained at different times by different experimenters may need to be combined in some way. If the sample rates of the two signals are different, then a resampling will be necessary in order to proceed.

The Scilab function **intdec** is based on the work of [52] and accomplishes a sampling rate change through a process of interpolation followed by decimation. The interpolation takes the input signal and produces an output signal that is sampled at an integer rate L times more frequently than the input. Then decimation takes the input signal and produces an output signal that is sampled at an integer rate M times less frequently than the input. To change the sampling rate of a signal by a noninteger quantity it suffices to perform a combination of the interpolation and decimation operations. The function **intdec** works on both 1-D and 2-D signals.

As an example, a signal at a 1 kHz sample rate can be converted to one at a 400 Hz sample rate:

```
dt=.001;                              ← sample period is 1 ms
t=(0:999)*dt;                         ← sample rate is 1000 Hz
f=sin(%pi*t).*sin(16*%pi*t);          ← modulated test signal
g=intdec(f,2/5);                      ← signal resampled to 400 Hz

xsetech([0,0,1,.5])
plot(f)
xtitle('Signal at 1000 Hz Sample Rate')
xsetech([0,0.5,1,.5])
plot(g)
xtitle('Signal at 400 Hz Sample Rate')
```

The results of the above Scilab session are illustrated in Figure 7.1. As can be seen in the figure, the signal has been converted from 1000 samples to 400 samples (i.e., a sample rate of 1 kHz to one of only 400 Hz) without any apparent loss of signal character.

7.1.2   The DFT and the FFT

The FFT ("fast Fourier transform") is a computationally efficient algorithm for calculating the DFT ("discrete Fourier transform") of finite-length, discrete-time sequences. Calculation of the DFT from its definition requires order N^2 multiplications, whereas the FFT requires order $N \log_2 N$ multiplications. For a finite-length sequence $x(n)$, where $n = 0, 1, \ldots, N - 1$, the

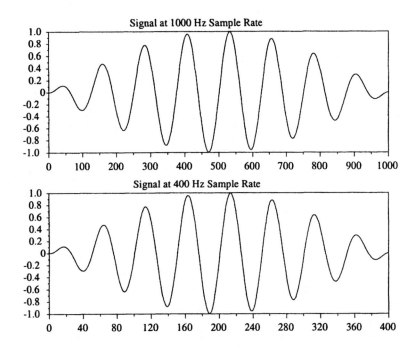

Figure 7.1: *Resampling a signal using* `intdec`

definition of the DFT of $x(n)$ is the finite-length sequence $X(k)$:

$$X(k) = \sum_{n=0}^{N-1} x(n)e^{-j\frac{2\pi}{N}nk} \quad k = 0, 1, \ldots, N-1. \tag{7.1}$$

The N points of the sequence $x(n)$ can be recovered from N points of $X(k)$. This recovery is called the inverse DFT and takes the form

$$x(n) = \frac{1}{N} \sum_{k=0}^{N-1} X(k)e^{j\frac{2\pi}{N}nk}. \tag{7.2}$$

It is equally possible to compute a multidimensional DFT. For a multidimensional sequence $x(n_1, n_2, \ldots, n_M)$ the multidimensional DFT is defined by

$$
\begin{aligned}
X(k_1, k_2, \ldots, k_M) &= \sum_{n_1=1}^{N_1-1} e^{-j\frac{2\pi}{N_1}n_1k_1} \sum_{n_2=1}^{N_2-1} e^{-j\frac{2\pi}{N_2}n_2k_2} \times \cdots \\
&\quad \times \sum_{n_M=1}^{N_M-1} e^{-j\frac{2\pi}{N_M}n_Mk_M} x(n_1, n_2, \ldots, n_M).
\end{aligned}
$$

The inverse multidimensional DFT is analogous to the calculation above with a change of sign for the complex exponentials and with a factor of $1/(N_1 N_2 \cdots N_M)$. The FFT algorithm is a computationally efficient way of calculating the DFT.

An example is now presented illustrating how to use the **fft** function in Scilab. A simple cosine signal is chosen for the one-dimensional signal. The following Scilab session illustrates the example, and the results are plotted in Figure 7.2.

```
dx=1/100;                           ←  sample period is 10 ms
x=(0:99)*dx;                        ←  sample rate is 100 Hz
y=cos((2*%pi)*(10*x));              ←  cosine test signal
yf=fft(y,-1);                       ←  FFT of test signal

xsetech([0,0,1,.5])
plot2d(x,y)
xtitle('Time Domain 1-D signal','time','amplitude')
xsetech([0,0.5,1,.5])
plot2d(x/dx,abs(yf))
xtitle('Magnitude of the Fourier Transform','frequency',..
        'magnitude')
```

As can be seen in the figure, the magnitude of the Fourier transform of the cosine signal gives rise, as it should, to two spikes. The first spike is located at 10 Hz, the frequency of the cosine, and the second spike is at 90 Hz, which when wrapped around the Nyquist frequency of 50 Hz is actually -10 Hz.

As specified in (7.2), the original signal may be recovered by computing the inverse DFT. The inverse DFT is also computed using the **fft** function but where the second argument is +1 instead of −1. These arguments come from the definition of the sign of the argument of the exponential in (7.2) and (7.1), respectively.

A two-dimensional DFT can be computed using the same **fft** function. The syntax is not different from that for the one-dimensional problem. For signals of dimension higher than 2, which are nevertheless stored in Scilab as 1-D or 2-D arrays, the DFT is computed using the mulidimensional function for computing the FFT: **mfft**.

7.1.3   Transfer Function Representation of Signals

Scilab can be used to represent signals in much more sophisticated ways than as just sampled signals in the time or frequency domain. This section gives a brief description of how continuous or discrete signals can be

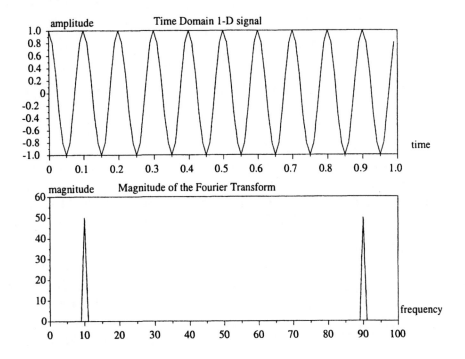

Figure 7.2: *Computing the DFT using* fft

represented as rational functions in the Laplace or z-transform domains, respectively.

Matrix polynomials and matrix rational functions are defined types in Scilab. Scalar polynomials are easily created and manipulated. The **poly** primitive in Scilab can be used to define a polynomial either by its coefficients or by its roots. These features of Scilab are more fully described in Section 2.4.1. Any signal that can be represented as the solution of a constant-coefficient differential or difference equation can be represented in Scilab as a rational transfer function. For example, the differential equation

$$\sum_{n=1}^{N} a_n \frac{d^n}{dt^n} y(t) = \sum_{m=1}^{M} b_m \frac{d^m}{dt^m} x(t) \tag{7.3}$$

can be Laplace transformed and arranged in transfer function form to yield

$$H(s) = \frac{\sum_{n=1}^{N} a_n s^n}{\sum_{m=1}^{M} b_m s^m}, \tag{7.4}$$

where s is the Laplace transform variable. The difference equation

$$\sum_{n=1}^{N} a_n y(n\Delta t) = \sum_{m=1}^{M} b_m x(m\Delta t) \qquad (7.5)$$

can be z-transformed to give

$$H(z) = \frac{\sum_{n=1}^{N} a_n z^n}{\sum_{m=1}^{M} b_m z^m}, \qquad (7.6)$$

where z is the z-transform variable.

Scilab has two built-in polynomial variables, **%s** and **%z**, which can be used for the construction of rational transfer functions. As an example, a single-input single-output system with one zero and two poles can be constructed by

```
-->H = (1+%s)/((10+%s)*(100+%s))      ← 1 zero, 2 pole transfer function
 H  =

         1 + s
    ---------------
                    2
    1000 + 110s + s
```

Multi-input multi-output systems are created in Scilab using matrix rational functions. An example of a simple 2×2 rational transfer matrix is

```
-->H=[1/%s %s/(1+%s^2);1/(1+%s) 1/%s] ← 2 × 2 rational transfer matrix
 H  =
!   1              s     !
!   -            -----   !
!                     2  !
!   s            1 + s   !
!                        !
!       1          1     !
!     -----        -     !
!     1 + s        s     !
```

For signal processing purposes, the evaluation of rational transfer functions is accomplished using the function **freq**. For example, a discrete filter can be evaluated on the unit circle in the z-plane as follows:

```
p=2*%pi;                          ←  define a pole in radians
H=1/(1+%s/p);                     ←  define a transfer function
f=.01:.01:10;                     ←  define a vector of frequencies
Hf=freq(H(2),H(3),2*%pi*%i*f);    ←  evaluate H on the jw axis
bode(f,Hf)                        ←  plot magnitude and phase
```

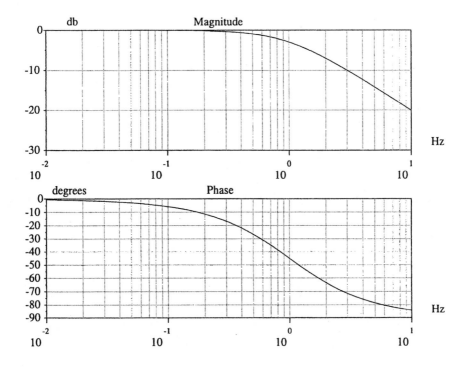

Figure 7.3: *Bode plot obtained using* **freq**

This gives rise to the Bode plot shown in Figure 7.3, which as expected has a magnitude response which downs −3 dB at the pole and then drops at a rate of −20 dB/decade, whereas the phase goes from 0° through −45° at the pole and finishes at −90°.

Furthermore, the impulse response of discrete linear systems represented by rational transfer matrices can be had using **rtitr**. As an example:

```
H=1/(1+.9/%z);                          ← create a single pole filter
impulse=[1,zeros(1,64)];                ← create an impulse
Ht=rtitr(H(2),H(3),impulse);            ← the discrete time response to
                                          the impulse
plot(Ht)
xtitle('Impulse Response of a Discrete Single Pole Filter')
```

which yields the plot shown in Figure 7.4.

When rational matrices in Scilab are intended to represent linear time-invariant systems, there is a very useful data abstraction tool called **syslin**, whose syntax is as follows:

```
sl=syslin(domain,num,den)
```

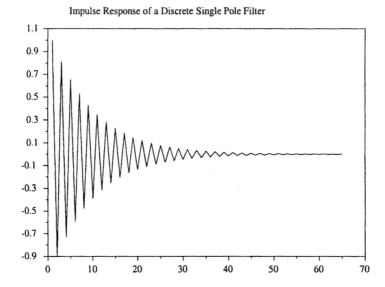

Figure 7.4: *Impulse response obtained with* **rtitr**

where **domain** is the character chain **'c'** or **'d'** for continuous or discrete systems, **num** and **den** are the numerator and denominator polynomial matrices of the rational function, and **sl** is the resulting abstract linear system data object. The usefulness of the **syslin** linear system abstraction will be clearer in the discussion that follows.

7.1.4   State-Space Representation

Another useful way of representing signals is as the output of a state-space representation of a linear system. The classical state-space description of a continuous time linear system is

$$
\begin{aligned}
\dot{x}(t) &= Ax(t) + Bu(t), \\
y(t) &= Cx(t) + Du(t), \\
x(0) &= x_0,
\end{aligned}
$$

where A, B, C, and D are matrices and x_0 is a vector, and for a discrete-time system it takes the form

$$\begin{aligned} x(n+1) &= Ax(n) + Bu(n), \\ y(n) &= Cx(n) + Du(n), \\ x(0) &= x_0. \end{aligned}$$

State-space descriptions of systems in Scilab, like rational transfer functions, use the **syslin** function:

```
sl=syslin(domain,a,b,c [,d[,x0]])
```

where **domain** is the character chain **'c'** or **'d'** for continuous or discrete systems, and **a**, **b**, **c**, and **d** are corresponding state matrices. The returned value is a list, **sl=list('lss',a,b,c,d,x0,domain)**.

Similar to the case of rational transfer function matrices, state-space descriptions of linear systems can be used to examine the impulse response and frequency response of the system. The impulse response is obtained using **ltitr** and the frequency response using **repfreq**.

One of the values of having an abstract data object that represents a state-space description of a system is that operators and functions can be overloaded to correctly handle linear systems regardless of their format. For example, linear systems can be combined in parallel or in cascade, transformed from state-space descriptions to transfer function descriptions (or vice versa), and discretized or interpolated by continuous-time systems in a totally transparent fashion. The next section discusses changing between system representations.

7.1.5 Changing System Representation

Sometimes linear systems are described by their transfer function and sometimes by their state space equations. In the event where it is desirable to change the representation of a linear system, there exist two Scilab functions that are available for this task. The first function, **tf2ss**, converts a system described by a transfer function to a system described by state-space representation. The second function, **ss2tf**, works in the opposite sense.

The syntax of **tf2ss** is as follows:

```
sl=tf2ss(h)
```

where **h** is an abstract linear system data object created using **syslin**. An example of the use of **tf2ss** is:

```
-->H=syslin('c',1/(1-2*%s+%s^2))
  H  =
             1
        ---------
                  2
        1 - 2s + s
```

```
-->ssprint(tf2ss(H))
  .      | 0 -1 |     | 0 |
  x =    | 1  2 |  +  | 1 |u
```

```
  y =  |-1  0 |x
```

where the function **ssprint** is responsible for the pretty-print representation of the derived state-space system. The fact that the resulting state-space description corresponds to the original transfer function can be verified by computing $C(sI - A)^{-1}B$, which yields

```
-->s1=tf2ss(H);
```

```
-->a=s1(2); b=s1(3); c=s1(4);
```

```
-->c*inv(%s*eye-a)*b
  ans  =
             1
        ---------
                  2
        1 - 2s + s
```

where **eye** creates an identity matrix of the appropriate dimensions. The above computation corresponds, of course, to the function that converts state-space systems into transfer functions

```
-->ss2tf(s1)
  ans  =
             1
        ---------
                  2
        1 - 2s + s
```

7.1.6 Frequency-Response Evaluation

In the previous sections several methods have been discussed for obtaining the frequency representation of a signal. In this section three additional tools for evaluating the frequency response of a system are presented. The first, **bode**, is a sophisticated way of representing a signal magnitude and phase response, where the signal may be represented as a continuous or discrete linear system in either transfer function or state-space form. The second tool, **frmag**, is specifically for computing the magnitude response of FIR and IIR digital filters. Finally, the third tool, **group**, computes the group delay of a digital filter.

Bode Plots

The Bode plot is used to plot the phase and log-magnitude response of functions of a single complex variable. Before computers were widely used for engineering calculations, the log-scale characteristics of the Bode plot permitted rapid "back-of-the-envelope" calculations of a system's magnitude and phase response. For this the Bode plot was an extremely useful engineering tool.

Bode plots are still very useful, and in Scilab they are precisely computed. In the following discussion only real, causal systems are considered. Consequently, any poles and zeros of the system occur in complex conjugate pairs (or are strictly real), and the poles are all located in the left-half s-plane for continuous systems or within the unit circle for discrete systems.

The Bode plot of a transfer function, $H(s)$, of the complex variable s really is two plots: a log-magnitude plot and a phase plot. The log-magnitude of $H(s)$ is defined by

$$M(\omega) = 20 \log_{10} |H(s)_{s=j\omega}|, \tag{7.7}$$

and the phase of $H(s)$ is defined by

$$\Theta(\omega) = \tan^{-1} \left[\frac{\text{Im}(H(s)_{s=j\omega})}{\text{Re}(H(s)_{s=j\omega})} \right]. \tag{7.8}$$

The magnitude, $M(\omega)$, is plotted on a log-linear scale where the independent axis is marked in decades (sometimes in octaves) of degrees or radians and the dependent axis is marked in decibels. The phase, $\Theta(\omega)$, is also plotted on a log-linear scale where the independent axis is the same as for the magnitude plot and the dependent axis is marked in degrees (and sometimes radians). Several different calls to **bode** exist. In the following example the call is based on the linear system data type. Other call syntaxes can be obtained from Scilab's on-line help.

When using **bode** to plot the frequency response of a signal corresponding to a linear system, the syntax can take two different forms:

```
bode(sl,[fmin,fmax] [,step] [,comments])
bode(sl,frq [,comments])
```

For both, the input **sl** is a linear system that was created using **syslin**. In the first syntax the user may specify the scalar frequency limits, **fmin** and **fmax**, and/or the step, **step**, between frequencies for which **bode** will plot the response. If these are not specified, **bode** automatically determines relevant values by determining the poles and zeros of the system. The second syntax plots the frequency response at a set of frequencies specified by the vector **frq**.

The function **bode** demonstrates the power of data abstraction in Scilab, since both

```
H=syslin('c',1/(1-2*%s+%s^2));     ←   transfer function representation
bode(H)
```

and

```
sl=tf2ss(H);                                      ←   state-space representation
bode(sl)
```

yield exactly the same result even though the first is a transfer function and the second a state-space representation of the same system.

The Function frmag

Later in this chapter it will be very useful to have a general tool for evaluating the magnitude of the frequency response for FIR and IIR filters. The main tool that will be used is **frmag**. The function **frmag** evaluates the magnitude response of the transfer function on the unit circle in the z-plane. The function **frmag** has two syntaxes. The first one is

```
[xm,fr]=frmag(h,npts)
```

where **h** can be a vector, a polynomial, or a rational function specifying the digital filter, and **npts** is an integer giving the number of points to be evaluated around the unit circle in the z-plane. If **h** is a vector or a polynomial, then it is taken to be a representation of an FIR filter. If **h** is a rational function then it is presumed to be a representation for an IIR filter. The outputs are the magnitude responses, **xm**, at the frequencies, **fr**.

The second syntax for **frmag** is

```
[xm,fr]=frmag(num,den,npts)
```

where **num** and **den** can be either vectors or polynomials representing the numerator and denominator of a rational transfer function. In this syntax **frmag** computes the magnitude response of an IIR filter. The remaining arguments are the same as for the first syntax of **frmag**.

Group Delay

Let $H(\omega)$ denote the Fourier transform of a signal

$$H(\omega) = A(\omega)e^{j\theta(\omega)} \tag{7.9}$$

where $A(\omega)$ is the magnitude of $H(\omega)$, and $\theta(\omega)$ is the phase of $H(\omega)$. The group delay, $t_g(\omega)$, is defined by

$$t_g(\omega) = d\theta(\omega)/d\omega. \tag{7.10}$$

A nonconstant group delay tends to cause signal deformation. This is due to the fact that the different frequencies that compose the signal are time shifted by different amounts according to the value of the group delay at that frequency. Consequently, it is valuable to examine the group delay of filters during the design procedure. The Scilab function **group** computes the group delay of a digital filter.

The function **group** accepts filter parameters in several formats as input and returns the group delay as output. The main syntax of the function is as follows:

```
[tg,fr]=group(npts,h)
```

The group delay **tg** is evaluated in the interval $[0, 0.5)$ at equally spaced samples contained in **fr**. The number of samples is governed by the integer **npts**. Three formats can be used for the specification of the filter. The filter **h** can be specified by a vector of real numbers, by a rational polynomial representing the z-transform of the filter, or by a matrix polynomial representing a cascade decomposition of the filter.

As an example of the use of **group**:

```
hz=iir(3,'lp','ellip',[.15,0],.025*[1 1]);     ← low-pass elliptic
                                                        filter
[tg,fr]=group(256,hz);                          ←  group delay

plot(fr,tg)
xtitle('Group Delay of a Low-Pass Elliptic Filter',..
       'Frequency (Hz)','Group Delay')
```

which yields the results illustrated in Figure 7.5.

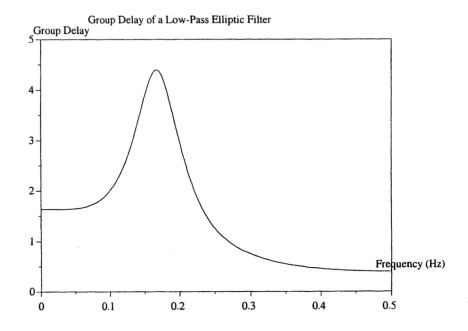

Figure 7.5: *Group delay computed with* **group**

7.1.7 The Chirp z-Transform

The discrete Fourier transform (DFT) of a finite-length, discrete-time signal $x(n)$ is defined by

$$X(k) \;=\; \sum_{n=0}^{N-1} x(n)e^{-j(2\pi nk)/N},$$
$$k \;=\; 0, 1, \ldots, N-1,$$

and the z-transform of $x(n)$ is given by

$$X(z) \;=\; \sum_{n=-\infty}^{\infty} x(n)z^{-n}$$
$$=\; \sum_{n=0}^{N-1} x(n)z^{-n}.$$

The $N-1$ points of the DFT of $x(n)$ are related to the z-transform of $x(n)$ in that they are samples of the z-transform taken at equally spaced intervals on the unit circle in the z-plane.

There are applications [50] where it is desired to calculate samples of the z-transform at locations either off the unit circle or at unequally spaced angles on the unit circle. The chirp z-transform (CZT) is an efficient algorithm that can be used for calculating samples of some of these z-transforms. In particular, the CZT can be used to efficiently calculate the values of the z-transform of a finite-length, discrete-time sequence if the z-transform points are of the form

$$z_k = AW^{-k}, \tag{7.11}$$

where

$$A = A_0 e^{j\theta},$$
$$W = W_0 e^{-j\phi},$$

and where A_0 and W_0 are real-valued constants and θ and ϕ are angles. The set of points $\{z_k\}$ lie on a spiral where z_0 is at distance A_0 from the

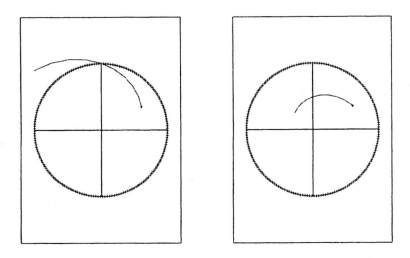

Figure 7.6: *The chirp z-transform samples the z-transform on spirals*

origin and at angle θ from the **x**-axis. The remaining points are located at equally spaced angles ϕ and approach the origin for $W_0 > 1$, move away from the origin for $W_0 < 1$, and remain on a circle of radius A_0 for $W_0 = 1$. Figure 7.6 shows the location of samples of the z-transform for $W_0 < 1$ on the left-hand side of the figure and of $W_0 < 1$ on the right-hand side of the figure. In both parts of the figure the position of z_0 is indicated by the sample connected to the origin by a straight line.

The syntax of **czt** is as follows:

```
[czx]=czt(x,m,w,phi,a,theta)
```

where **x** is a vector giving the input data sequence, **m** is an integer giving the number of points desired in the z-plane, **w** is the magnitude multiplier, **phi** is the phase increment, **a** is the initial magnitude, **theta** is the initial phase, and **czx** is a vector giving the resulting chirp z-transform output.

7.2 Filtering and Filter Design

Filtering and filter design are a core component of signal processing. Scilab has several built-in filtering tools and has an array of filter design functions.

7.2.1 Filtering

A linear time-invariant filter can be implemented as a convolution. The convolution operation is defined for discrete time functions as follows. Let $x(n)$ and $h(n)$ represent a discrete time signal and a discrete filter, respectively. The result of convolving these two sequences, $y(n)$, is defined by

$$y(n) = \sum_{k=-\infty}^{\infty} h(n-k)x(k). \tag{7.12}$$

Straight convolution between two sequences is accomplished in Scilab using **convol**. The **convol** function has two calling syntaxes. The first syntax calculates a convolution based on two discrete-length sequences that are passed in their entirety as arguments to the function. The second format performs updated convolutions using the overlap-add method described in [46]. This method is useful when the signal to be filtered is so long that it cannot be passed directly to the function because of memory limitations.

The first syntax for **convol** is

```
y=convol(h,x)
```

where both **h** and **x** are finite-length vectors and **y** is a vector representing the resulting convolution of the inputs. As an example:

```
-->x=1:3;

-->h=ones(1,2);

-->y=convol(h,x)
 y  =
!   1.    3.    5.    3. !
```

The second syntax for **convol** is

```
[y,e1]=convol(h,x,e0)
```

where **e0** and **e1** are required to update the calculations of the convolution at each iteration. Typically, use of this syntax for **convol** requires a small program of the following form:

```
x1=readx(1,Nx);                        ← user function to read first Nx points of data
[y,e]=convol(h,x1);                                ← 1st convolution
writex(y);                             ← user function to store convolved data
for m=2:M-2
    xm=readx(m*Nx+1,(m+1)*Nx);                  ← read more data
    [y,e]=convol(h,xm,e);                 ← convolve new data
    writex(y);                              ← store new data
end,
xM=readx((M-1)*Nx+1,M*Nx);                   ← read final data
y=convol(h,xM,e);                      ← convolve final data
writex(y);                          ← store final convolved data
```

where **readx** and **writex** are user-supplied functions that read and write the data, **Nx** is the length of a data chunk to be read, **h** is the filter, and **M** is the number of sections of data to be convolved.

Filtering of discrete signals by linear systems defined by **syslin** can be accomplished using the function **flts**. Using the function **flts** takes advantage of the data abstraction features of Scilab. Thus, the result is the same whether the linear system is defined as a transfer function or in a state-space representation. The syntax of the function is as follows:

```
[y[,x]]=flts(u,sl[,x0])
```

where **sl** represents a linear system defined using **syslin**, and **u** is a vector representing the data to be filtered. The optional input **x0** is either the initial state vector if **sl** is in state-space form or a matrix giving the necessary past input–output history if **sl** is in transfer function form (see Scilab's on-line help for **flts** for details). The input variable **x0** allows for filtering of lengthy signal sequences similar to the overlap-add method described above for **convol**. Finally, the optional output variable **x** gives the state sequence if the input linear system is in state-space form.

The following is an example of filtering with **flts**:

```
hz=iir(3,'lp','butt',[.1 0],[0 0]);          ← make a low-pass filter
hz=syslin('d',hz);                       ← make it a discrete linear system
t=1:200;                              ← make a signal with two components
x1=sin(2*%pi*t/20);                  ← ... a low-frequency component
x2=sin(2*%pi*t/3);               ← ... and a high-frequency component
x=x1+x2;
y=flts(x,hz);                                       ← filter and compare

xsetech([0,0,1,.5])
plot(x)
xtitle('Signal is Sum of Two Sinusoids','time','amplitude')
xsetech([0,0.5,1,.5])
plot(y)
xtitle('Filtered Signal','time','amplitude')
```

This example illustrates the use of **flts** to low-pass filter a signal consisting of two sinusoids using a third-order Butterworth filter. The resulting filtered output contains only a single sinusoid, as is illustrated in Figure 7.7.

7.2.2   Finite Impulse Response Filter Design

Filters whose z-transform can be represented by

$$H(z) = \sum_{k=0}^{N} h_k z^{-k} \qquad (7.13)$$

are known as finite impulse response (FIR) filters. Filters of this type are also known as all-zero filters and moving-average filters. One advantage of these types of filters is that they can usually be designed to have linear phase. Linear-phase filters do not change the overall phase delay of the filtered signal, only the group delay. Because of this characteristic of FIR filters, the parts of the signal that are not magnitude filtered retain their original shape. This is an advantage in some applications, especially those where signal shape plays a role in data interpretation. Scilab has two tools for designing these types of filters. The first is a suboptimal method known as the windowing technique. The second is an optimal method based on the Remez algorithm [19].

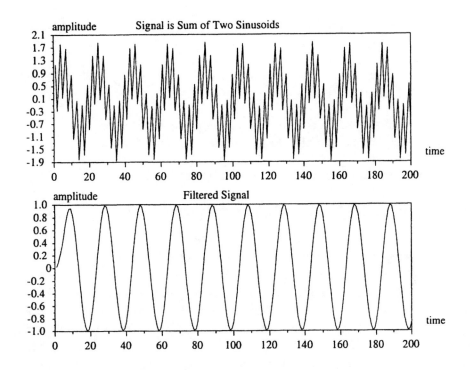

Figure 7.7: *Filtering using* **flts**

Windowing Techniques

The Scilab function **wfir** designs four different types of FIR linear phase filters: low-pass, high-pass, band-pass, and stop band filters using the windowing technique (see [46] or [49]). Four types of windows can be used: the triangular, generalized Hamming, Kaiser, and Chebyshev windows.

The syntax for the function **wfir** can take two formats. The first format is interactive:

```
[wft,wfm,fr]=wfir()
```

where the parentheses are a required part of the function call. This format of the function will prompt the user for required input parameters such as the filter type (**lp='low pass'**, **hp='high pass'**, **bp='band pass'**, **sb='stop band'**), filter length (an integer $n > 2$), window type (**re='rectangular'**, **tr='triangular'**, **hm='hamming'**, **kr='kaiser'**, **ch='chebyshev'**) and other special parameters such as α for the the generalized Hamming window ($0 < \alpha < 1$) and β for the Kaiser window ($\beta > 0$). The three returned arguments are:

wft A vector containing the windowed filter coefficients for a filter of length n.

wfm A vector of length 256 containing the frequency response of the windowed filter.

fr A vector of length 256 containing the frequency axis values ($0 \leq$ **fr** ≤ 0.5) associated to the values contained in **wfm**.

The second format of the function is as follows:

```
[wft,wfm,fr]=wfir(ftype,forder,cfreq,wtype,fpar)
```

This format of the function is not interactive and, consequently, all the input parameters must be passed as arguments to the function. The first argument, **ftype**, indicates the type of filter to be constructed and can take the values **'lp'**, **'hp'**, **'bp'**, and **'sb'** representing, respectively, the filters low-pass, high-pass, band-pass, and stop-band. The argument **forder** is a positive integer giving the order of the desired filter. The argument **cfreq** is a two-vector for which only the first element is used in the case of low-pass and high-pass filters. Under these circumstances **cfreq(1)** is the cut-off frequency (in normalized hertz) of the desired filter. For band-pass and stop-band filters both elements of **cfreq** are used, the first being the low-frequency cut-off and the second being the high-frequency cut-off of the filter. Both values of **cfreq** must be in the range $[0, 0.5)$ corresponding to the possible values of a discrete frequency response. The argument **wtype** indicates the type of window desired and can take the values **'re'**, **'tr'**, **'hm'**, **'hn'**, **'kr'**, and **'ch'** representing, respectively, the windows rectangular, triangular, Hamming, Hanning, Kaiser, and Chebyshev. Finally, the argument **fpar** is a two-vector for which only the first element is used in the case of a Kaiser window and for which both elements are used in the case of a Chebyshev window. In the case of a Kaiser window the first element of **fpar** indicates the relative trade-off between the main lobe of the frequency response window and the side-lobe height and must be a positive integer. For more on this parameter see [49]. For the case of the Chebyshev window one can specify either the width of the window's main lobe or the height of the window's side lobes. The first element of **fpar** indicates the side-lobe height and must take a value in the range $[0, 1)$, and the second element gives the main-lobe width and must take a value in the range $[0, 0.5)$. The unspecified element of the **fpar** vector is indicated by assigning it a negative value. Thus, **fpar=[.01,-1]** means that the Chebyshev window will have side lobes of height 0.01 and the main-lobe width is left unspecified.

Note: Because of the properties of FIR linear phase filters, it is not possible to design an even-length high-pass or stop-band filter.

The following code presents three examples using **wfir**:

```
[wft,wfm,fr]=wfir('lp',33,[.2 0],'kr',[5.6 0]);
xsetech([0,0,1,1/3])
plot2d(fr,20*log10(wfm))
xtitle('Low Pass Filter with Kaiser Window',..
        'freq (Hz)','Amplitude (dB)')
[wft,wfm,fr]=wfir('sb',127,[.2 .3],'hm',[0 0]);
xsetech([0,1/3,1,1/3])
plot2d(fr,20*log10(wfm))
xtitle('Stop Band Filter with Hamming Window',..
        'freq (Hz)','Amplitude (dB)')
[wft,wfm,fr]=wfir('bp',55,[.15 .35],'ch',[.001 -1]);
xsetech([0,2/3,1,1/3])
plot2d(fr,20*log10(wfm))
xtitle('Band Pass Filter with Chebyshev Window',..
        'freq (Hz)','Amplitude (dB)')
```

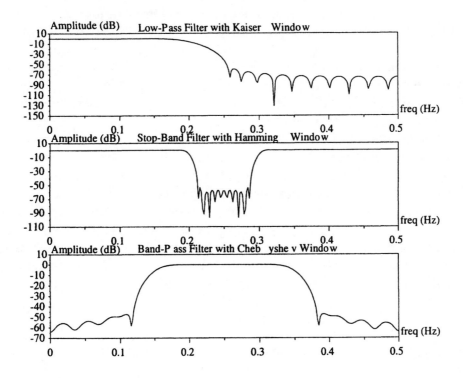

Figure 7.8: *Examples using* **wfir**

These three examples are illustrated in Figure 7.8. In the first example we choose a low-pass filter of length $n = 33$ using a Kaiser window with parameter $\beta = 5.6$. The resulting magnitude of the windowed filter is plotted in the top frame of Figure 7.8, where the magnitude axis is given on a log scale. The second example is a stop-band filter of length 127 using a Hamming window with parameter $\alpha = 0.54$. The resulting magnitude of the windowed filter is plotted in the middle frame of Figure 7.8, where the magnitude is given on a log scale. The third example is a band-pass filter of length 55 using a Chebyshev window with a main lobe width of 0.001 (the side lobe height is automatically determined, since the second element of the vector is negative). The resulting magnitude of the windowed filter is plotted in the lower frame of Figure 7.8, where the magnitude is given on a log scale.

Minimax FIR Filter Design

In the previous section the design of FIR linear phase filters using the windowing technique was described. The windowing technique tries to obtain an optimal, minimum mean-squared error filter of various types while keeping pass-band ripple, stop-band ripple, and transition bands to a minimum. The process is iterative, since the trade-offs between ripple, bandwidth, and filter length cannot be determined analytically. Furthermore, the final design is not optimum in the mean-square sense. In this section a technique is described that does yield an optimum filter, not in the mean-square sense but in the minimax sense.

For the design of FIR linear phase filters that are optimal in the minimax sense Scilab has the function **eqfir**. This function yields better-performing filters for the same length filter than does **wfir** and is more general, since it can be used to design error-weighted multiband filters.

The design philosophy for **eqfir** is different from that for **wfir**. For **wfir** the pass-band ripple, stop-band ripple, and transition bandwidth change as a function of filter length and window type. For **eqfir** the transition band(s) and relative pass-band and stop-band ripples are specified, and then the overall weighted ripple is optimally found as a function of filter length.

The function **eqfir** has the following syntax:

```
[hn]=eqfir(nf,bedge,des,wate)
```

where **nf** is an integer giving the filter length, **bedge** is an $M \times 2$ matrix defining pairs of edges for each of M bands (all in the range $[0, 0.5]$), **des** is an *M*-vector giving the desired magnitude in each band, **wate** is an *M*-vector that gives the weight of the error in each band, and **hn** is a vector of length **nf** giving the resulting filter.

Two examples are presented here. The first example compares the performance of the minimax filter design method to the first windowed filter technique presented in Figure 7.8. The second example illustrates the design of a multiband filter.

In order to compare the result of **wfir** with that for **eqfir** it was necessary to iteratively change the calling arguments of the two functions. The resulting comparison is giving in the following code:

```
[wft,wfm,fr]=wfir('lp',33,[.23 0],'kr',[5.6 0]);     ← low-pass
                                                        filter with wfir
bedge=[0 .2;.2875 .5];              ← pass-band and stop-band
des=[1 0];
wate=[.025 1];                       ← relative error weights
hn=eqfir(33,bedge,des,wate);         ← low-pass filter with eqfir

xsetech([0 0 1 .5])
plot2d(fr,20*log10(wfm)),xgrid()
xtitle('Low-Pass Filter Using wfir','Frequency (Hz)',..
       'Amplitude (dB)')
xsetech([0 .5 1 .5])
[hm,fr]=frmag(hn,256);
plot2d(fr,20*log10(hm)),xgrid()
xtitle('Low-Pass Filter Using eqfir','Frequency (Hz)',..
       'Amplitude (dB)')
```

The results are illustrated in Figure 7.9. As shown in the figure, the cut-off frequency (0.2 Hz) and the transition band of both filters are identical. The filter designed by **eqfir**, however, has its first side lobe at −70 dB, whereas the first side lobe of the design from **wfir** is at −60 dB.

The sidelobes for the filter from **wfir** do not fall below −70 dB until 0.34 Hz on the normalized frequency scale. This example shows that although the filter design required an iterative procedure, the final result obtained from **eqfir** should have better overall performance than that obtained using **wfir**.

The second example shows the result of the design of a multiband filter:

```
bedge=[0 .1;.15 .2;.25 .3;.35 .4;.45 .5];      ← design of 5-band
                                                  filter
des=[0 1 0 .5 0];               ← desired response in each bands
wate=[1 .025 1 1 1];            ← relative error weights in each band
```

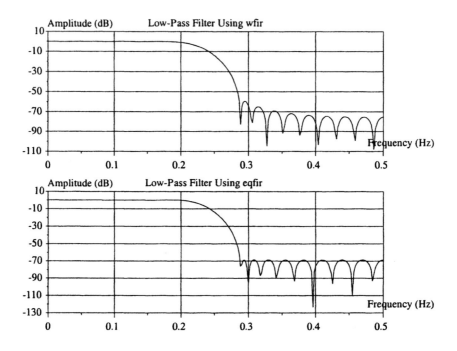

Figure 7.9: *Comparing* **eqfir** *with* **wfir**

```
hn=eqfir(64,bedge,des,wate);                    ← design with eqfir
[hm,fr]=frmag(hn,256);
plot2d(fr,hm)
xtitle('5-band Filter Designed with eqfir',..
      'Frequency (Hz)','Amplitude')
```

The filter is illustrated in Figure 7.10. Note that the filter was designed with pass-bands of different magnitudes (1 and 0.5), and the relative error in the first pass-band is greater than that in the other bands.

Finally, it is important to note that it is theoretically impossible to design high-pass and stop-band filters of even length. Neither **wfir** nor **eqfir** should be used to design such filters, since the results will not be correct.

There exists in Scilab a function that designs optimal FIR filters in the minimax sense that are much more general than those described in the previous section. This function is called **remezb** and can be used to find the optimum minimax filter for any filter magnitude response. The method works best, in fact, for magnitude responses that are continuous. The syntax for this function is

```
[an]=remezb(nc,fg,ds,wt)
```

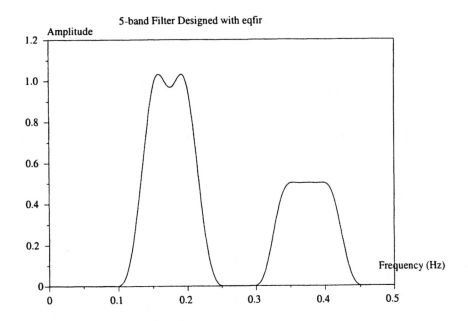

Figure 7.10: *Design of a multiband filter using* **eqfir**

where **nc** is an integer giving the half filter length, **fg** is a vector giving a dense grid of normalized frequencies, **ds** is a vector of desired magnitude values on this grid, and **wt** is the error weighting vector. The vector **an** is a canonical representation for the filter coefficients. The actual filter is obtained by applying

```
hn(1:nc-1)=an(nc:-1:2)/2;
hn(nc)=an(1);
hn(nc+1:2*nc-1)=an(2:nc)/2;
```

An example using **remezb** is given here to design a filter with a triangular magnitude response:

```
nc=21;                                    ←  half filter length
ngrid=nc*16;                              ←  number of points in dense grid
fg=.5*(0:(ngrid-1))/(ngrid-1);           ←  frequency grid
ds(1:ngrid/2)=(0:-1+ngrid/2)*2/(ngrid-2);  ←  desired function
                                                  (a triangle)
ds(ngrid/2+1:ngrid)=ds(ngrid/2:-1:1);
wt=ones(fg);                              ←  weighting function
an=remezb(nc,fg,ds,wt);                   ←  optimal coefficients canonical form
h=an(nc:-1:2)/2;                          ←  obtain filter coefficients
h(nc)=an(1);
```

```
h(nc+1:2*nc-1)=h(nc-1:-1:1);

[hm,fr]=frmag(h,256);
plot2d(fr,hm)
xtitle('Minimax Approximation to Triangular Filter',..
        'Frequency (Hz)','Magnitude')
```

The filter is illustrated in Figure 7.11. Note that the filter error is equiripple.

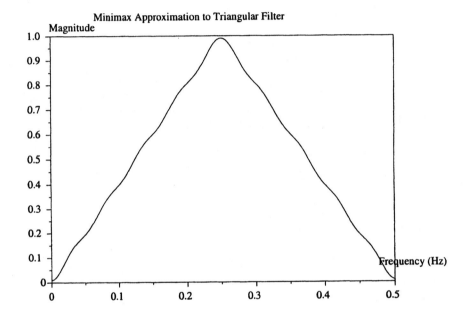

Figure 7.11: *Design of a triangular magnitude response with* **remezb**

7.2.3   Infinite Impulse Response Filter Design

Discrete infinite impulse response filters can be designed in Scilab using two functions: **iir** and **eqiir**.

The **iir** Function

The algorithm used by **iir** is based on the bilinear transform applied to an analog filter. The function designs a canonical analog low-pass filter of the appropriate design (Butterworth, Chebyshev type I, Chebyshev type II, or elliptic), which it then transforms into a digital low-pass filter using the bi-linear transform. The bilinear transform is then applied again to obtain the specific low-pass, high-pass, band-pass, or stop-band filter that is desired.

The function **iir** has the following syntax:

```
[hz]=iir(n,ftype,fdesign,frq,delta)
```

The argument **n** is the filter order. For low-pass and high-pass filters **n** is the order of the numerator and denominator polynomials of **hz**. For band-pass and stop-band filters the numerator and denominator polynomials will be of order **2*n** due to the way in which discrete IIR filters are transformed from continuous ones. The argument **ftype** is the filter type and can take the value **'lp'** for a low-pass filter, **'hp'** for a high-pass filter, **'bp'** for a band-pass filter, or **'sb'** for a stop-band filter. The argument **fdesign** specifies the type of IIR filter and can be **'butt'** for a Butterworth filter design, **'cheb1'** for a Chebyshev filter design of the first type, **'cheb2'** for a Chebyshev filter design of the second type, or **'ellip'** for an elliptic filter design. The argument **frq** is a two-vector that contains cut-off frequencies of the desired filter. For low-pass and high-pass filters only the first element of this vector is used ,to specify the cut-off frequency of the desired filter. The second element of this vector is used for band-pass and stop-band fil-ters, and it gives the upper band edge of the band-pass or stop-band filter, whereas the first element gives the lower band edge. The argument **delta** is a two-vector of ripple values. In the case of the Butterworth filter, which yields a filter that is maximally flat in the pass band, **delta** is not used. For Chebyshev filters of the first type, only the first element of this vector is used, and it serves as the value of the ripple in the pass band. Consequently, the magnitude of a Chebyshev filter of the first type ripples between 1 and 1-**delta(1)** in the pass band. For a Chebyshev filter of the second type only the second element of **delta** is used. This value of **delta** is the ripple in the stop band of the filter. Consequently, the magnitude of a Chebyshev filter of the second type ripples between 0 and **delta(2)** in the stop band. Finally, for the elliptic filter, both the values of the first and second elements of the vector **delta** are used, and they are the ripple errors in the pass and stop bands, respectively. The output of the function, **hz**, is a rational function giving the coefficients of the desired filter.

The following example of the use of **iir** illustrates the four design types:

```
hz1=iir(5,'lp','butt',[.15 0],[0 0]);        ← Butterworth low-pass
                                                filter

[hf1,fr1]=frmag(hz1,256);
```

```
hz2=iir(5,'hp','cheb1',[.35 0],[.05 0]);   ← Chebyshev I high-pass
                                                            filter
[hf2,fr2]=frmag(hz2,256);
hz3=iir(5,'bp','cheb2',[.15 .35],[0 .05]);        ← Chebyshev II
                                                        band-pass filter
[hf3,fr3]=frmag(hz3,256);
hz4=iir(5,'sb','ellip',[.15 .35],[.05 .025]); ← elliptic stop-band
                                                            filter
[hf4,fr4]=frmag(hz4,256);

xsetech([0,0,.5,.5])
plot2d(fr1,hf1)
xtitle('Butterworth Low-Pass Filter','Frequency','Amplitude')
xsetech([0,0.5,.5,.5])
plot2d(fr2,hf2)
xtitle('Chebyshev Type I High-Pass Filter','Frequency',..
       'Amplitude')
xsetech([.5,0,.5,.5])
plot2d(fr3,hf3)
xtitle('Chebyshev Type II Band-Pass Filter','Frequency',..
       'Amplitude')
xsetech([.5,0.5,.5,.5])
plot2d(fr4,hf4)
xtitle('Elliptic Stop-Band Filter','Frequency','Amplitude')
```

The results are illustrated in Figure 7.12. The upper left part of the figure shows the low-pass Butterworth filter with a cut-off frequency of 0.15 normalized Hz. The lower left shows the Chebyshev type I high-pass filter. Notice how the filter has ripple only in the pass band. The upper right shows a Chebyshev II band-pass filter, which has ripple only in the stop bands. Finally, the lower right shows an elliptic band-pass filter, which has ripple in the pass and stop bands.

The eqiir Function

The eqiir function is similar to iir in that it designs low-pass, high-pass, symmetric stop-band, and symmetric pass-band filters. The main difference between iir and eqiir is that for eqiir the filter order is not specified. Rather, the cut-off frequencies, pass-band, and stop-band ripples are specified, and the filter order is automatically determined to meet these specifications. Furthermore, the function eqiir returns a vector of numerically stable second-and first-order rational functions whose product gives the filter transfer function. These cells can then be used to easily obtain a hardware implementation of the filter.

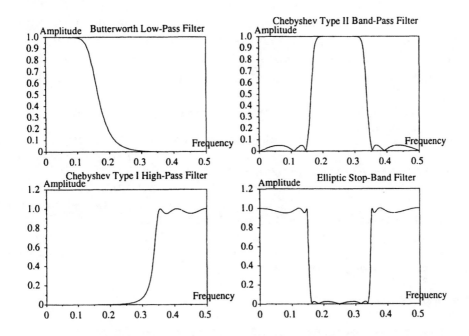

Figure 7.12: *Design of IIR filters using* **iir**

The syntax for **eqiir** is as follows:

```
[cells,fact,zzeros,zpoles]=eqiir(ftype,approx,om,deltap,
                                  deltas)
```

Here **ftype** can take the values **'lp'**, **'hp'**, **bp'**, **'sb'** for a low-pass, high-pass, band-pass, or stop-band filter, respectively. The input variable **approx** indicates the filter approximation and can take the values **'butt'**, **'cheb1'**, **'cheb2'**, or **'ellip'** for filter designs that are Butterworth, Chebyshev type I, Chebyshev type II, or elliptic, respectively. The input **om** is a 2-vector for low-pass and high-pass filters and a 4-vector for band-pass and stop-band filters. For a low-pass filter the first element of the vector gives the pass-band edge and the second element the stop-bend edge, and for a high-pass filter the first value is the stop-band edge, whereas the second value is the pass-band edge. For a band-pass filter the first two values give the first pass-band and stop-band edges, whereas the second two values give the second stop-band and pass-band edges. Finally, for stop-band filters the first two values give the first stop-band and pass-band edges, whereas the second two values give the second pass-band and stop-band edges. In contrast to **iir**, where frequencies are specified in normalized hertz in $[0, 0.5]$, **eqiir** specifies frequencies in the range $[0, 2\pi]$ for similar results. Finally, the inputs **deltap** and **deltas** are the allowable ripples in the pass and stop bands, respectively. These inputs are taken into account only for the appropriate filter. Thus, **deltap** is used with the Chebyshev type I and elliptic filters, whereas **deltas** is used with the Chebyshev type II and elliptic filters.

As an example of using the function **eqiir** the following code is used:

```
[c,g]=eqiir('sb','butt',2*%pi*[.1 .2 .3 .4],.05,.025);     ←
                                              Butterworth low-pass filter
[hf1,fr1]=frmag(g*prod(c(2))/prod(c(3)),256);
[c,g]=eqiir('hp','cheb2',2*%pi*[.3 .4],.05,.025);  ← Chebyshev
                                                 II band-pass filter
[hf2,fr2]=frmag(g*prod(c(2))/prod(c(3)),256);
[c,g]=eqiir('lp','ellip',2*%pi*[.1 .2],.05,.025);     ← elliptic
                                              stop-band filter
[hf3,fr3]=frmag(g*prod(c(2))/prod(c(3)),256);

xsetech([0,0,1,1/3])
plot2d(fr1,hf1)
xtitle('Butterworth Stop-Band Filter','Frequency',..
       'Amplitude')
xsetech([0,1/3,1,1/3])
plot2d(fr2,hf2)
xtitle('Chebyshev Type II Band-Pass Filter','Frequency',..
       'Amplitude')
xsetech([0,2/3,1,1/3])
plot2d(fr3,hf3)
xtitle('Elliptic Low-Pass Filter','Frequency','Amplitude')
```

The results are illustrated in Figure 7.13. The top part of the figure shows a stop-band Butterworth filter with cut-off frequencies of 0.1 and 0.4 normalized Hz. The middle shows a Chebyshev type II high-pass filter. Notice how the filter has ripple only in the stop band. Finally, the bottom shows an elliptic low-pass filter that has ripple in the pass and stop bands.

7.3   Spectral Estimation

The power spectrum of a deterministic, finite-length, discrete-time signal $x(n)$ is defined to be the magnitude squared of the signal's Fourier transform

$$S_x(\omega) = \frac{1}{N} \left| \sum_{n=0}^{N-1} x(n)e^{-j\omega n} \right|^2 . \tag{7.14}$$

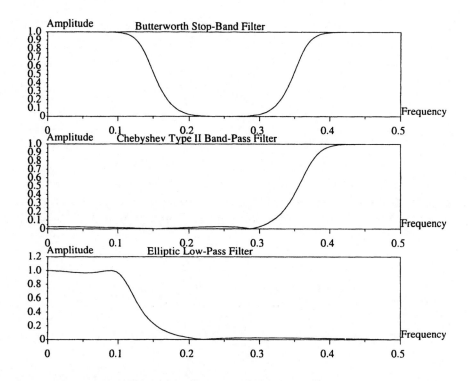

Figure 7.13: *Design of IIR filters using* `eqiir`

In an analogous fashion the cross-spectrum of two signals $x(n)$ and $y(n)$ is defined to be

$$S_{xy}(\omega) = \frac{1}{N}(\sum_{n=0}^{N-1} x(n)e^{-j\omega n})(\sum_{n=0}^{N-1} y(n)e^{-j\omega n})^*. \qquad (7.15)$$

The power spectra of random, zero-mean, wide-sense stationary signals are obtained from the Fourier transform of the correlation functions of these signals. Thus, for R_x representing the autocorrelation function of x, and R_{xy} representing the cross-correlation function of x with y, we have by definition that

$$
\begin{aligned}
R_x(m) &= E\{x(n+m)x^*(n)\}, \\
R_{xy}(m) &= E\{x(n+m)y^*(n)\}.
\end{aligned}
$$

Consequently, the power spectrum and cross-power spectrum of $x(n)$ and of $x(n)$ with $y(n)$ are, respectively,

$$S_x(\omega) \;=\; \sum_{m=-\infty}^{\infty} R_x(m)e^{-j\omega m}, \tag{7.16}$$

$$S_{xy}(\omega) \;=\; \sum_{m=-\infty}^{\infty} R_{xy}(m)e^{-j\omega m}. \tag{7.17}$$

The formulas (7.16) and (7.17) require estimates of the correlation functions. Possible candidates for the estimates of the auto-and cross-correlation functions of finite-length random signals (i.e., $x(n) \neq 0$ and $y(n) \neq 0$ for $n = 0, 1, \ldots, N-1$) are

$$\hat{R}_x(m) \;=\; \frac{1}{N} \sum_{n=0}^{N-1-m} x(n+m)x^*(n), \tag{7.18}$$

$$\hat{R}_{xy}(m) \;=\; \frac{1}{N} \sum_{n=0}^{N-1-m} x(n+m)y^*(n). \tag{7.19}$$

The estimates in (7.18) and (7.19) are unbiased, consistent estimators in the limit as $N \to \infty$. Furthermore, in the case where the random signals are jointly Gaussian, these estimators are the maximum likelihood estimates of the correlation functions. Another interesting property of the estimators in (7.18) and (7.19) is that when substituted, respectively, into the expressions in (7.16) and (7.17), after some algebraic manipulation they yield exactly the expressions in (7.14) and (7.15).

Unfortunately, there is a serious problem with the above power spectrum estimation scheme in that the resulting power spectral estimates are not consistent. That is, the error variance of the estimate does not decrease with increasing data. Consequently, power spectral estimates obtained by taking the magnitude squared of the Fourier transform are high-variance, poor performance estimates.

In the sections that follow two techniques are discussed that yield improved spectral estimates. These techniques are both based on averaging spectral estimates obtained from the classical approach just described. This averaging, although introducing some biasing, yields greatly improved estimates in that in the limit, these estimates become consistent.

The first averaging technique sections the data into overlapping segments. In this case the magnitude squared of the Fourier transform is calculated from each segment, and then these transforms are averaged together

to yield the spectral estimate. This technique is called the modified periodogram method of spectral estimation.

The second averaging technique also sections the data into overlapping segments. For each segment an estimate of the correlation function
is calculated. These estimates are then averaged, and the estimated power
spectral density is the Fourier transform of the average. This technique is
known as the correlation method for spectral estimation.

7.3.1   The Modified Periodogram Method

The modified periodogram method of spectral estimation repeatedly calculates the periodogram of windowed subsections of the data. These periodograms are then averaged together and normalized by an appropriate
constant to obtain the final spectral estimate. It is the averaging process
that reduces the variance in the estimate.

The periodogram of a finite data sequence is defined by

$$I(\omega) = \frac{1}{U} \left| \sum_{n=0}^{N-1} w(n)x(n)e^{-j\omega n} \right|^2, \tag{7.20}$$

where $w(n)$ is a window function with energy U. Consequently, if K sub-
segments of length N are used to calculate the spectrum of the signal, then
the modified periodogram spectral estimate $\hat{S}_x$ is just the average of the K
periodograms

$$\hat{S}_x(\omega) = \frac{1}{K} \sum_{k=0}^{K-1} I_k, \tag{7.21}$$

where each of the I_k is the periodogram (calculated as in (7.20)) of the kth
segment of data.

The variance of the spectral estimate decreases proportionally to the
number of independent data segments. Of course, with a limited quantity of data the number of data segments must also be limited. Thus, there
is often a trade-off between the roughness (i.e., variance) of the estimated
spectrum and the bias due to the truncation (i.e., windowing) of the correlation length. It is possible to diminish the effects of bias in shorter data
sequences by using overlapping segments. This, of course, will affect the
overall reduction in the variance of the estimate. It can be shown that an
overlap of fifty percent results in an approximately fifty percent reduction
in the variance of the estimate of the power spectrum [49].

The Function pspect

The function **pspect** calculates an estimate of the power spectrum using the modified periodogram method. The syntax for pspect is as follows:

```
[sm,cwp]=pspect(sec_step,sec_leng,wtype,x[,y],wpar)
```

Here the input **sec_step** is an integer giving the relative offset of consecutive data windows. The input **sec_leng** is an integer giving the length of each window. The input **wtype** is a string specifying the window to apply to each data segment and can take the values **'re'**, **'tr'**, **'hm'**, **'hn'**, **'ch'**, or **'kr'** for a rectangular, triangular, Hamming, Hanning, Chebyshev, or Kaiser window, respectively. The input **x** is the data sequence whose spectrum is to be estimated. If **y** is specified, then the cross-spectrum of **x** and **y** is computed. Finally, the input **wpar** is used to specify properties of the Chebyshev and Kaiser windows. For a Chebyshev window, type **wpar** is a 2-vector, but only one of the elements is used for the window. If the first element is in $(0, 0.5)$ and the second is negative, then the first element is used to specify the main-lobe width of the window's frequency response. Otherwise, if the first element is negative and the second element is greater than zero, then the second element specifies the side-lobe height of the window. For the Chebyshev window there is a trade-off between main-lobe width and side-lobe height. If one of these parameters is specified, the other can be computed, and the output **cwp** gives the value of the computed parameter. Finally, for a Kaiser window, **wpar** is a scalar, which must be greater than 0. The output **sm** is a vector of length **sec_leng** that contains the two-sided spectral estimate (i.e., for the positive and negative frequencies).

As an example of the use of **pspect** a random signal is generated by passing zero-mean white noise of unit variance through a low-pass filter. The spectrum of the data will be the magnitude squared of the filter frequency response. The low-pass filter is an FIR filter of length 33 generated using the function **eqfir**:

```
rand('normal');
rand('seed',0);
x=rand(1:4096-33+1);          ←  generate zero-mean, white, Gaussian noise
nf=33;
bedge=[0 .1;.125 .5];
des=[1 0];
wate=[1 1];
h=eqfir(nf,bedge,des,wate);   ←  make an equiripple low-pass filter
y=convol(h,x);                ←  convolve the white noise with the filter
sm=pspect(100,200,'tr',y);    ←  compute the estimate of the spectrum
[hf,fr]=frmag(h,100);         ←  compute the magnitude of the filter
```

```
xsetech([0 0 1 .5])
plot2d(fr,20*log10(hf))
xtitle('Squared Magnitude of the Filter Response',..
       'Frequency (Hz)','Magnitude (dB)')
xgrid()
xsetech([0 .5 1 .5])
plot2d(fr,10*log10(sm(1:100)))
xtitle('Estimate of the Spectrum Using pspect',..
       'Frequency (Hz)','Magnitude (dB)')
xgrid()
```

A total of 4096 data points are available for the estimation of the spectrum. Note that a triangular window is used to calculate the spectral estimate. The logarithm of the magnitude squared of the filter frequency response is shown in the top part of Figure 7.14. The log-magnitude of the estimate of the spectrum, **sm**, is plotted in the lower part of Figure 7.14. Note that the correlation length of the data cannot be greater than the filter length, and so the value of **sec_leng** need not be greater than 33. Nevertheless, the value of **sec_leng** was chosen to be 200, so that the spectral representation would have a smooth, interpolated response that could be compared nicely with that of the filter magnitude.

As can be seen from Figure 7.14, the estimated spectrum matches well with the theoretical spectrum represented by the magnitude squared of the filter response. In particular, the magnitudes of the pass and stop bands are accurate, and even the ripples in the stop band line up well between the two plots.

7.3.2 The Correlation Method

The correlation method for power spectral estimation calculates the Fourier transform of an averaged estimate of the autocorrelation function. This consists in repeatedly calculating estimates of the autocorrelation function as in (7.18) from possibly overlapping segments of the data and then averaging them.

For each N-point segment of the data, $x_k(n)$, the estimate of $2M$ points of the autocorrelation function is calculated by

$$\hat{R}_k(m) = \sum_{n=0}^{N-1-m} x(n+m)x^*(n) \tag{7.22}$$

for $m = 0, \pm1, \pm2, \ldots, \pm M$. For K estimates of the autocorrelation function

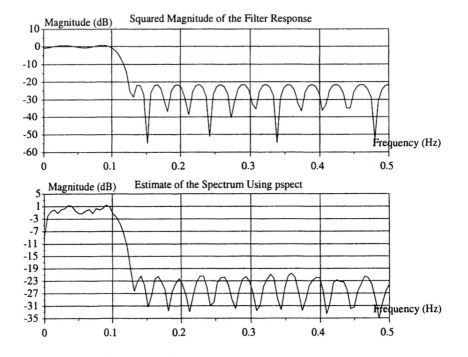

Figure 7.14: *Example of spectral estimation using* pspect

calculated as in (7.22) the power spectral estimate is obtained from

$$\hat{S}_x(\omega) = \mathcal{F}\{\tilde{R}_x(m)w(m)\}, \tag{7.23}$$

where $\mathcal{F}\{\cdot\}$ represents the Fourier transform operation, $w(m)$ is a window function, and $\tilde{R}_x(m)$ is the average of the K estimates,

$$\tilde{R}_x = \frac{1}{K}\sum_{k=1}^{K}\hat{R}_k. \tag{7.24}$$

The Function cspect

The function **cspect** calculates an estimate of the power spectrum using the correlation method. The syntax for **cspect** is as follows:

```
[sm,cwp]=cspect(nlags,ntp,wtype,x[,y],wpar)
```

Here, **nlags** is an integer giving the total number of correlation lags to compute for each segment. The input **ntp** is an integer giving the number of transform points to compute. The input **wtype** is a character string taking the values, **'re'**, **'tr'**, **'hm'**, **'hn'**, **'ch'**, or **'kr'** for a rectangular, triangular, Hamming, Hanning, Chebyshev, or Kaiser window, respectively. The input **x** is the data sequence whose spectrum is to be estimated. If **y** is specified then, the cross-spectrum of **x** and **y** is computed. Finally, the input **wpar** is used to specify properties of the Chebyshev and Kaiser windows. For a Chebyshev window, type **wpar** is a 2-vector, but only one of the elements is used for the window. If the first element is in $(0, 0.5)$ and the second is negative, then the first element is used to specify the main-lobe width of the window's frequency response. Otherwise, if the first element is negative and the second element is greater than zero, then the second element specifies the side-lobe height of the window. For the Chebyshev window there is a trade-off between main-lobe width and side-lobe height. If one of these parameters is specified, the other can be computed, and the output **cwp** gives the value of the computed parameter. Finally, for a Kaiser window, **wpar** is a scalar, which must be greater than 0. The output **sm** is a vector of length **ntp** that contains the two-sided spectral estimate (i.e., for the positive and negative frequencies).

As an example of the use of **cspect**, the same random signal used in the above example for **pspect** is used again here:

```
rand('normal');
rand('seed',0);
x=rand(1:4096-33+1);        ← generate zero-mean, white, Gaussian noise
nf=33;
bedge=[0 .1;.125 .5];
des=[1 0];
wate=[1 1];
h=eqfir(nf,bedge,des,wate);    ← make an equiripple low-pass filter
y=convol(h,x);                 ← convolve the white noise with the filter
sm=cspect(33,200,'ch',y,[.02,-1]);     ← compute the estimate
                                          of the spectrum
[hf,fr]=frmag(h,100);          ← compute the magnitude of the filter

xsetech([0 0 1 .5])
plot2d(fr,20*log10(hf))
xtitle('Squared Magnitude of the Filter Response',..
       'Frequency (Hz)','Magnitude (dB)')
xgrid()
xsetech([0 .5 1 .5])
plot2d(fr,10*log10(sm(1:100)))
xtitle('Estimate of the Spectrum Using cspect',..
       'Frequency (Hz)','Magnitude (dB)')
xgrid()
```

A total of 4096 data points are available for the estimation of the spectrum. Here, a Chebyshev window with a main-lobe width of 0.02 normalized hertz is used in the computation of the spectral estimate. The logarithm of the magnitude squared of the filter frequency response is shown in the top part of Figure 7.15. The log-magnitude of the estimate of the spectrum **sm** is plotted in the lower part of Figure 7.15.

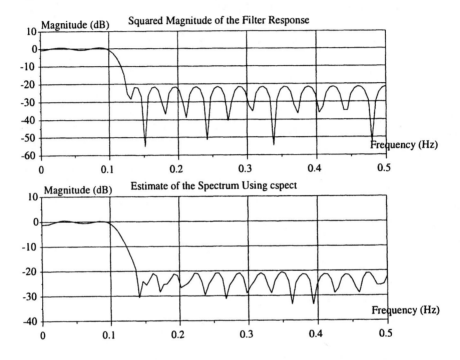

Figure 7.15: *Example of spectral estimation using* **cspect**

As can be seen from Figure 7.15, the estimated spectrum also matches well with the theoretical spectrum represented by the magnitude squared of the filter response.

Chapter 8

Simulation and Optimization Tools

This chapter is devoted to simulation and optimization tools in Scilab. The simulation and optimization of systems is an important part of engineering work. For example, in control design it is often necessary to simulate the behavior of systems that can be modeled by differential equations as well as to find an optimal control for the system.

In this chapter we begin by describing various models used for simulation and optimization problems in Scilab. We then describe Scilab's numerical differential equation solver (which is effective for both for ODE and DAE systems). Finally, we describe Scilab's nonlinear optimization tools. Examples are given. Particular emphasis is given to mechanical problems that can be described using DAEs.

8.1 Models

In this section we describe the various types of equations that arise in solving differential systems or optimization problems.

Simulation problems can often be modeled by first-order explicit systems of ordinary differential equations (ODEs):

$$
\begin{aligned}
\dot{y} &= f(t, y), \\
y(t_0) &= y_0,
\end{aligned}
\tag{8.1}
$$

where y is the vector $(y_1, \ldots, y_n)^T$, t is the independent variable, $f(t, y)$ is the vector function $(f_1(t, y), \ldots, f_n(t, y))^T$, t_0 is the initial time, and y_0 is the vector of initial conditions $y_0 = y(t_0)$. We have adopted the preceding notation, since the independent variable often denotes time, and consequently, we employ the "dot" notation for derivatives with respect to time.

Simulation problems can also be modeled by implicit first-order systems of differential algebraic equations (DAEs):

$$\begin{aligned} f(t, y, \dot{y}) &= 0, \\ y(t_0) &= y_0, \end{aligned} \qquad (8.2)$$

where the notation is as above. Here it is no longer possible to express $\dot{y}$ explicitly, and the system (8.2) cannot be simply manipulated to obtain system (8.1).

The function **ode** is used to solve ODEs and is described in Section 8.2. When the vector function f in equation (8.2) is linear in $\dot{y}$, the system is implicitly linear, and the function **impl**, described in Section 8.3.1, can be used to solve it. Otherwise, the **dassl** and **dasrt** functions are used to solve general DAEs and they are described in Section 8.3.

Optimization problems can be expressed as

$$\min_x f(x),$$

where x belongs to $\mathbb{R}^n$ and f is a function from $\mathbb{R}^n$ to $\mathbb{R}$. The problem is to find the value of $\bar{x}$ for which $f(\bar{x})$ is minimum.

The **linpro** function is used to solve linear optimization problems. The **quapro** and **optim** functions are used to solve nonlinear optimization problems and are described in Section 8.4. In this chapter only nonlinear optimization is discussed.

In the following sections we will describe all the above-mentioned functions (except for **linpro**). For each function there are many parameters and options that can be set by the user. We do not describe all of them, but we will show how to use them in the problems we solve and with respect to the type of result we want to obtain.

For each problem type, ODEs, DAEs, and optimization, we give examples.

8.2   Integrating ODEs

The Scilab function for solving explicit ODEs described by equation (8.1) is **ode**. This function is mainly an interface to the ODEPACK package [35] and to a set of programs implementing Runge-Kutta methods. Scilab can solve stiff and nonstiff systems. The notion of "stiffness" is not well-defined (for instance, see [33]). Practically speaking, stiffness means that implicit solvers are needed and that the Jacobian of the system needs to be employed.

According to the optional argument, **type**, of the **ode** function, different solvers are used and/or different types of problems are solved. If given, **type** must be the first argument to **ode**.

If **type** is not given, the **lsoda** solver of the package ODEPACK is called by default. It automatically selects between the nonstiff predictor–corrector Adams method and the stiff backward differentiation formula (BDF) method. It uses nonstiff methods initially and dynamically monitors the results in order finally to decide which method to use.

type='adams' This is for nonstiff problems. The **lsode** solver of package ODEPACK is called, and it uses the Adams method.

type='stiff' This is for stiff problems. The **lsode** solver of package ODEPACK is called, and it uses the BDF method.

type='rk' Adaptive Runge-Kutta of order 4 (RK4) method [48].

type='rkf' or 'fix' The Shampine and Watts program based on Fehlberg's Runge-Kutta pair of order 4 and 5 (RKF45) method [26] is used. This is for nonstiff and mildly stiff problems where derivative evaluations are inexpensive. This method generally should not be used when the user requires high accuracy.

When the type is **fix**, the same solver is used, but the user function call is simpler, i.e., only **rtol** and **atol** need to be be passed to the solver (see page 259).

type='root' This is useful when the user needs a root-finding capability built into the solution. This is particularly useful for finding stopping conditions at which the definition of the function $f(t, y)$ changes. The **lsodar** solver from the package ODEPACK is used here. It is a variant of the **lsoda** solver with the addition that it finds the roots of a given vector function $h(t, y) = 0$.

The function **ode** has a number of optional arguments. It is also possible to specify additional parameters for the ODEPACK solvers. This is done by giving a value to the variable **%ODEOPTIONS**. The **odeoptions** function is used to assign values to this variable.

We do not describe all optional arguments and ODEPACK parameters. These can be found in the on-line help for **ode** and **odeoptions**. Instead, we show practical applications of **ode**, explaining only the most important parameters.

8.2.1 Calling ode

Simple Call

The simplest call of the **ode** function for solving system (8.1) is

```
y=ode(y0,t0,t,f)
```

where **y0** is the vector of initial conditions y_0 (normally a column n-vector), **t0** is the initial time, **t** is the vector of instants for which the solution is to be computed, and **f** describes the function $f(t, y)$. The output **y** is an $n \times T$ matrix, where T is the size of t and where each column j of **y** is the value of y at time t_j.

The function **f** can be written in Scilab or C or as a Fortran subroutine. If written in C or Fortran, it must be linked into Scilab. If it is a Scilab function, its calling sequence must be **ydot=f(t,y)**. If it is an external C function or Fortran subroutine, **f** is a string, and the arguments of the corresponding C function or Fortran subroutine must be **(n,t,y,ydot)**, where **n** is the dimension of **y**, **t** is the time, **y** is the vector y, and **ydot** is the value of $f(t, y)$ (i.e., the output of the program). For a Fortran subroutine we have

```
subroutine f(n,t,y,ydot)
doubleprecision t,y(n),ydot(n)
...
```

For a C function we have

```
f(n,t,y,ydot)
int *n;                                    ← Fortran variables are pointers
double *t,*y,*ydot;
{
...
}
```

Note that the preceding C function is called by Scilab, which is a Fortran program, so all input and output variables of this C function belong to the Fortran world; in particular, variables are pointers.

Examples of Passing $f(t, y)$ to ode

- **Using a Scilab function.** This is convenient when the function uses mainly matrix or vector operations, for which Scilab is efficient. This function can be defined on-line or in a `.sci` file and loaded with the `getf` function.

As a first example, let us consider the linear series RLC circuit where a resistor, an inductor, and a capacitor are interconnected. The input is the voltage $u(t)$ applied to the circuit, and the output is the charge of the capacitor $q(t)$.

If $i(t)$ denotes the current and $q(t)$ denotes the charge of the capacitor, the second Kirchhoff law gives

$$L\frac{di}{dt} + Ri + \frac{1}{C}q = u,$$

$$q = \int_0^t i(s)ds.$$

From this we get the linear ODE

$$\frac{d}{dt}\begin{pmatrix} q \\ i \end{pmatrix} = \begin{pmatrix} i \\ -\frac{R}{L}i - \frac{1}{LC}q + \frac{1}{L}u \end{pmatrix}.$$

The corresponding Scilab session using the function **ode** is

```
R=1.5;L=0.25;C=0.5;
F=[0 1;-1/(L*C) -R/L]; G=[0; 1/L];
deff('[ydot]=RLC(t,y)','ydot=F*y+G*u(t)');     ← RLC ODE
deff('[ut]=u(t)','ut=sin(5*t)');               ← sinusoidal input
y0=[0;0];t0=0;T=[0:0.05:5];
Y=ode(y0,t0,T,RLC);                            ← Y is vector(i,q)
plot2d("onn",T',Y',[-1 1],'061');
xtitle('','Time','Intensity-Charge')
```

Figure 8.1 shows the time response of variables $i(t)$ and $\dot{q}(t)$, where $q(t)$ is represented by the curve denoted by "+" signs.

We now treat a simple nonlinear example:

$$\begin{aligned} \dot{y}_1 &= -y_2 + (a - \|y\|_2)y_1, \\ \dot{y}_2 &= y_1 + (a - \|y\|_2)y_2. \end{aligned} \tag{8.3}$$

The construction of **f** corresponding to $a = 1$ is

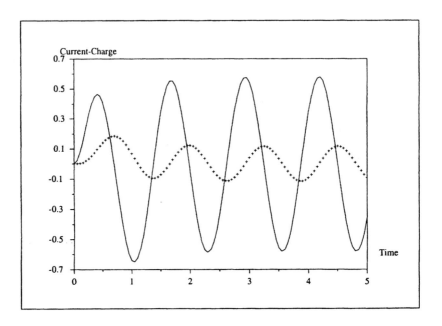

Figure 8.1: *RLC circuit simulation*

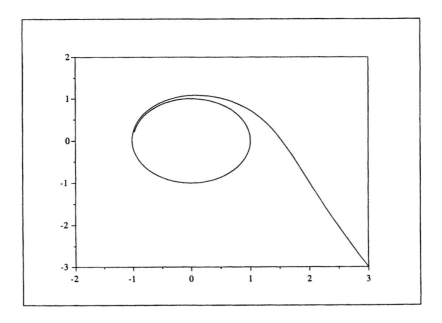

Figure 8.2: *Simple ode plot*

```
J=[0,-1;1,0]; a=1;
deff('ydot=f(t,y)','ydot=J*y+(a-norm(y))*y')
t0=0;y0=[3;-3];T=0:0.1:10;
y=ode(y0,t0,T,f);
plot(y(1,:),y(2,:))
```

Figure 8.2 shows the trajectory of points corresponding to $(y_1(t), y_2(t))$.

- **Using a function $f(t, y)$ written in Fortran or C and to be linked using `link` function.** This approach is used when the function already exists as a Fortran or C subroutine or when the use of Fortran or C is necessitated for reasons of efficiency (see Chapter 10).

Let us consider the following, which is known as Robertson's chemical reaction:

$$\begin{aligned}
\dot{y}_1 &= -0.04\, y_1 + 10^4\, y_2 y_3, \\
\dot{y}_2 &= -\dot{y}_1 - \dot{y}_3, \\
\dot{y}_3 &= 3.10^7\, y_2^2.
\end{aligned} \tag{8.4}$$

The system can be modeled with the following C function:

```
f(n,t,y,ydot)
int *n;
double *t, *y, *ydot;
{
  ydot[0] = -0.04 * y[0] + 1.0e4 * y[1] * y[2];
  ydot[2] = 3.0e7 * y[1] * y[1];
  ydot[1] = - ydot[0] - ydot[2];
}
```

We compile this function, link it to Scilab, and then make use of it with the function `ode`:

```
-->link("f.o","f","C");
linking files f.o  to create a shared executable
shared archive loaded
Linking f (in fact f)
Link done

-->y0=[1;0;0];t0=0;T=logspace(0,5);

-->y=ode(y0,t0,T,'f');
```

Note that it is also possible to have the function f returning a $p \times q$ matrix instead of a vector. Using the usual matrix notation, we then solve the $n = p + q$ ODE $\dot{Y} = F(t, Y)$, where $\dot{Y}$ is the element-wise derivative of Y. The initial conditions **Y0** must also be a $p \times q$ matrix, and the output of **ode** is the $p \times q\,(T + 1)$ matrix $[Y(t_0), Y(t_1), \ldots, Y(t_T)]$ (where T is the length of t). As an example, we use this formulation for the well-known Riccati equations $\dot{Y} = A^T Y + Y A - Y^T B Y + C$:

```
-->deff('Ydot=ric(t,Y)','Ydot=A''*Y+Y*A-Y''*B*Y+C')
```

```
-->A=[1,1;0,2];B=[1,0;0,1];C=[1,0;0,1];
```

```
-->Y0=eye(A);                          ← initial matrix is identity
```

```
-->Y=ode(Y0,0,[1,2],ric)               ← find solution at times 1 and 2
 Y  =
!    2.1259363    0.6246094    2.2613257    0.6141092  !
!    0.6246094    4.3123604    0.6141092    4.4180621  !
```

Passing Parameters to $f(t, y)$

Often, parameters appear in an ODE. For instance, if we want to solve the simple ODE $\dot{y} = \sin(y + kt)$, it should be convenient to have the parameter k as an argument of the function f. The way to accomplish this is to pass a list **list(f,u1,...,un)** in the position of the argument **f** to **ode**. In this case, the calling sequence of the function that computes $f(t, y)$ must be **ydot=f(t,y,u1,...,un)**. This is simple to implement for Scilab functions. For example, for the above function we would write:

```
deff('[ydot]=f(t,y,k)','ydot=sin(y+k*t)')
k=1; y=ode(0,0,1:10,list(f,k));
```

Note that this could have been accomplished simply with the standard calling sequence $f(t, y)$ and without having used a list, since parameters as such as **k** above are global variables and as such available to the environment inside **f**. However, for the sake of clarity it is preferable to pass parameters as arguments to f instead of relying on global variables.

Passing parameters is a bit trickier when dealing with C or Fortran programs. An easy way to accomplish this is to create a Scilab function that subsequently calls the C function or Fortran subroutine. Then we can use the list approach described above in conjunction with the Scilab function **f** as the argument to **ode**. The parameter is then passed to the C or Fortran program inside the Scilab function.

Another way to pass parameters to Fortran or C programs is to use the **creadmat** Fortran function or the **ReadMatrix** C procedure to make the parameter values available to the Fortran subroutine or the C function of interest. For instance, we can use the example of system (8.4), where the numerical values have been replaced by the 3-vector p. This system becomes

$$
\begin{aligned}
\dot{y}_1 &= p_1 y_1 + p_2 y_2 y_3, \\
\dot{y}_2 &= -\dot{y}_1 - \dot{y}_3, \\
\dot{y}_3 &= 3 p_3 y_2^2.
\end{aligned}
$$

Now we define the vector p as a Scilab variable **p**, and we desire to access it in a Fortran program defining an ODE system. We do this in the following way:

```
      subroutine fp(neq, t, y, ydot)
      double precision t, y, ydot, p
      logical creadmat
      dimension y(3), ydot(3), p(3)
c     If "p" does not exist return else load p
      if(.not.creadmat('p'//char(0),m,n,p)) return
c     ***********************************
      ydot(1) = -p(1)*y(1) + p(2)*y(2)*y(3)
      ydot(3) = p(3)*y(2)*y(2)
      ydot(2) = -ydot(1) - ydot(3)
      end
```

where **m** and **n** are the row and column dimensions of the variable **p** (i.e., outputs from the **creadmat** Fortran subroutine). The function **creadmat** searches for the **p** variable in the Scilab environment and puts it into the Fortran variable **p**.

Similarly, a C function for the same system is

```
int fp(neq, t, y, ydot)
    int *neq;
    double *t, *y, *ydot;
{
    static int m, n;
    static double p[3];
    ReadMatrix("p", &m, &n, p);
    ydot[0] = -p[0] * y[0] + p[1] * y[1] * y[2];
    ydot[2] = p[2] * y[1] * y[1];
    ydot[1] = -ydot[0] - ydot[2];
    return(0);
}
```

where the C procedure **ReadMatrix** is used to access the value of the Scilab variable **p**.

We compile this function, link it into Scilab, define the vector **p** and use **ode**:

```
-->link("fp.o","fp","C");
linking files fp.o  to create a shared executable
shared archive loaded
Linking fp (in fact fp)
Link done

-->p=[0.04,1.e4,3.e7];

-->y0=[1;0;0];t0=0;T=logspace(0,5);

-->y=ode(y0,t0,T,'fp');
```

The drawback of using the **creadmat** subroutine or the **ReadMatrix** procedure is that a copy of the parameter variable is made, and this can be computationally time-consuming if it is a large matrix. Thus, there exists a more complicated way to pass parameters to C functions or Fortran subroutines. This can be done by using the **cmatptr** Fortran subroutine or the **GetMatrixptr** C procedure, which directly accesses internal data in Scilab. We will not describe these functions here.

See the files **EX-ode.f** and **EX-ode-more.f** in the Scilab directory **SCIDIR/routines/defaults/** for more examples on how to use all the above-mentioned possibilities. Some simple examples using Fortran or C are also provided in the directory **SCIDIR/examples/link-example**. Table 8.1 lists the names of files and the corresponding features of **ode** they illustrate.

File name	Illustrated feature
ext9*	Simple **ode** example
ext10*	Transmitting additional parameters using **list('xxx',param)**
ext11*	Using **creadmat** or **ReadMatrix**
ext12*	Using **cmatptr** or **GetMatrixptr**

Table 8.1: *Coding ODE in Fortran or C*

Stiff Problems

Methods for solving stiff ODEs are usually integration methods that use the Jacobian of $f(t, y)$ with respect to y, i.e., the $n \times n$ matrix f_y, where the element (i, j) is equal to $\partial f_i / \partial y_j$. This is the case for the backward differentiation formula (BDF) method used by the **lsoda** and **lsode** subroutines of the package ODEPACK used by Scilab.

Giving the Jacobian is optional, but when the problem is expected to be stiff, efficiency will be improved if the user supplies it. An approximation of the Jacobian can also be given. If it is not given, it is approximated by finite differences.

Note that giving the Jacobian is useful only for methods that make use of it, i.e., when the value of the optional argument, **type**, is chosen to be **'stiff'** or when the function **ode** determines automatically that the problem is stiff.

If the Jacobian is given, it must come just after **f** in the calling sequence of **ode**: **y=ode(y0,t0,t,f,jac)**. It can be a Scilab function with the following calling sequence: **j=jac(t,y)**. It can also be an external C function or Fortran subroutine. Then **jac** must be a string, and the arguments of the corresponding C function or Fortran subroutine must be **(n,t,y,ml,mu,pd, nrowpd)**, where **n** is the dimension of the variable **y**, **t** is the time, **y** is the value of y, and **pd** is the value of f_y, (output of the program). There are more arguments, because the Jacobian matrix can be given as a full or banded matrix. By default, it is supposed that the Jacobian is a full matrix, and the **ml** and **mu** arguments are ignored and **nrowpd** has the same value as **n**. So, for a Fortran subroutine we have

```
subroutine jac(n,t,y,ml,mu,pd,nrowpd)
doubleprecision t,y(n),pd(n,nrowpd)
...
```

For a C function we have

```
jac(n,t,y,ml,mu,pd,nrowpd)
int *n,*ml,*mu;                              ← Fortran variables are pointers
double *t,*y,*pd;←  pd  is a vector where the Jacobian is stored columnwise
...
}
```

In the C function **jac**, **pd** is a one-dimensional array that contains entries of the Jacobian matrix stored columnwise. Note that the C function is called by Scilab, which is a Fortran program, so all input and output variables of this C function belong to the Fortran world; in particular, variables are pointers, and matrices are stored columnwise.

As an example, using the example of system (8.4), we can give the Jacobian as the following C function:

```
jac(n,t,y,ml,mu,pd,nrowpd)
int *n,*ml,*mu;
double *t,*y,*pd;
```

```
{
  pd[0] = -0.04;
  pd[1] = -pd[0];
  pd[2] = 0;

  pd[3] = 10000 * y[2];
  pd[5] = 6.0e7 * y[1];
  pd[4] = -pd[3] - pd[5];

  pd[6] = 10000 * y[1];
  pd[7] = -pd[6];
  pd[8] = 0;
}
```

We compile this function, link it into Scilab, and use **ode**:

```
-->link("f.o","f","C");
linking files f.o  to create a shared executable
shared archive loaded
Linking f (in fact f)
Link done

-->link("jac.o","jac","C");
linking files jac.o  to create a shared executable
shared archive loaded
Linking jac (in fact jac)
Link done

-->y0=[1;0;0];t0=0;T=[0.4,4];

-->y=ode(y0,t0,T,'f','jac');
```

For large problems where the Jacobian is a sparse banded matrix, it is better to give the solver only the nonzero elements of the matrix. For that, we use the half-bandwidth parameters **ml** and **mu**. They are, respectively, the widths of the lower and upper parts of the band, excluding the main diagonal (the full bandwidth is **ml+mu+1**).

The diagonals of the Jacobian f_y are assembled into the rows of the banded Jacobian matrix from the top down, right justified for upper diagonal lines and left justified for lower diagonal lines. Thus if **J** is the Jacobian matrix and **pd** is the corresponding banded Jacobian matrix, we have **pd(i-j+mu+1,j)=J(i,j)** when **J(i,j)** is not 0. As an example, if f_y is the following 5×5 matrix where mu=1 and ml=2,

$$\begin{pmatrix} 1 & 1 & 0 & 0 & 0 \\ 1 & 2 & 1 & 0 & 0 \\ 2 & 1 & 3 & 1 & 0 \\ 0 & 2 & 1 & 4 & 1 \\ 0 & 0 & 2 & 1 & 5 \end{pmatrix},$$

then the corresponding 4×5 banded matrix is

$$\begin{pmatrix} 0 & 1 & 1 & 1 & 1 \\ 1 & 2 & 3 & 4 & 5 \\ 1 & 1 & 1 & 1 & 0 \\ 2 & 2 & 2 & 0 & 0 \end{pmatrix}.$$

The solver must know that we are giving a banded matrix for the Jacobian. This is accomplished by specifying a few parameters. These are the parameters jactyp, ml, and mu in the variable %ODEOPTIONS. By default the value of this variable is [1,0,0,%inf,0,2,500,12,5,0,-1,-1], where jactyp is the 6th element, and ml and mu are respectively the next-to-last and the last elements. To specify a banded Jacobian, jactyp must have the value 4 and ml and mu the correct values. For instance, for the preceding Jacobian, we have

```
jactyp=4; ml=2; mu=1;
%ODEOPTIONS=[1,0,0,%inf,0,jactype,500,12,5,0,ml,mu];
```

Note that it is also possible to use the odeoptions function to interactively give %ODEOPTIONS a value.

When we use a Scilab function for the computation of the banded Jacobian, there is nothing more to do than setting %ODEOPTIONS. But when the banded Jacobian is a C function or a Fortran subroutine, in addition to giving %ODEOPTIONS correct values, we also need to specify the number of rows of pd in nrowpd, and this must be equal to ml+mu+1. Examples of using banded Jacobians are given in the directory
SCIDIR/examples/misc-examples/odeoptions.sce.

Tuning the ODE Solver

We have seen in the preceding section that it is possible to specify parameters to ode using %ODEOPTIONS. It is also possible to give ode a number of arguments in order to tune its performance. We are going to describe the most important of these.

The `rtol` and `atol` arguments of `ode` control the relative and absolute error tolerances of the integration. If they are given, they must be positive and they must be positioned in this order in the calling list after the time parameter `t`. They can be scalars or vectors of length n (the dimension of the solution y). They determine the error control performed by the solver: the estimated local error in `y(i)` will be controlled so as to be less than `rtol(i)*abs(y(i))+atol(i)`; i.e., the local error test passes if in each component, either the absolute error is less than `atol(i)` or the relative error is less than `rtol(i)`.

If `rtol` and/or `atol` is a constant, `rtol(i)` and/or `atol(i)` are set to this constant value.

Note that according to the method specified by the optional argument `type` of `ode` function, only scalar values of `atol` and `rtol` are significant: This is the case for `type='rkf'` or `type='fix'`. For `type='rk'` only a scalar value of `atol` is meaningful.

The default values for `rtol` and `atol` are respectively 10^{-7} and 10^{-9} for most solvers. They are respectively 10^{-3} and 10^{-4} for the `type='rkf'` and `type='fix'` solvers.

An example using `atol` and `rtol` can be seen in the example of the pendulum Section 8.3.2.

Other options for tuning solvers can be made using the `%ODEOPTIONS` variable. An explanation of these can be had by looking at the on-line help of `odeoptions`, which instructs the user how to specify these options interactively.

Hot Start of the ODE Solver

When using `ode`, it is possible to stop the integration of the ODE at a given point to do some other work and to subsequently restart the integration at the stopped point: This is called *hot start of the solver*.

Hot starting is possible for all solvers with the exceptions of `type='rk'` and `type='fix'`. To perform a hot start the values of two real vectors `w` and `iw` must be obtained on the first call to `ode`: `[y,w,iw]=ode(...)`. These vectors store internal information from the solver needed for the hot start.

To restart the integration from the state at the stopping point it is only necessary to pass the two real vectors `w` and `iw` to `ode` in the final two positions of the argument list: `[y,w,iw]=ode(...,w,iw)`.

8.2.2 Choosing Between Methods

The precision of the Adams method (**type='adams'** and the different variants) is high, but it cannot be applied to stiff problems because of a relative lack of stability. For stiff problems the BDF method (**type='stiff'**) is recommended; its precision is less than the Adams method, but its stability is ensured for a very large class of problems.

The BDF method gives better results if we have an explicit formula for the Jacobian instead of computing a numerical approximation. For this reason it is highly recommended that the user supply the Jacobian. A detailed procedure has been defined for supplying the Jacobian in different ways: as a full or banded matrix. Furthermore, it is not necessary to provide a precise calculation of the Jacobian; even a coarse approximation is preferable to the internal numerical approximation.

If maximum efficiency is necessary, some experimentation with the various methods may be required. To this point in the presentation the more subtle points of calls to **ode** have not been developed. Many powerful options can be employed, such as how to choose the starting procedure, how to change the integration step, how to change the order. These are all that is needed to obtain an efficient global method.

To learn the more sophisticated options of **ode** it is suggested that the user look at the source files of the ODEPACK programs (which are distributed with the Scilab source). The files are well documented and can be found in the directory **SCIDIR/routines/integ/**.

8.2.3 ODE Integration with Stopping Times

In implementing ODE integration with stopping times, the solver integrates either a stiff or nonstiff ODE system (8.1) and at the same time locates the roots of a set of m functions $h_i(t, y)$:

$$
\begin{aligned}
\dot{y} &= f(t, y), \\
y(t_0) &= y_0,
\end{aligned}
\tag{8.5}
$$

and find τ such that for some i, $h_i(\tau, y(\tau)) = 0$.

The solver tries to find the first time τ for which at least one of the set of constraint functions $h_i(t, y)$ is equal to zero. It then returns the solution $y(\tau)$ at this time.

You can interpret this problem as computing the solution of the ODE until the solution $y(t)$ crosses the surface $h(t, y) = 0$, where $h(t, y)$ denotes the vector function $(h_1(t, y), \dots, h_m(t, y))^T$. Note that if a stop condition (given by the integration time or by using **%ODEOPTIONS** described on page

259) occurs sooner than the next root, the solver returns the solution according to the specified stop condition.

The simplest calling sequence of **ode** with root finding is

[y,rd]=ode('root',y0,t0,t,f,nh,h)

where **y0, t0, t, f,** and **y** have the standard meaning. The output variable, **rd**, is a vector of size k. The first entry is the stopping time of the integration. The other entries indicate which components of $h(t, y)$ have changed sign, so if k is greater than 2, then more than one surface ($k - 1$ exactly) have been simultaneously crossed. The input, **h**, can be a Scilab function or an external C function or Fortran subroutine (which must be linked into Scilab). The manner in which these functions are passed to **ode** is described on page 253.

If **h** is a Scilab function, its calling sequence must be **z=h(t,y)**, where **z** is a vector of size **nh** corresponding to the number of constraints. If it is an external C function or Fortran subroutine, then **h** must be a string, and the corresponding arguments must be **(n,t,y,nh,hout)**, where **n** is the dimension of **y**, **t** is the time, **y** is the vector y, **nh** is the number of constraints, and **hout** is the value of $h(t, y)$ (output of the program).

For a Fortran subroutine we have

```
subroutine h(n,t,y,nh,hout)
doubleprecision t,y(n),hout(nh)
...
```

and for a C function we have

```
h(n,t,y,nh,hout)
  int *n,*nh;                    ← Fortran variables are pointers
  double *t,*y,*hout;
{
...
}
```

Note that the preceding C function is called by Scilab, which is a Fortran program, so all input and output variables of this C function belong to the Fortran world; in particular, variables are pointers.

It is possible to use all the tools with root finding that were previously described for the solver. It is possible to pass parameters to the function $h(t, y)$. This is accomplished in the same way as for $f(t, y)$ and is described on page 254. The optional parameters **rtol** and **atol** and the variable **%ODEOPTIONS** can be used as described on page 259. We repeat that for stiff problems it is better to provide a computation of the Jacobian, and the syntax for passing the Jacobian to **ode** is described on page 256. Finally, it is possible to hot start the solver as described on page 260.

As an example let us consider the system (8.3) where the initial state, y_0, is close to zero and where we are seeking the root corresponding to when the surface $\|y\| = b$ is crossed. This is done as follows:

```
J=[0,-1;1,0]; a=1;
deff('ydot=f(t,y)','ydot=J*y+(a-norm(y))*y')
deff('s=surf(t,y)','s=norm(y)-b');                    ← one surface
t0=0;y0=[0.01;0.02];T=0:0.1:10;b=0.99;
[y,rd]=ode('roots',y0,t0,T,f,1,surf);               ← crossing a surface
isoview(-1,1,-1,1);plot2d(y(1,:),y(2,:),1,'001',' @ @ ');
xarc(-b,b,2*b,2*b,0,360*64)
xset('mark',1,5)
plot2d(-b,0,-2,'001',' @ @ ');
```

The plot is shown in Figure 8.3, and the cross shows the point where the trajectory reaches the surface defined by the plotted circle.

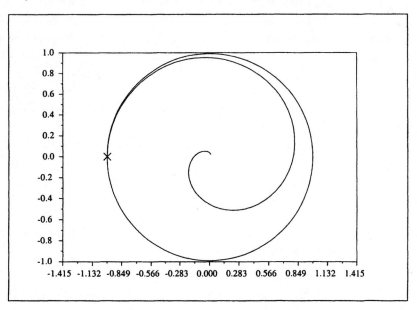

Figure 8.3: *Ode with surface crossing*

8.2.4   Sampled Systems

We describe here the case where we have to solve a family of ODEs depending on a parameter that switches at given sampling times. Note that such systems can be conveniently described and simulated with Scicos (see Chapter 9).

To be more specific, consider the sampling instants ($t_0 = \Delta, t_1 = \Delta + h, \ldots, t_k = \Delta + kh, \ldots$) and a dynamical system with state $y = (y_c, y_d)$ partitioned in such a way that the "continuous" state y_c and the "discrete" state y_d evolve according to

$$\dot{y}_c = f_c(t, y_c, y_d);$$
$$y_d(t_i^+) = f_d(y_c(t_i), y_d(t_{i-1})).$$

The function **odedc** is devoted to the simulation of such systems. Its syntax is

```
yt=odedc([y0c;y0d],nd,[h,delta],t0,t,f)
```

where **y0c** and **y0d** are the initial conditions, **nd** is the dimension of **y0d**, **h** is step h, **delta** gives the delay $\Delta = h \times$ **delta**, **t0** is the initial time, and **t** is the vector of instants where **yt** is calculated. Functions f_c and f_d are defined in a single Scilab function **f** passed to **odedc** with calling sequence **ycd=f(t,yc,yd,flag)**. If **flag** is 0, this function should compute f_c and return $\dot{y}_c$; if **flag** is 1, this function should compute f_d and update y_d. As usual, **f** can represent an external C function or Fortran subroutine (which must be linked into Scilab). The manner in which these functions are passed to **odedc** is described on page 253.

To illustrate **odedc** we define the simplest one dimensional case: The t_i's are integers ($0, 1, \ldots$), and $y(t)$ is such that $\dot{y} = a$ for t in odd intervals and $\dot{y} = -a$ for t in even intervals. At integer values of time, y_d switches from $+a$ to $-a$ or conversely.

```
-->deff('ycd=f(t,yc,yd,flag)',..          ← switching function
        'if flag==0 then ycd=yd;..
        else ycd=-yd;end');
```

```
-->a=1;y0c=0;y0d=a;nd=size(y0d,2);..      ← defining constants of the
    h=1;delta=0;t0=0;t=0:0.1:10;                              problem
```

```
-->ycd=odedc([y0c;y0d],nd,[h,delta],t0,t,f);
```

```
-->rect=[0,-2,10,2];
```

```
-->plot2d2('onn',t',ycd(2,:)',[1],'051',' ',rect);
```

```
-->plot2d1('onn',t',ycd(1,:)',[2],'000');
```

Figure 8.4 shows the switching trajectories.

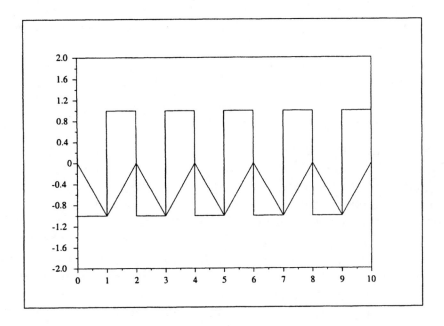

Figure 8.4: *ODE with sampling*

8.3 Integrating DAEs

DAEs are implicit ODEs described by equation (8.2). Such equations usually arise in attempts at solving nonlinear optimal control problems or mechanical problems such as multibody mechanisms with constraints. DAEs also naturally appear as limits of systems having both slow and fast modes. System (10.6) in Section 10.4 and system (10.9) in Section 10.5 are examples of mechanical DAEs. We will see in this section the main Scilab functions for solving such types of problems. Implicit ODEs arise in most areas of engineering, and we give a few examples showing how Scilab can be used to solve them. First, we treat the special case where the function to be integrated, f, is linear, followed by the general nonlinear case. For the general case, we will see that only a limited class of implicit ODEs can be solved, depending on the index of the system.

8.3.1 Implicit Linear ODEs

Implicit linear ODEs are a special case of DAEs described by the equation

$$
\begin{aligned}
A(t,y)\dot{y} - g(t,y) &= 0, \\
y(t_0) &= y_0,
\end{aligned}
\tag{8.6}
$$

where $A(t, y)$ is a square matrix of dimension n and $g(t, y)$ is a vector function of the same dimension.

The Scilab function to solve this type of problem is `impl`, which uses the `lsodi` solver of the package ODEPACK. The simplest call of this function is

```
y=impl(y0,ydot0,t0,t,res,adda)
```

where `y0` is the vector of initial conditions y_0, `ydot0` is the vector of initial time derivatives $\dot{y}_0$, `t` is the vector of the instants where the solution is computed, `res` is a function computing the residual $g(t, y) - A(t, y)\dot{y}$, and `adda` is a function for computing $A(t, y)$. The output, `y`, a matrix where each column j is of dimension n, is the value of y at time t_j.

It is usually necessary to provide the value of $\dot{y}(t_0)$ for DAE solvers to find the solution efficiently. In fact, if $A(t, y)$ is invertible for all t, this system is not a true DAE, since it could be solved in order to obtain an explicit ODE. Under these conditions it would no longer be necessary to provide $\dot{y}(t_0)$. If $\dot{y}(t_0)$ is unknown, it can be computed by solving the equation $A(t_0, y_0)\dot{y}(t_0) - g(t_0, y_0) = 0$ using either the $\backslash$ operator for solving linear systems or by using the function `fsolve` (see Section 8.4.3).

The functions `res` and `adda` can be, as usual, Scilab functions or external C functions or Fortran subroutines that are subsequently linked into Scilab. These functions are passed to `impl` in the same way as described on page 253. If `res` is a Scilab function, `r=res(t,y,ydot)` must be its calling sequence. If it is an external C function or Fortran subroutine, `res` must be a string, and the arguments of the corresponding C function or Fortran subroutine must be `(n,t,y,ydot,r,ires)`, where `n` is the number of equations, `t` is the time, `y` is the vector y, `ydot` is the vector $\dot{y}$, and `r` is the value of $g(t, y) - A(t, y)\dot{y}$ (output of the program). The argument `ires` can be ignored here.

If `adda` is a Scilab function, its calling sequence must be `r=adda(t,y,p)` and it must return $A(t, y) + p$. This function adds the matrix $A(t, y)$ to another matrix p represented by the Scilab variable `p`. The matrix p is only a formal parameter. If `adda` is an external C function or Fortran subroutine, `adda` must be a string giving the name of the linked program, and the arguments of the corresponding C function or Fortran subroutine must be `(n,t,y,ml,mu,p,nrowp)`, where `n` is the dimension of `y`, `t` is the time, `y` is the vector y, and `p` is an $n \times n$ matrix used both as input, $A(t, y)$, and as output, $A(t, y) + p$. The variable `nrowp` is the number of rows of `p`, which here is equal to `n`, and `ml` and `mu` are unused (eventually, when implemented, they will be used for banded matrices in a way similar to that for passing the Jacobian, as described on page 256).

Parameters can be passed to both **adda** and **res**. The method for this is using lists, as described on page 254.

As in **ode**, the first argument of **impl** can be the type of method we want to use, and it can be **'adams'** (Adams method for nonstiff ODEs) or **'stiff'** (BDF method for stiff ODEs). If **type** is not given, **'stiff'** is assumed. If the matrix $A(t, y)$ is nonsingular, then stiffness means that the explicit ODE $\dot{y} = A(t, y)^{-1}g(t, y)$ is stiff. If $A(t, y)$ is singular, the concept of stiffness is not well-defined. As usual, if we do not know whether the problem is stiff, it is recommended to use **'stiff'** (or no type at all), because if the problem is stiff, the BDF method will work better. However, if the problem is not stiff, the advantage of using the Adams method is only nominal.

The optional parameters **rtol** and **atol** can be used in the same way as for standard ODE problems, and this is described on page 259. The position of these arguments in the calling list to **impl** is just after **t**. Default values for **rtol** and **atol** are respectively 10^{-7} and 10^{-9}.

For stiff problems it is, as usual, better to provide a computation of the Jacobian of $g(t, y) - A(t, y)\dot{y}$ with respect to y. If the Jacobian is given, it must come just after **adda** in the calling sequence of **impl**. Giving the Jacobian can be done in the same way as for standard ODE problems, and it is described on page 256. The only difference here is that $\dot{y}$ is given as an argument, so that the calling sequence of the Scilab function calculating the Jacobian is **j=jac(t,y,ydot)**. In the case of using an external C function or Fortran subroutine to perform the computation, the calling sequence is **(n,t,y,ydot,ml,mu,pd,nrowpd)**, where **ydot** is the value of $\dot{y}$.

The file **EX-impl.f** in the directory **SCIDIR/routines/defaults/** illustrates an example of using Fortran subroutines in conjunction with **impl**.

It is also possible to hot start the solver for **impl**, and the procedure is identical to that described on page 260.

8.3.2   General DAEs

General DAEs are implicit ODEs. They are described by the system (8.2) where the function f is a function of $\dot{y}$:

$$\begin{aligned} f(t, y, \dot{y}) &= 0, \\ y(t_0) &= y_0. \end{aligned} \tag{8.7}$$

Index of DAEs

In general, not all DAEs can be solved. We can class DAEs by using their index. There are at least two definitions of the index: the differentiation index and the perturbation index. The simplest is the differentiation index, d, which expresses the fact that we have to differentiate at most d times the DAE system (8.7) in order to obtain an explicit ODE system. For more complete definitions of the index, see [33].

Note: In what follows we write $f(y, u)$ instead of $f(t, y, u)$ to simplify notation, and we do not explicitly specify associated initial conditions.

Index 1 DAEs   A more straightforward, but less general, definition of a DAE is

$$\begin{aligned} \dot{y} &= f(y, u), \\ 0 &= g(y, u), \end{aligned} \tag{8.8}$$

where $g(y, u)$ is a vector function of dimension p, and u is a vector of dimension m. This system consists of a differential equation $\dot{y} = f(y, u)$ together with an algebraic equation $0 = g(y, u)$, which consists of a set of constraints. The input u can be considered to be a control input variable.

If we differentiate the constraints, $g(y, u) = 0$, once with respect to time and substitute $f(y, u)$ for $\dot{y}$, we obtain the following system with (y, u) as the new unknown:

$$\begin{aligned} \dot{y} &= f(y, u), \\ g_u(y, u)\dot{u} &= -g_y(y, u)f(y, u). \end{aligned} \tag{8.9}$$

Remember that we are using vector notation for functions, so $g_u(y, u)$ is the matrix of size $p \times m$ where the element (i, j) is equal to $(g_i)_{u_j} = \partial g_i / \partial u_j$, and $g_y(y, u)$ is the matrix of size $p \times n$ where the element (i, j) is equal to $(g_i)_{y_j} = \partial g_i / \partial y_j$, and thus $g_u(y, u)\dot{u}$ is the product of a matrix by a vector.

If $g_u(y, u)$ is an invertible square matrix, the preceding system can be inverted to give

$$\begin{aligned} \dot{y} &= f(y, u), \\ \dot{u} &= -g_u^{-1}(y, u)g_y(y, u)f(y, u), \end{aligned} \tag{8.10}$$

which is an explicit system, so the index of the original system is 1.

Index 2 DAEs   Now suppose that in equation (8.8), g does not depend on u. If we differentiate $0 = g(y)$ with respect to time, we obtain $0 = g_y(y)\dot{y} = g_y(y)f(y, u)$, which leads us to the same type of constraint as in equation (8.8).

So we need to differentiate again and substitute $f(y, u)$ for $\dot{y}$ in order to obtain

$$
\begin{aligned}
\dot{y} &= f(y, u), \\
g_y(y) f_u(y, u) \dot{u} &= -(g_{yy}(y)(f, f) + g_y(u) f_y(y, u) f(y, u)) .
\end{aligned}
\tag{8.11}
$$

With the same matrix notation as for (8.9), $g_y(y)$ is the matrix of size $p \times n$ where the element (i, j) is equal to $(g_i)_{y_j} = \partial g_i / \partial y_j$, and $f_u(y, u)$ is the matrix of size $n \times m$ where the element (i, j) is equal to $(f_i)_{u_j} = \partial f_i / \partial u_j$. We have that $g_{yy}(y)(f, f)$ is the p-vector

$$
\begin{pmatrix} \cdots \\ f^T g_{i_{yy}}(y) f \\ \cdots \end{pmatrix},
$$

where $g_{i_{yy}}(y)$ is the Hessian matrix of g_i, i.e., the matrix of size $n \times n$ where the element (i, j) is equal to $\partial^2 g_i / \partial y_i \partial y_j$. The matrix $f_y(y, u)$ is the matrix of size $n \times n$ where the element (i, j) is equal to $(f_i)_{y_j} = \partial f_i / \partial y_j$.

If $g_y(y) f_u(y, u)$ is an invertible square matrix, the preceding system can be inverted to give

$$
\begin{aligned}
\dot{y} &= f(y, u), \\
\dot{u} &= -(g_y(y) f_u(y, u))^{-1} (g_{yy}(y)(f, f) + g_y(u) f_y(y, u) f(y, u)),
\end{aligned}
\tag{8.12}
$$

which is an explicit system, so the index of the system is 2.

Examples of various index systems are given by a class of mechanical problems that can be modeled by the Lagrange equations

$$
\begin{aligned}
\frac{d}{dt}\left(\frac{\partial L}{\partial \dot{q}_k}\right) - \frac{\partial L}{\partial q_k} &= 0, \quad k = 1, \dots, n, \\
0 &= g(q),
\end{aligned}
\tag{8.13}
$$

where $q = (q_1, \dots, q_n)^T$ is the vector of generalized coordinates of the system, $L(q, \dot{q})$ is the Lagrangian of the system and $g(q) = (g_1(q), \dots, g_m(q))^T$ are m constraints defining the geometry of the system and are independent of the generalized velocities $\dot{q}$. If T denotes the kinetic energy, U the potential energy, and $\lambda = (\lambda_1, \dots, \lambda_m)^T$ are the Lagrange multipliers corresponding to the constraints, then the Lagrangian L is defined by

$$
L(q, \dot{q}) = T(q, \dot{q}) - U(q, \dot{q}) - \sum_{i=1}^{m} \lambda_i g_i(q).
$$

It can be shown that upon differentiating the above system, we obtain the following:

$$\begin{aligned} \dot{q} &= u, \\ M(q)\dot{u} &= f(q,u) - g_q^T(q)\lambda, \\ 0 &= g(q), \end{aligned} \qquad (8.14)$$

where $M(q)$ is a positive definite matrix and $u = (\dot{q}_1, \ldots, \dot{q}_n)^T$. From this point a few various transformations are possible, leading to DAEs with index 1, 2, or 3. We will treat an example later in this section (see Section 8.3.2).

To solve DAE problems, we must find consistent initial conditions, i.e., $y(t_0)$ and $\dot{y}(t_0)$ satisfying $f(t_0, y(t_0), \dot{y}(t_0))$. Usually, some components of $y(t_0)$ and $\dot{y}(t_0)$ are given. The remaining components can be computed using the function **fsolve** (see Section 8.4.3). Note that the index reduction process can sometimes lead to systems in which it is difficult to find consistent initial conditions.

The **dassl** Function

In order to solve DAE systems described by equation (8.7), it is necessary to use the **dassl** function, which makes use of the DASSL solver [47]. This solver is able to solve systems of index 1 and a few systems of index 2. If you use it for DAEs with index greater than 2 (it is possible that the system's index is indeterminate), the solver is likely not to converge. Consequently, it is strongly encouraged that the user take steps to reduce the index of the system. Many index reduction methods exist (for instance, see [33]).

The DASSL solver uses the backward differentiation formulas (BDF) of orders one through five to solve the system. Values for y and $\dot{y}$ at t_0 must be given as input. It is important that these values be consistent. If $\dot{y}(t_0)$ is unknown, it can be computed by solving the equation $f(t_0, y_0, \dot{y}_0) = 0$ using the function **fsolve** (see Section 8.4.3).

The simplest call of the function **dassl** takes the form

```
yyt=dassl([y0,ydot0],t0,t,res,info)
```

where **y0** is the vector of initial conditions y_0, **ydot0** is the vector of initial time derivatives $\dot{y}_0$, **t0** is the initial time, **t** is the vector of the instants for which the solution is computed, **res** represents the function $f(t, y, \dot{y})$, and **info** is explained later.

The output **yyt** is a matrix where each column j represents the vector $(t_j, y(t_j), \dot{y}(t_y))^T$ of size $2n + 1$. Thus, the matrix of the values of y for all times is **yyt(2:n+1,:)**, and the matrix of the values of $\dot{y}$ for all times is **yyt(n+2:2*n+1,:)**.

The initial conditions are specified in the $n \times 2$ matrix **[y0,ydot0]**. If $\dot{y}_0$ is unknown, it is possible to provide only **y0**. In this case the solver tries to find a value for **ydot0**. If an approximate but inconsistent value of $\dot{y}_0$ is known (i.e., $f(t_0, y_0, \dot{y}_0,) \neq 0$), this must be indicated to the solver by setting **info(7)=1**.

The function **res** can be either a Scilab function or an external C function or Fortran subroutine to be subsequently linked into Scilab. The syntax for passing this external to **dass1** is the same as described on page 253.

If **res** is a Scilab function, its calling sequence must be

```
[r,ires]=res(t,y,ydot)
```

where the flag **ires** should be set to 0 when the computation of $f(t, y, \dot{y})$ is successful, -1 if the value of $f(t, y, \dot{y})$ is not defined, and -2 if parameters are out of admissible range. If **res** is an external C function or Fortran subroutine, it must be a string giving the function name, and the arguments of the corresponding C function or Fortran subroutine must be **(t,y,ydot,delta,ires,rpar,ipar)**, where **t** is the time, **y** is the vector y, **ydot** is the vector $\dot{y}$, **delta** is the value of $g(t, y, \dot{y})$ (the output of the program), and **ires** is a flag (as described above). The inputs **rpar** and **ipar** are respectively double-precision and integer arrays, which can be ignored here.

If necessary, parameters can be passed to **res**. The method for accomplishing this is described on page 254.

Finally, **info** is a list with 7 elements. A complete description of its meaning is given on page 279. For the moment it suffices to say that it can take the default values **list([],0,[],[],[],0,0)**.

Example Using **dass1**

Here an example of a simulation of a simple mechanical system using **dass1** is given. The problem is that of solving the motion of a free pendulum moving in three-dimensional space. The example is solved by classical methods for constrained mechanical systems. Although it is possible to eliminate the constraint by an appropriate choice of coordinates, the purpose of this example is to illustrate the use of **dass1** for a simple nontrivial problem. Importantly, it is shown in this example how to reduce the index of the system and how to stabilize the solution of an index 1 system.

We use the Lagrange equation defined by system (8.13) where the kinetic energy of the pendulum is $T = m(\dot{x_1}^2 + \dot{x_2}^2 + \dot{x_3}^2)/2$ and the potential energy of the pendulum is $U = -m\gamma x_3$, where $q = (x_1, x_2, x_3)$ and m are respectively the coordinates and the mass of the end point of the pendulum and γ is the acceleration due to gravity. This end point of the pendulum must satisfy the constraint $x_1^2 + x_2^2 + x_3^2 = l^2$ where l is the length of the pendulum, so we have $g(q) = x_1^2 + x_2^2 + x_3^2 - l^2$. Denoting by λ the Lagrange multiplier associated with the constraint, the Lagrange equations (8.13) for the pendulum are

$$
\begin{aligned}
m\ddot{x_1} &= -2x_1\lambda, \\
m\ddot{x_2} &= -2x_2\lambda, \\
m\ddot{x_3} &= -m\gamma - 2x_3\lambda, \\
0 &= x_1^2 + x_2^2 + x_3^2 - l^2.
\end{aligned}
$$

Introducing the velocities as new state variables $u = (u_1, u_2, u_3)$, the following first-order system, corresponding to system (8.14), is obtained:

$$
\begin{aligned}
\dot{x_1} &= u_1, \\
\dot{x_2} &= u_2, \\
\dot{x_3} &= u_3, \\
\dot{u_1} &= -2x_1\lambda/m, \\
\dot{u_2} &= -2x_2\lambda/m, \\
\dot{u_3} &= -\gamma - 2x_3\lambda/m, \\
0 &= x_1^2 + x_2^2 + x_3^2 - l^2.
\end{aligned}
$$

Defining $\varphi(q, u) = (0, 0, -\gamma)^T$, $M(q) = I$, and $g_q(q) = \partial g(q)/\partial q = 2q^T$, we obtain the classical system

$$
\begin{bmatrix} I & 0 & 0 \\ 0 & M(q) & 0 \\ 0 & 0 & 0 \end{bmatrix} \begin{bmatrix} \dot{q} \\ \dot{u} \\ \dot{\lambda} \end{bmatrix} - \begin{bmatrix} 0 & I & 0 \\ 0 & 0 & -g_q^T(q) \\ 0 & 0 & 0 \end{bmatrix} \begin{bmatrix} q \\ u \\ \lambda \end{bmatrix} - \begin{bmatrix} 0 \\ \varphi(q, u) \\ g(q) \end{bmatrix} = 0
$$

$$(8.15)$$

which is similar to equation (8.7), where the state is $y = (q, u, \lambda)$.

To solve this problem, we must find consistent initial conditions, i.e., $y(t_0)$ and $\dot{y}(t_0)$ such that $f(t_0, y(t_0), \dot{y}(t_0)) = 0$, which implies in particular that $g(q(t_0)) = 0$. Also, system (8.15) is an index 3 DAE, which cannot be solved directly by **dassl**, even for small integration time and large tolerance.

An equivalent and simpler problem is obtained by replacing the constraint equation $g(q) = 0$ by its derivative $\dot{g}(q) = 0$. This yields the following index 2 system:

$$
\begin{bmatrix} I & 0 & 0 \\ 0 & M(q) & 0 \\ 0 & 0 & 0 \end{bmatrix} \begin{bmatrix} \dot{q} \\ \dot{u} \\ \dot{\lambda} \end{bmatrix} - \begin{bmatrix} 0 & I & 0 \\ 0 & 0 & -g_q^T(q) \\ 0 & g_q(q) & 0 \end{bmatrix} \begin{bmatrix} q \\ u \\ \lambda \end{bmatrix} - \begin{bmatrix} 0 \\ \varphi(q,u) \\ 0 \end{bmatrix} = 0.
$$
$$(8.16)$$

This index 2 DAE can be solved with **dassl** by taking large tolerances for λ. We first define functions corresponding to the preceding system:

```
function res=M(q)                                        ← matrix M(q)
res=eye(n,n);

function res=phi(q,u)                                    ← vector φ(q,u)
res=[0;0;-gama];

function res=g(q)                                        ← constraint g(q)
res=q'*q-l^2;

function res=g_q(q)                                      ← constraint g_q(q)
res=2*q';

function [res,ires]=index2(t,y,yd)                       ← index 2 function
q=y(1:n);u=y(n+1:2*n);
res=[I,      Z,  Z0;
     Z,  M(q),  Z0;
             Zb        ]*yd -..
  [ Z,         I,          Z0;
    Z,         Z,  -g_q(q)';
   Z2, g_q(q),          Z3]*y -..
  [Z1;phi(q,u);Z4];
ires=0;
```

Then we set some initial conditions and compute the other components. First the initial position of the end point of the pendulum is given using spherical coordinates. Then the velocity components and $\lambda(t_0)$ are chosen such that $f(t_0, y(t_0), \dot{y}(t_0)) = 0$ is satisfied, implying that

$$
\dot{g}(q(t))|_{t=t_0} = g_q(q(t_0))u(t_0) = 0
$$

and

$$
\begin{bmatrix} M(q(t_0)) & g_q(t_0)^T \\ g_q(t_0) & 0 \end{bmatrix} \begin{bmatrix} (\dot{u})(t_0) \\ \lambda(t_0) \end{bmatrix} = \begin{bmatrix} \varphi(q(t_0), u(t_0)) \\ 0 \end{bmatrix}.
$$

The value of $\lambda(t_0)$ is obtained by solving this linear system:

```
n=3;nc=1;N=n+n+nc;
Z=zeros(n,n);I=eye(n,n);Zb=zeros(nc,N);Z0=zeros(n,nc);
Z1=zeros(n,1);Z2=zeros(nc,n);Z3=zeros(nc,nc);Z4=zeros(nc,1);
```

```
m=1;gama=9.8;l=1;                                          ← constants
```

```
theta0=0.2;phi0=%pi/4;
q0=l*[sin(phi0)*cos(theta0);..                             ← initial position
     cos(phi0)*cos(theta0);sin(theta0)];
u0=[1;-1;0];                                               ← initial velocity
w=[M(q0),g_q(q0)';g_q(q0),Z3]\[phi(q0,u0);Z4];
ud0=w(1:n);
lambda0=w(n+1:$);                                          ← $\lambda(t_0)$
```

Finally, we call **dassl** using a large tolerance for λ because the value of λ does not interest us:

```
-->y0=[q0;u0;lambda0];

-->yd0=[u0;ud0;0];

-->info=list([],0,[],[],[],0,0);

-->atol=[0.0001;0.0001;0.0001;0.001;0.001;0.001;1];

-->rtol=atol;

-->t0=0;T=0:0.01:15;

-->y2=dassl([y0,yd0],t0,T,rtol,atol,index2,info);

-->q2=y2(1+1:1+n,:);

-->norm(q2(:,$),2)                                      ← constraint is satisfied
 ans  =
    0.9998415

-->param3d(q2(1,:),q2(2,:),q2(3,:),31,78," @ @ ",[2,3]);
```

Figure 8.5 illustrates the 3-D trajectory of the end point of the pendulum obtained from the simulation using **dassl**.

It is also possible to transform the system (8.15) into an index 1 system by replacing the constraint equation $g(q) = 0$ by its second derivative $\ddot{g}(q) =$

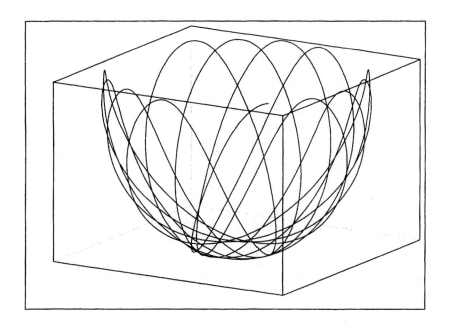

Figure 8.5: *3-D trajectory of the pendulum*

0. The resulting system is

$$
\begin{bmatrix} I & 0 & 0 \\ 0 & M(q) & 0 \\ 0 & g_q(q) & 0 \end{bmatrix} \begin{bmatrix} \dot{q} \\ \dot{u} \\ \dot{\lambda} \end{bmatrix} - \begin{bmatrix} 0 & I & 0 \\ 0 & 0 & -g_q^T(q) \\ 0 & 0 & 0 \end{bmatrix} \begin{bmatrix} q \\ u \\ \lambda \end{bmatrix} - \begin{bmatrix} 0 \\ \varphi(q,u) \\ -g_{qq}(q,u) \end{bmatrix} = 0,
$$
(8.17)

where $g_{qq}(q,u)$ denotes the second derivative of $g(q)$ with respect to q (see the definition of g_{qq} on page 269).

We define two new functions for this system:

```
function res=g_qq(q,u)
res=2*u'*eye(n,n)*u;

function [res,ires]=index1(t,y,yd)
q=y(1:n);u=y(n+1:2*n);
res=[ I,       Z, Z0;
      Z,    M(q), Z0;
     Z2, g_q(q), Z3]*yd-..
   [ Z, I,        Z0;
      Z, Z, -g_q(q)';
            Zb         ]*y-..
   [Z1;phi(q,u);-g_qq(q,u)];
```

```
ires=0;
```

Now, solving this DAE requires finding an initial condition that satisfies $f(t_0, y(t_0), \dot{y}(t_0)) = 0$, which implies $\dot{g}(q(t))|_{t=t_0} = g_q(q(t_0))u(t_0) = 0$ and

$$\begin{bmatrix} M(q(t_0)) & g_q(t_0)^T \\ g_q(t_0) & 0 \end{bmatrix} \begin{bmatrix} (\dot{u})(t_0) \\ \lambda(t_0) \end{bmatrix} = \begin{bmatrix} \varphi(q(t_0), u(t_0)) \\ -g_{qq}(q(t_0), u(t_0)) \end{bmatrix}.$$

The value of $\lambda(t_0)$ is obtained by solving this linear system:

```
theta0=0.2;phi0=%pi/4;
q0=1*[sin(phi0)*cos(theta0);cos(phi0)*cos(theta0);sin(theta0)];
u0=[1;-1;0];

w=[M(q0),g_q(q0)';g_q(q0),Z3]\[phi(q0,u0);-g_qq(q0,u0)];
ud0=w(1:n);
lambda0=w(n+1:$);
```

Finally, the solution is obtained as above for an index 2 system. Employing the same parameters as above we obtain

```
-->y1=dassl([y0,yd0],t0,T,rtol,atol,index1,info);

-->q1=y1(1+1:1+n,:);

-->norm(q1(:,$),2)
 ans  =
     0.9905839
```

Note that the constraint does not hold exactly. This is common for systems obtained by index reduction. A simple way to improve the solution is to stabilize the constraint. This is done by replacing the constraints $g(q) = 0$ by

$$\beta^2 g(q) + 2\alpha\dot{g}(q) + \ddot{g}(q) = 0,$$

where α and β are constants and the polynomial $s^2 + 2\alpha s + \beta^2$ is stable, i.e., has roots with negative real parts.

The stabilized index 1 system becomes

$$\begin{bmatrix} I & 0 & 0 \\ 0 & M(q) & 0 \\ 0 & g_q(q) & 0 \end{bmatrix} \begin{bmatrix} \dot{q} \\ \dot{u} \\ \dot{\lambda} \end{bmatrix} - \begin{bmatrix} 0 & I & 0 \\ 0 & 0 & -g_q^T(q) \\ 0 & -2\alpha g_q(q) & 0 \end{bmatrix} \begin{bmatrix} q \\ u \\ \lambda \end{bmatrix}$$

$$-\begin{bmatrix} 0 \\ \varphi(q,u) \\ -g_{qq}(q,u) - \beta^2 g(q) \end{bmatrix} = 0. \qquad (8.18)$$

We define the new function for this system:

```
function [res,ires]=index1_s(t,y,yd)
q=y(1:n);u=y(n+1:2*n);
res=[ I,        Z, Z0;
      Z,     M(q), Z0;
     Z2, g_q(q), Z3]*yd-..
    [ Z,             I,          Z0;
      Z,             Z, -g_q(q)';
     Z2, -2*alfa*g_q(q),         Z3]*y-..
   [Z1;phi(q,u);-beta^2*g(q)-g_qq(q,u)];
ires=0;
```

The initial condition can be taken as for the unstabilized index 1 system (8.17). We take $\alpha = \beta = 20$, and the solution of the stabilized DAE system is

```
-->alfa=20;beta=20;

-->y1_s=dassl([y0,yd0],t0,T,rtol,atol,index1_s,info);

-->q1_s=y1_s(1+1:1+n,:);

-->norm(q1_s(:,$),2)
 ans   =
    0.9999284
```

We can see that the constraint is now well satisfied. However, if α and β are chosen too large, the problem becomes stiff, and if they are too small, the constraint is poorly stabilized.

Note that $\dot{\lambda}$ is not constrained by the DAE equation (8.18). Introducing $\dot{\mu} = \lambda$, we obtain the index 0 DAE

$$\begin{bmatrix} I & 0 & 0 \\ 0 & M(q) & g_q^T(q) \\ 0 & g_q(q) & 0 \end{bmatrix} \begin{bmatrix} \dot{q} \\ \dot{u} \\ \dot{\mu} \end{bmatrix} - \begin{bmatrix} 0 & I & 0 \\ 0 & 0 & 0 \\ 0 & -2\alpha g_q(q) & 0 \end{bmatrix} \begin{bmatrix} q \\ u \\ \mu \end{bmatrix}$$

$$-\begin{bmatrix} 0 \\ \varphi(q,u) \\ -g_{qq}(q,u) - \beta^2 g(q) \end{bmatrix} = 0 \qquad (8.19)$$

We define the new function for this system:

```
function [res,ires]=index0_s(t,z,zd)
q=z(1:n);u=z(n+1:2*n);
res=[I,Z,Z0;
 Z,M(q),g_q(q)';
 Z2,g_q(q),Z3]*zd-..
[Z,I,Z0;
 Z,Z,Z0;
 Z2,-2*alfa*g_q(q),Z3]*z-..
 [Z1;phi(q,u);-beta^2*g(q)-g_qq(q,u)];
ires=0;
```

The initial condition is taken to be the same as for the unstabilized index 1 system, with $\dot{\mu}(t_0) = \lambda(t_0)$ and $\mu(t_0) = 0$ (μ is defined up to an additive constant). We take $\alpha = \beta = 100$, and the solution of the stabilized DAE system (8.19) is

```
-->alfa=100;beta=100;

-->z0_s=dassl([z0,zd0],t0,T,rtol,atol,index0_s,info);

-->q0_s=z0_s(1+1:1+n,:);

-->norm(q0_s(:,$),2)
 ans  =
    0.9999823
```

We can see that as for the stabilized index 1 case, the constraint is still well satisfied. In addition, the computation time is significantly shorter even though we have chosen bigger values for α and β. The large values were chosen for better stabilization of the constraint, which induced some stiffness in the system.

Stiff Problems

For stiff problems, it is as usual better to give the Jacobian of $g(t, y, \dot{y})$ with respect to y. If the Jacobian, **jac**, is given, it must come just after **res** in the calling sequence of **dassl**. It can be a Scilab function or an external C function or Fortran subroutine to be subsequently linked into Scilab. It is passed to **dassl** in the same way as described on page 253.

   If **jac** is a Scilab function, **j=jac(t,y,ydot,c)** must be its calling sequence, and it must return $g_y + cg_{\dot{y}}$, where c is a vector of the same size as y. If it is an external C function or Fortran subroutine, **jac** must be a string, and the arguments of the corresponding C function or Fortran subroutine must be **(t,y,ydot,pd,c,rpar,ipar)**, where **t** is the time, **y** is the vector

y, **ydot** is the vector $\dot{y}$, **pd** is the value of $g_y + cg_{\dot{y}}$ (output of the program), and **c** has the same meaning as above. The variables **rpar** and **ipar** are respectively double-precision and integer arrays, which can be ignored here.

By default, **pd** is a matrix of size $n \times n$. But if the problem is large and the matrix is banded, then it is better that the matrix be passed in a more efficient manner. The matrix can be more efficiently passed using the banded matrix procedure described on page 259. Here **pd** is a matrix of size **(2*ml+mu+1)** $\times n$, and the input **info(3)=[ml,mu]** is used to tell the solver that a banded matrix is being used. The values of **ml** and **mu** are the widths of the lower and upper parts of the band, excluding the main diagonal, and if **J** is the Jacobian matrix, we have **pd(i-j+ml+mu+1,j)=J(i,j)** when **J(i,j)** is different from 0. The matrix is packed in the same way as on page 259, but here there are **ml** more rows used internally for the solver.

See the file **EX-dassl.f** in the directory **SCIDIR/routines/defaults/** for an example using Fortran subroutines in **dassl**.

It is also possible to hot start the solver as described on page 260, but now only one vector, **hd**, passed as the last argument of **dassl** is used: **[yyt,hd]=dassl(...,hd)**.

Tuning the DASSL Solver

The argument **info** is passed to **dassl** as a list of length 7 and is used mainly for fine tuning. The meaning of **info(3)** has already been described for the banded Jacobian and **info(7)** for information concerning $\dot{y}(t_0)$. Other informational elements can be employed using **info**. These are described in the on-line help of **dassl**.

The optional parameters **rtol** and **atol** can also be used in the same way as for standard ODE problems, and this is described on page 259. These parameters must come just after **t** in the calling sequence of **dassl**. Default values for **rtol** and **atol** are respectively 10^{-7} and 10^{-9}

8.3.3 DAEs with Stopping Time

In the same way as for ODEs, DAEs can be solved with stopping times (see Section 8.2.3). For DAEs the following problem is solved:

$$\begin{aligned} f(t, y, \dot{y}) &= 0, \\ y(t_0) &= y_0, \end{aligned} \tag{8.20}$$

and find τ such that $h_i(\tau, y(\tau)) = 0$ for some i.

This is like system (8.7), but at the same time, the solver finds the roots of at least one of the set of constraint functions $h_i(t, y)$ as a function of t.

This problem can be interpreted as computing the solution of the DAE until the value of $y(t)$ crosses the surface $h(t, y) = 0$, where $h(t, y)$ denotes the vector $(h_1(t, y), \ldots , h_m(t, y))^T$. The problem is solved using a variation of the DASSL solver called DASRT, for which the corresponding function in Scilab is **dasrt**.

The simplest calling sequence of **dasrt** is

```
[yyt,nt]=dasrt([y0,ydot0],t0,t,res,nh,h,info)
```

where **y0**, **ydot0**, **t0**, **t**, **res**, **info**, and **yyt** have the same meaning as for **dassl**. The output, **nt**, is a 2-vector that indicates the time at which the surface is crossed and the number of the corresponding constraint surface.

The function **h** can be a Scilab function or an external C function or Fortran subroutine to be subsequently linked into Scilab. It is passed to **dasrt** in the same manner as described on page 253. If the constraint is implemented as a Scilab function, its calling sequence must be **z=h(t,y)**, where **z** is a vector of size **nh** corresponding to the **nh** constraints. If the constraints are implemented as an external C function or Fortran subroutine, **h** is a string giving the name of the function. The arguments of the corresponding C function or Fortran subroutine must be **(n,t,y,nh,hout,rpar,ipar)**, where **n** is the number of equations, **t** is the time, **y** is the vector y, **nh** is the number of constraints, and **hout** is the value of $h(t, y)$ (output of the program). The arguments **rpar** and **ipar** are respectively double-precision and integer arrays, which can be ignored here.

Parameters can be passed to $h(t, y)$. This is done in the same way as described on page 254.

The optional parameters **rtol** and **atol**, using the Jacobian **jac**, hotstarting the solver, and tuning **dasrt** using the parameter **info** follow the same rules as described for the function **dassl**.

8.4   Solving Optimization Problems

Optimization problems arise in all engineering fields. In physics it is common to formulate problems according to a variational principle that requires the minimization of a function (e.g., the energy of a system). Similar problems exist in economics and many other fields. Furthermore, many optimization problems arise from attempting to minimize some function constrained by a set of differential equations. Such problems also give rise to a variational approach that can be solved using Lagrange multipliers. Such is the case for many problems in automatic control. In this section we describe how Scilab can be used to solve mainly nonlinear optimization problems with and without constraints (linear programming problems can be solved using the function `linpro`).

The standard optimization problem consists in the minimization of a function f from $\mathbb{R}^n$ to $\mathbb{R}$:

$$\min_x f(x), \tag{8.21}$$

i.e., we are looking for the value $\bar{x}$ for which $f(\bar{x})$ is minimum. Usually f represents a cost to be minimized.

If f is convex and continuously differentiable and if ∇f denotes the gradient of f with respect to x, i.e., the vector $(\partial f / \partial x_i)$, then problem 8.21 has a solution, and the minimum is reached when $\nabla f = 0$. Most optimization algorithms use the gradient ∇f to solve the optimization problem. Some algorithms also use the Hessian (or an approximation to the Hessian), denoted by $\nabla^2 f$. The Hessian is an $n \times n$ matrix where the element (i, j) is $\partial^2 f / \partial x_i \partial x_j$.

Optimization problems with inequality constraints are also of interest. These types of problems can be formulated as follows:

$$\begin{aligned} \min_x f(x), \\ l \leq x \leq u, \end{aligned} \tag{8.22}$$

where l and u are respectively the lower and upper constraint vectors having the same dimensions as x. Scilab can handle all of these types of problems, of which a good number are described in what follows.

8.4.1 Quadratic Optimization

First, we consider a very important class of problems: those that consist in quadratic optimization. Many practical problems can be modeled in such a way, but this approach is also useful for problems that are not strictly quadratic. For such cases the quadratic algorithm is used as part of a more general iterative algorithm.

The standard quadratic optimization problem without constraints is formulated as

$$\min_x \frac{1}{2} x^T Q x - p^T x, \tag{8.23}$$

where Q is an $n \times n$ symmetric positive definite matrix and p is a vector of dimension n. This quadratic optimization problem corresponds to the solution of the linear system $Qx = p$, which is the simplest case of quadratic programming.

The function for solving quadratic optimization problems in Scilab is called **quapro**. In fact, this function can also solve quadratic optimization problems with constraints that are formulated in the following way:

$$\begin{aligned}
&\min_x \frac{1}{2}x^T Q x - p^T x, \\
&(Cx)_i = b_i \quad i = 1, \ldots, m_e, \\
&(Cx)_i \leq b_i \quad i = m_e + 1, \ldots, m, \\
&l \leq x \leq u,
\end{aligned} \tag{8.24}$$

where C is an $m \times n$ matrix, b is a vector of dimension m, and l and u are respectively the lower and upper constraint vectors with the same dimensions as b. Here Q need not be positive definite and may also have negative eigenvalues.

The simplest call to the function **quapro** is

```
x=quapro(Q,p,C,b[,x0])
```

where **Q** is the matrix Q, **p** is the vector p, **C** is the matrix C, and **b** is the vector b. The output, **x**, is the optimum point found by the function. With this calling sequence, **quapro** takes all the constraints to be inequalities, i.e., it is supposed that $Cx \leq b$. The optional argument **x0** specifies an initial guess for the value of **x**. If no initial point is given, or if **x0** is equal to the string **'v'**, the program computes a feasible initial point that is a vertex of the region of feasible points. If **x0** is equal to the string **'g'**, the program computes a feasible initial point that is not necessarily a vertex. This mode is advisable when the quadratic form is positive definite and there are few constraints in the problem, or when there are large bounds on the variables that are just security bounds and that are likely to be inactive at the optimal solution.

Other syntaxes for specifying all the constraints described in (8.24) are available and are described in the on-line help of **quapro**.

An example of a quadratic optimization problem is the following quadratic network flow problem:

$$\min \sum_{(i,j)\in\mathcal{A}} \frac{1}{2}w_{i,j}(x_{i,j})^2,$$

$$\sum_{j:(i,j)\in\mathcal{A}} x_{i,j} - \sum_{j:(j,i)\in\mathcal{A}} x_{j,i} = 0 \qquad \forall i \in \mathcal{N}, \tag{8.25}$$

$$l_{i,j} \leq x_{i,j} \leq u_{i,j} \qquad \forall (i,j) \in \mathcal{A},$$

where the graph of the network is a directed graph defined by the set of n nodes $\mathcal{N}$ and the set of m arcs $\mathcal{A}$. The vector w represents the cost of the transport on the arcs, and the vectors l and u are the lower and upper constraints on the arcs (minimum and maximum capacities).

This problem corresponds to problem (8.24) where $p = 0$, $b = 0$, and $m_e = n$ (only equality constraints for Cx). We have to find the Q and C matrices corresponding to problem (8.25). Here Q is the diagonal matrix of size m, and an easy way to find C is to start from the description of the graph and to use Metanet functions (see Chapter 11). The Scilab session using the **quapro** function is

```
-->n=15; m=31;

-->ta=[1 1 2 2 2 3 4 4 5 6 6 6 7 7 7 8 9..
        10 12 12 13 13 13 14 15 14 9 11 10 1 8];

-->he=[2 6 3 4 5 1 3 5 1 7 10 11 5 8 9 5 8..
        11 10 11 9 11 15 13 14 4 6 9 1 12 14];

-->l=[5 4 4 3 3 0 3 1 5 5 5 4 4 5 2 0 3 2 4 1..
       2 4 4 1 1 1 1 3 1 0 2];

-->u=[38 39 35 47 45 46 47 46 47 36 38 35 43 43 41 33 45..
       44 32 34 35 40 33 37 48 43 31 31 39 32 45];

-->g=make_graph('foo',1,15,ta,he);

-->C=full(graph_2_mat(g));            ← obtain matrix C  from graph g

-->Q=diag(ones(1,31)); p=zeros(31,1); b=zeros(15,1);

--> [x,lagr,f]=quapro(Q,p,C,b,l',u');
```

In fact, the standard way to solve this quadratic network flow problem is to use the **min_qcost_flow** function (see Section 11.5.6):

```
g('edge_min_cap')=l;
g('edge_max_cap')=u;
g('edge_q_orig')=0*ones(1,31);
g('edge_q_weight')=ones(1,31);
[cost,phi,flag]=min_qcost_flow(0.001,g);
```

We can see the difference between the two methods:

```
-->cost, f
  cost  =
```

```
     575.
 f  =
     558.5
```

```
-->(C*phi')'                          ← value of constraints with min_qcost_flow
  ans  =
          column  1 to 11
 !   0.     0.     0.     0.     0.     0.     0.     0.     0. !
          column 12 to 15
 !   0.     0.     0.     0.     0.     0 !
```

```
-->format(3); (C*x)'                   ← value of constraints with quapro
  ans  =
    1.0D-14 *
          column  1 to 11
 ! - 0.   - 1.   - 0.     0.   - 0.     0.   - 0.     0.     0. !
          column 12 to 15
 !   2.     0.   - 1.     0.     0.     0. !
```

```
-->format(1); phi-x'                   ← difference between flows
  ans  =
          column  1 to 14
 ! -0.   -1.     0.     0.     0.     0.    -0.     0.    -1.     0.     0. !
          column 15 to 28
 !  0.     0.     0.     0.    -1.     0.     0.     0.     0.     1.     0. !
          column 29 to 31
 !  0.     1.    -0.     0.     1.     0.     0.     0.     1. !
```

This example illustrates some differences with the optimization algorithms; of course in this case the second method using **min_qcost_flow** is the right specific one and this problem is only an example for the first method using **quapro**. The **min_qcost_flow** function strictly respects the flow constraint $Cx = 0$ and with the **quapro** function we have a better cost (3% better) but the flow constraint is not exactly observed (10^{-14}).

8.4.2   General Optimization

Scilab is able to solve problem (8.22) where the function $f(x)$ need not be continuously differentiable. The function that does this is called **optim**.

The main technique for solving general optimization problems is Newton's algorithm, which corresponds to solving a Taylor series expansion of the problem to order 2. However, the direct use of Newton's algorithm requires the computation of the Hessian as well as its inverse, and this is very costly in numerical computation. Fortunately, there are many algorithms that can efficiently compute an approximation of the inverse of the Hessian. Many of these algorithms are implemented in the function **optim**. These are taken from the MODULOPT library developed at INRIA [36], [13].

The exact algorithm used by **optim** is specified using the optional argument **algo**. The following is a description of the different values this argument can take:

algo='qn' This is the default value when **algo** is not explicitly given. A quasi-newton algorithm is used that consists in iteratively constructing a symmetric positive definite matrix approximating the inverse of the Hessian. For this algorithm, the BFGS (Broyden, Fletcher, Goldfarb, Shanno) update formula is used.

algo='gc' This is the conjugate gradient method, which uses conjugate directions with respect to the quadratic approximation of the function. Note that this algorithm converges in n steps for a quadratic function in $\mathbb{R}^n$.

algo='nd' Here a bundle method for nondifferentiable optimization is used. The descent direction is obtained by projection of the origin on a set of already computed gradients. For this option bounding constraints are not allowed.

Simple Call to **optim**

The simplest call to **optim**, corresponding to the quasi-Newton method, is

```
[f,xopt]=optim(costf,x0)
```

where **costf** is the cost function and **x0** is an initial starting point for the algorithm. The cost function, **costf**, to be written by the user, returns both the cost and the gradient, ∇f, given an input value for **x**. The output of **optim** is the best cost found for **f** as well as the argument, **xopt**, yielding the optimal value.

The function **costf** can be a Scilab function or an external C function or Fortran subroutine to be subsequently linked into Scilab. If **costf** is a Scilab function, its calling sequence is **[f,g,ind]=costf(x,ind)**, where **f** and **g** are respectively the cost function and the gradient of the cost with respect to the argument. The variable **ind** is used both as input and as output. If **ind** is equal to 2 (respectively 3 or 4), **costf** must provide **f** (respectively **g**, **f**, and **g**). If **ind** is equal to 1, nothing is computed (it is used for display purposes only). On output, if **ind** is negative, it means that $f(x)$ cannot be evaluated at **x**, and if **ind** is equal to 0, the optimization is interrupted.

If `costf` is an external C function or Fortran subroutine, `costf` is a string giving the name of the program, and the arguments of the corresponding C function or Fortran subroutine must be `(ind,n,x,f,g,ti,tr, td)`, where the parameters `ind`, `n`, `x`, `f`, and `g` have the same meaning as above and `ti`, `tr`, and `td` are working arrays. The use of these external programs is identical to that described on page 253.

See the file `EX-optim.f` in the directory `SCIDIR/routines/defaults/` for an example using Fortran subroutines in `optim`.

Simple Example

Here a simple example using `optim` is presented. The coefficients of a polynomial are found that yield the best fit to given values at given points. The function to be found is the polynomial $P(x) = \sum_{i=0}^{N} i * x^i$ of order $N = 5$, and we compute its values at $N + 1$ given points $x_k = k/N$.

```
-->N=5;

-->aref=[1:N];

-->X=0.2*[1:N+1];

-->Pref=poly(aref,'x','coef')          ← reference polynomial
 Pref  =
                2    3    4
   1 + 2x + 3x + 4x + 5x

-->Y=horner(Pref,X);                    ← Y_i = P(X_i)
```

We use a Scilab function for the cost function. For a vector of N coefficients this function returns the cost, the sum of the square deviations at the $N + 1$ given points, and also the corresponding gradient. The given points are global variables, which can be accessed inside the function `costf`.

```
function [cost,grad,ind]=costf(a,ind)
grad=0*a;
P=poly(a,'x','coef');
ya=horner(P,X);
cost=(ya-Y)*(ya-Y)';                    ← computing the cost
grad(1)= 2*(ya-Y)*ones(X)'
for i=2:size(grad,'*'),
  grad(i)=2*(ya-Y)*(X.^(i-1))'          ← computing the gradient
end;
```

Then we choose an initial value for the coefficient vector to be optimized, e.g., 0, and then we call the **optim** function:

```
-->[f,xopt]=optim(costf,0*ones(a))
 end of optimization
 xopt  =                                          ← best value found
 !   1.     2.     3.     4.     5. !
   f  =                                            ← value of optimal cost
     3.402E-30

-->poly(xopt,'x','coef')
 ans  =
                    2     3     4
     1 + 2x + 3x + 4x + 5x
```

Of course, the optimum value is close to the solution; as the maximum number of calls has been reached, we can simply change the initial value to the value given by **optim** and call it again; in fact, the improvement will be generally very poor.

Example of an Assignment Problem

We present here another example of the use of the function **optim**. The model leads to an integer programming problem, and it can be considered in two ways: very simple if we only want an approximate solution, and very complicated if we consider the generic integer modeling. The purpose of this example is limited to the illustration, for practical problems, of the benefit of a short set of commands with the use of **optim** in order to quickly improve a naive solution.

The example is described in what follows. A company wants to perform a survey about the water consumption, Q, of customers living in a specific geographic zone including $N = 80$ towns. The global number of surveyed people is given, or more precisely determined, by the global funding for this task. The problem is to determine the optimal number, n_h, of families to be interviewed in each town. Here optimal means that the variance of the estimated value of Q is to be minimized. We want to have the information about at least one family in every town because other information about the towns such as funding or taxes are asked in the same questionnaire and so we add the following constraint: $n_h \geq 1$.

The variance of the estimated value of the water consumption is

$$\text{Var}(Q) = \sum_{h=1}^{N} \left(\frac{N_h}{N}\right)^2 \left(1 - \frac{n_h}{N_h}\right) \frac{S_h^2}{n_h},$$

where N_h is the total number of families in the town h and S_h is the variance of the corresponding water consumption. As we do not have additive information about the variances S_h^2, we suppose that they are all the same; so the problem becomes

$$\min_{n_h} \left(\sum_{h=1}^{N} \left(\frac{N_h}{N} \right)^2 \left(1 - \frac{n_h}{N_h} \right) \frac{1}{n_h} \right).$$

We add the constraints, we simplify the term to minimize, and we obtain the following assignment problem:

$$\min_{n_h} \sum_{h=1}^{N} ; \ \frac{N_h^2}{n_h} ; \ n_h \geq 1 ; \ \sum_{h=1}^{N} n_h = 400.$$

In fact, for this particular problem, it is possible to find by hand a continuous optimal solution. We call it the "hand solution," and it is equal to $400 N_h / \sum N_h$.

But for sake of generality, and in the case where the data of the problem could change, we can obtain a simple solution with a few lines of code. The method used is the penalization of the constraint $\sum n_h = 400$. So we obtain the following penalized problem:

$$\min_{n_h} \left(\sum_{h=1}^{N} \frac{N_h^2}{n_h} + c \left(\sum_{h=1}^{N} n_h - 400 \right)^2 \right).$$

The main part of the code is very simple, with the definition of a criterion function and the call to the Scilab function **optim**.

```
N2=N.*N;     ← N is the vector of towns sorted by increasing number of families
N2max=maxi(N2);
N2R=N2./N2max;
deff('[f,g,ind]=cost(x,ind)',          ← definition of the criterion
     ['f=sum(N2R./x)+c*(sum(x)-400)**2';
      'x2=x.*x;g=-N2R./x2+2*c*(sum(x)-400)']);
x0=400*N/sum(N);                        ← initial point (hand solution)
c=1000;                                 ← penalization rate
[f,xopt]=optim(cost,'b',ones(x0),400*ones(x0),x0,'gc');
```

We have given **'b'** as the second argument of **optim** to specify that we are solving a problem with lower and upper bound constraints given respectively by the two following arguments. By specifying the **'gc'** argument we indicate that the conjugate gradient algorithm should be used.

We get the following results: The cities are sorted according to increasing values of n_h. The 32 smallest values are equal to 1 (the constraint), and the 48 biggest are

1.039 1.077 1.18 1.455 1.463 1.473 1.5 1.502 1.507 1.528 1.561 1.657 1.666 1.733 1.853 2.012 2.043 2.078 2.091 2.208 2.373 2.442 2.83 2.958 3.115 3.216 3.521 4.603 4.715 5.86 5.975 7.254 7.802 9.405 9.646 10.513 10.613 10.665 15.601 18.234 26.156 27.47 29.841 33.784 37.274 37.5.

The constraint $\sum_{h=1}^{N} n_h = 400$ is enforced up to a gap less than 10^{-6}.

Of course, the values of n_h are reals, and we want integers; so we have to round the previous values. In this operation we have to keep the global constraint to be satisfied, i.e., the sum of the rounded values must be equal to 400. The way to do that is to increase or decrease a few values. It is easy to do this with Scilab commands, and Figure 8.6 shows the distribution of the sample sizes. On the x-axis we have the index of the cities sorted according to their increasing populations. On the y-axis we have the corresponding number of families from the survey for the "hand" solution as well as for the optimal solution.

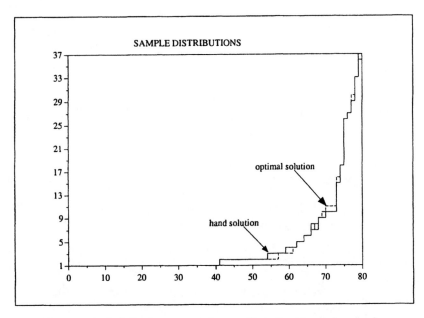

Figure 8.6: *Minimum variance distribution of samples*

The last thing to do is to measure the quality of the rounded integer solution with respect to the continuous solution. For this purpose we compute the values obtained by a perturbation around the continuous solution: We add a random sequence of -1, 0, or 1 values (with its sum equal to 0) to the continuous solution, and we plot the corresponding variation with the integer rounded solution. In Figure 8.7 we have represented the variation of the criterion with unitary perturbations of the continuous solution and the new integer rounded values.

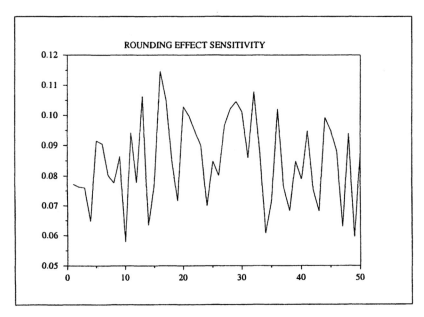

Figure 8.7: *Criterion variation (percentage) around the solution*

Complete Call to optim

A more sophisticated use can be made of **optim** by passing it additional arguments. Among the features that can be had are:

- Bound constraints can be specified (if the method is not '**nd**').

- Stopping procedures can be specified. For instance, the maximum number of iterations, a threshold on the gradient norm, a threshold on the decreasing value of the cost function $f(x)$, and a threshold on the variation of x can all be specified.

- Hot restart for the quasi-Newton method can be used. This is to allow stopping and restarting of an optimization without losing values critical to the estimation of the inverse of the Hessian.

- Communication of Scilab variables to the cost function when it is a C function or a Fortran subroutine.

Other features are available for tuning the **optim** function. The complete call syntax is described in the on-line help.

8.4.3 Solving Systems of Equations

Solving optimization problem 8.21 is related to solving the corresponding generally nonlinear system of equations $\nabla f(x) = 0$. The Scilab function for solving such a system of equations is **fsolve**.

The function **fsolve** is based on Minpack routines **hybrj** and **hybrd**. These functions find a zero of a system of n nonlinear functions in n variables by a modification of the Powell hybrid method. The user can provide the n-valued function as a Scilab function or as an external routine written in C or Fortran. The gradient of the function can be optionally given to reduce the computation time and for better accuracy of the solution. If the gradient is not provided, a finite difference approximation is internally calculated.

Let us consider, for instance, the problem of finding the intersection of an ellipsoid and a line in $\mathbb{R}^n$. The ellipsoid is defined by the equation $x^T Q x = 1$ where Q is a $n \times n$ symmetric positive matrix, and the line is defined by the equation $Ax = 0$ where A is a full-rank $(n-1) \times n$ matrix. Clearly, this problem has two solutions, x and $-x$. (Note that this simple example could be easily solved by a few Newton iterations:

```
x=x_init;for k=1:K, x=x-g(x)\f(x);end).
```

Illustrating the solution using the function **fsolve**, we have

```
-->deff y=f(x) 'y=[x''*Q*x-1;A*x]'

-->deff y=g(x) 'y=[2*x''*Q;A]'

-->n=8;Q=rand(n,n);Q=Q*Q';A=rand(n-1,n);x_init=rand(n,1);

-->[x_found,v,info]=fsolve(x_init,f);        ← without gradient

-->norm(f(x_found))
```

```
ans  =
   3.192D-16
```

`-->[x_found,v,info]=fsolve(x_init,f,g);`   ← using gradient

`-->norm(f(x_found))`
```
 ans  =
   2.776D-16
```

The function **fsolve** is useful for calculating steady-state equilibrium in differential equations. For instance, the trajectory $y(t)$ of Robertson's chemical reaction described above on page 253 converges for large values of t to an equilibrium that can be found as follows:

```
deff('[res]=equ(y)',..
   'res(1)=-0.04*y(1)+1e4*y(2)*y(3),..
   res(2)=0.04*y(1)-1d4*y(2)*y(3)-3d7*y(2)*y(2);..
   res(3)=y(1)+y(2)+y(3)-1;')
yeq=fsolve(y0,f)
```

The function **fsolve** is also useful for finding consistent initial conditions for DAE equations or to periodically "project" the solution of a DAE onto its constraint set.

Chapter 9

SCICOS – A Dynamical System Builder and Simulator

Even though it is possible to simulate mixed discrete and continuous (hybrid) dynamical systems in Scilab using Scilab's ordinary differential equation solver (the **ode** function, see Section 8.2), implementing the discrete recursions and the logic for interfacing the discrete and the continuous parts usually require a great deal of programming. These programs are often complex, difficult to debug, and slow.

Scicos (Scilab Connected Object Simulator) is a Scilab package for modeling and simulation of hybrid dynamical systems. Scicos is a simulation environment in which both continuous and discrete systems can coexist. It includes a graphical editor for building complex models by interconnecting blocks that represent either predefined basic functions defined in Scicos libraries (palettes) or user-defined functions. A large class of hybrid systems can be conveniently modeled this way, including many interesting problems in systems, control, and signal processing applications.

9.1 Hybrid System Formalism

For modeling dynamical systems encountered in engineering applications, different types of signals should be accounted for. Traditionally, two different types of signals have been used: continuous-time signals and discrete-time signals. Physical quantities such as position, speed, temperature, voltage and are naturally modeled by continuous-time signals. On the other hand, discrete-time signals are better suited for coding data flow in digital systems, where often all signals follow a unique rhythm set by a central clock (see Figure 9.1).

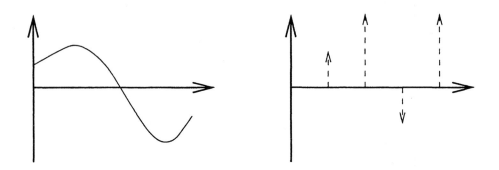

Figure 9.1: *Continuous and discrete-time signals*

The formalism used in Scicos is inspired in part by works on synchronous languages, in particular SIGNAL[11] and its extension to continuous time [9]. The basic idea is to associate to each signal a set of times (called activation times) over which the signal can evolve; this is similar to the notion of presence in SIGNAL. In Scicos though, signals are always present, even outside their activation times, where they simply remain constant (see Figure 9.2).

In general, a signal X is defined by an n-vector function of time $x(t)$ defined on $[0, \infty)$ associated with an activation time set T_x composed of the union of intervals and isolated points in time. The set T_x is called the set of activation times, and when $n = 0$ (i.e., X is pure timing information), we say that X is an activation signal. Activation signals at activation times corresponding to isolated points in time are called *events*.

Signals are generated by operators, which, because of their graphical representation in the Scicos editor, are referred to as *basic blocks*. These operators are driven by activation signals (in case of multiple-input activation signals, the operator is active if at least one of these input signals is active) and compute their outputs as functions of their inputs and internal states. The output signals inherit their activation times from the operator. Various types of basic blocks are described in Section 9.3.

A simulation model is obtained by interconnection of basic blocks. Such a model is properly defined when all the inputs of its blocks are connected, in which case it can be simulated by Scicos. The Scicos simulator advances the time by ensuring that the constraints imposed by the basic blocks are satisfied during their activation times. The simulation contains two distinct phases: a point phase due to events, and an interval phase. During this latter phase, the Scicos simulator uses the **ode** solver *lsodar* (see Section 8.2) to advance the time. In this phase, the internal states of the

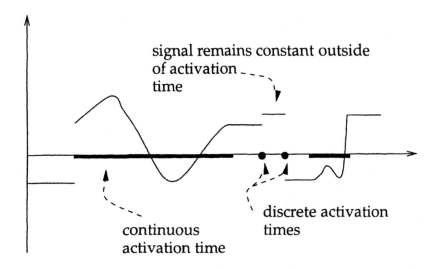

Figure 9.2: *A hybrid signal and its activation time set*

blocks evolve continuously, as opposed to the point phase, where the internal states can "jump" (this corresponds to the discrete behavior of the model).

9.2 Getting Started

This section presents the steps required for modeling and simulating a simple dynamical system in Scicos. In Scicos, systems are modeled in a modular way by interconnecting subsystems referred to as *blocks*. The model we construct here uses only existing blocks (available in various palettes); the procedure for creating new blocks will be discussed later.

9.2.1 Constructing a Simple Model

In Scilab's main window, type **scicos()**. This opens up an empty Scilab graphics window, which we call the Scicos main window. By default, this window is named **Untitled** (Figure 9.3).

Open up a palette by selecting the **Palettes** button in the **Edit** menu. You are then presented with a choice of palettes (see Figure 9.4). Click on **Inputs/ Outputs**; this opens up a palette, which is a new Scilab graphics window containing a number of blocks (Figure 9.5). To copy a block, select first the **copy** button in the **Edit** menu on top of the Scicos main window,

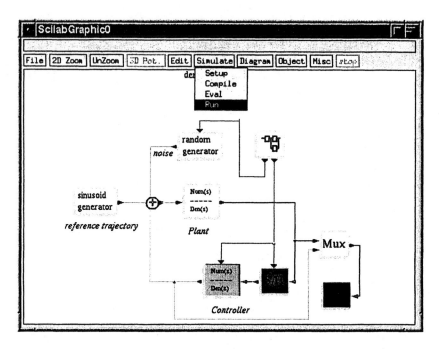

Figure 9.3: *Scicos main window*

then on the block to be copied in the palette, and finally somewhere in the Scicos main window, where the block is to be placed.

Now, copy the **MScope**, **sinusoid generator**, and the **Clock** (the clock with an output port on the bottom) blocks into the Scicos main window. Note that you do not need to select the **copy** button again; the selection remains in effect until you select a new operation. The result should look like Figure 9.6.

Open the **Linear** palette and copy the block **1/s** (integrator) into the Scicos main window. Connect the input and output ports of the blocks by selecting first the **link** button (**Edit** menu). Then click on the output port and then on the input port (or on intermediary points if you don't want just a straight-line connection) for each new link. Connect the blocks as illustrated in Figure 9.7. To make a link originate from another link (to split a link), click on the **Link** button and then on an existing link where the split is to be placed, and finally on an input port (or intermediary points before that). If in the process of creating a link you decide to stop and delete the link, click on the right mouse button.

The **MScope** is used to display the sine signal, which comes directly from the generator as well, as the cosine signal, which is the output of the

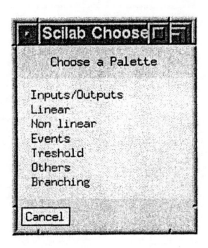

Figure 9.4: *Choice of palettes*

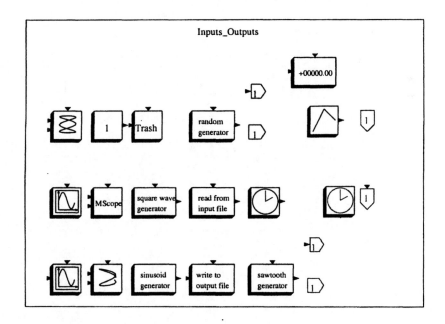

Figure 9.5: *Inputs/Outputs palette*

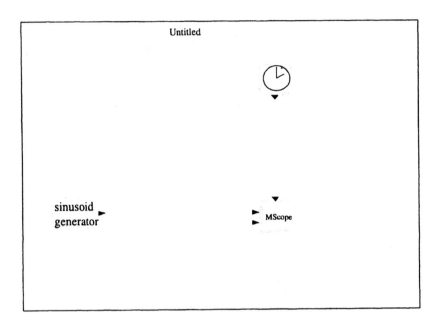

Figure 9.6: *Blocks copied from the Inputs/Outputs palette*

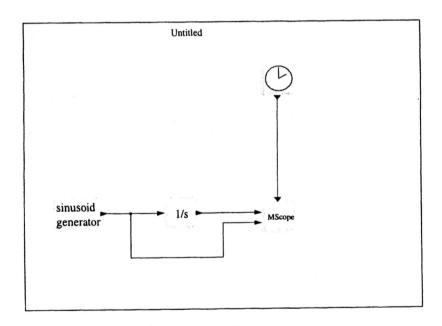

Figure 9.7: *Complete model*

integrator. The **Clock** generates an activation signal composed of a train of regularly spaced events in time telling the **MScope** block at what times the value of its inputs must be displayed (ports placed at the bottom of the blocks are the sources of activation signals). To inspect (and if needed change) the **Clock** parameters, select the **Open/Set** button (in the **Object** menu) and click on the **Clock** block. This opens up a dialogue panel as illustrated in Figure 9.8. At this point the period, the time period between events and the time of the first event can be changed. Let us leave them unchanged: Click on **Cancel** or **OK**. Similarly, you can inspect the parameters of other blocks. You can now save your diagram by clicking on the **Save** button. This saves your diagram in a file called **Untitled.cos** in the directory where Scilab was launched.

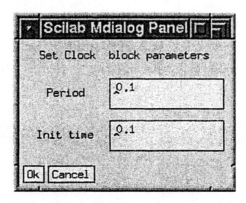

Figure 9.8: **Clock**'s *dialogue panel*

9.2.2   Model Simulation

It is now time to do a simulation. For select, select **Run** in the **Simulate** menu. After a short pause (time of compilation), the simulation starts; see Figure 9.9.

The simulation can be stopped by clicking on the **stop** button on top of the Scicos main window. It is clear that the **MScope**'s parameters need to be adjusted. So, click on the **MScope** block after selecting **Open/Set** in the **Object** menu; this opens up a dialogue box (see Figure 9.10).

To improve the display, we must change **Ymin** and **Ymax** of the two plots.

The first input ranges from 0 to 2 and the second from −1 to 1; **Ymin** and **Ymax** can thus be set to [0 −1] and [2 1], respectively. Increasing the **Buffer size** can speed up the simulation but can make the display "jerky."

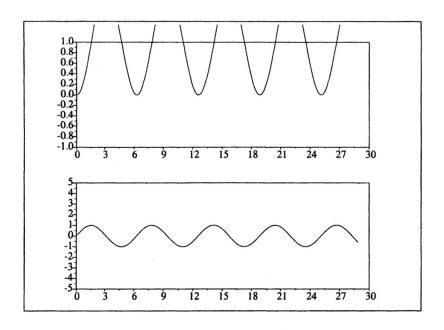

Figure 9.9: *Simulation result*

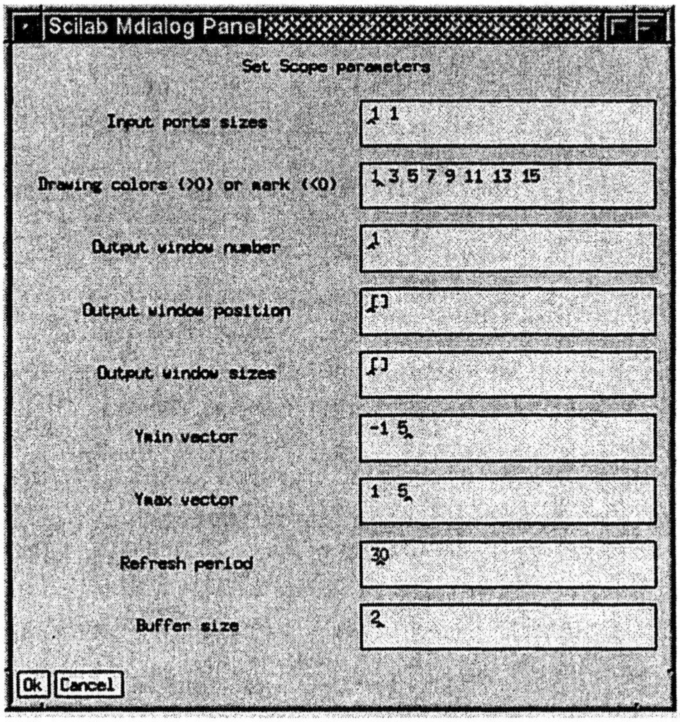

Figure 9.10: **MScope** *original dialogue box*

The refresh period is the maximum time displayed in a single window; let us change it to 50. We can also change the colors of the two curves by changing the values of the vector **Drawing colors**.

Click now on **OK** to register the new parameters and then get back to the Scicos main window. You can now restart the simulation by selecting **Run** and then **Restart** among the proposed choices. The result is depicted in Figure 9.9. Note that this time, the simulation starts right off, with no need for recompilation.

To make your diagram look good, you can use the operations in the **Object** menu. There you can change the colors of the blocks and links (**Color**), the text or pictures displayed inside blocks (**Icon**), and their sizes (**Resize**). For example, to color the block MScope, select the **Color** button, then click on the block. This opens up a color palette; select the desired color by clicking on it, confirming by **OK**.

You can also place text on your diagram by copying the block **Text** in the **Others** palette into your diagram and changing the text, the font, and the size by clicking on it.

Finally, you can print and export, in various formats, your diagram using the **File** menu on top of the corresponding graphics window. Make sure to do a replot (using the **Replot** button in the **Edit** menu) first, to clean up the diagram. The result will look like Scicos diagrams presented in this document.

9.2.3 Symbolic Parameters and "Context"

In the above example, the block parameters seem to be defined numerically. But in fact, even when a number is entered in a block's dialogue, it is first treated and memorized symbolically, and then evaluated. For example, in the block **sinusoid generator**, we can enter **2-1** instead of **1** for **Magnitude**, and the result would be the same (see Figure 9.11); opening up the dialogue the next time around would display **2-1**. We can go further and, for example, use a symbolic expression such as **sin(cos(3)-.1)**. In fact, we can use any valid Scilab expression, including Scilab variables. But these variables (symbolic parameters) should be defined in the current environment. For that select the **context** button in the **edit** menu. You are then presented with a "Dialogue Panel" (see Figure 9.12) in which you can define symbolic parameters (in this case **ampl**). Once you click on **OK**, the variable **ampl** can be used in all the blocks of the diagram. It can ,for example, be used to change, in the **sinusoid generator** block, the magnitude of the sine wave (see Figure 9.13).

Figure 9.11: *Symbolic expression as parameter*

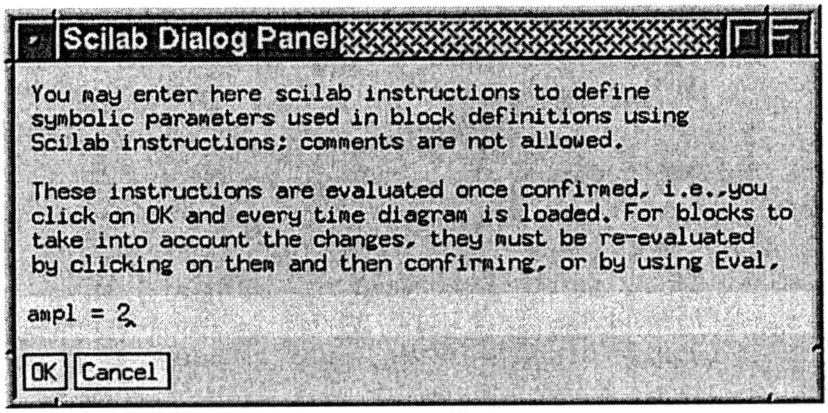

Figure 9.12: *Context is used to give numerical values to symbolic expressions*

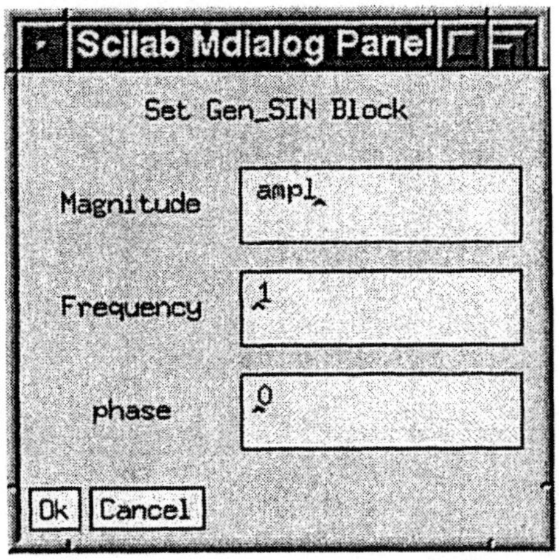

Figure 9.13: *Use of symbolic expressions in block parameter definition*

The context of the diagram is saved with the diagram, and the expressions are reevaluated when the diagram is reloaded. Note, however, that if you change the context, you must reevaluate the diagram if you want the blocks to take into account the changes. This can be done either by using the **eval** button in the **Simulate** menu or by selecting the **Open/Set** button, clicking on the concerned blocks and confirming by clicking on **Ok**.

9.2.4 Use of Super Block

It would be very difficult to model a complex system with hundreds of components in one diagram. For that, Scicos provides the possibility of grouping blocks together and defining subdiagrams called super blocks. These blocks have the appearance of regular blocks but can contain an unlimited number of blocks, even other super blocks. A super block can be duplicated using the **copy** button in the **Edit** menu and used any number of times.

One way to define a super block is to copy **Super Block** from the **Others** palette into your diagram (see Figure 9.14) and click on it, after selecting the **Open/Set** button. This opens up a new Scicos panel with an empty diagram. Edit this diagram by copying and connecting an **S/H** block (sample and hold), a discrete linear system from the **Linear** palette, and input and output Super Block ports from the **Inputs/Outputs** palette, as in Figure 9.15.

Once you exit (using the **Exit** button in the **Diagram** menu) from the super block panel, you get back the original diagram. Note that the super block now has the right number of inputs and outputs. To activate this

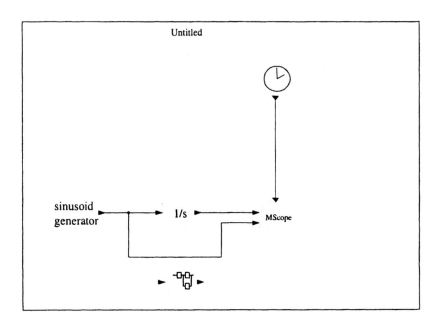

Figure 9.14: *Super block in the diagram*

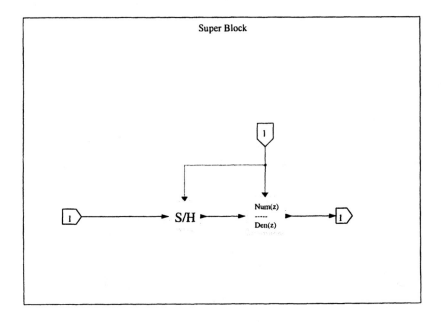

Figure 9.15: *Super block content*

discrete system, we need a **Clock**; just copy the **Clock** already in the diagram (see Figure 9.16). Note that there is no need to disconnect the links to **MScope** to change the number of its inputs. Simply update its parameters as shown in Figure 9.17; the input ports adjust automatically.

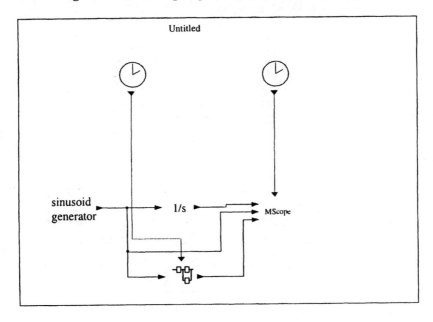

Figure 9.16: *Complete diagram with super block*

Before simulating, set the period of this new **Clock** to 2 (to view the discrete behavior, the period of discretization must be larger than that of the scope). You can now **Run** the diagram.

The other way to define a super block is to convert a whole region in a Scicos diagram into a super block using the **Region to Super Block** button in the **Diagram** menu. This operation creates a super block containing the subdiagram in the selected region, deletes this subdiagram from the main diagram, and replaces it with the super block.

9.2.5   Simulation Outside the Scicos Environment

When a file is compiled and saved as a **.cos** file, the next time it is loaded, it need not be recompiled. The saved file contains, in addition to all the properties and graphical information concerning the blocks and the diagram, the results of the compilation. In some cases it is useful to be able to extract just the data needed to do simulation, and do the simulation without hav-

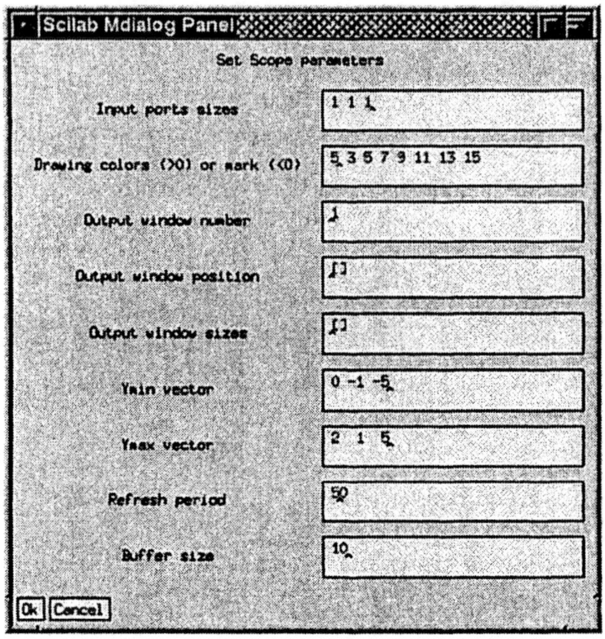

Figure 9.17: **MScope** *updated dialogue box*

ing to enter the Scicos environment. This can be done using the **scicosim** function.

To use **scicosim**, first compile and save your diagram. If you do not specify any name, your diagram will be saved in the file **Untitled.cos**. Then leave Scicos and, in Scilab, load this file:

load Untitled.cos

This will define three variables: **scicos_ver**, **scs_m**, **cpr**. The compilation results are in **cpr**, which is a list: **list(state,sim,cor,corinv)**. All the information needed to run the simulation can be found in **state** and **sim**; **cor** and **corinv** contain correspondence tables for block numbers in the original graphical structure and block numbers in the compiled structure.

To run **Untitled.cos**, you can proceed as follows:

```
state=cpr(1);sim=cpr(2);
tf=100;                                  ← final time of simulation
tol=[1d-5,1d-7,1d-9,tf];        ← error tolerances and max step size
[state,t]=scicosim(state,0,tf,sim,'start',tol)
[state,t]=scicosim(state,0,tf,sim,'run',tol)
[state,t]=scicosim(state,0,tf,sim,'finish',tol)
```

9.3 Basic Concepts

Models in Scicos are constructed by interconnection of basic blocks. There exist three types of basic blocks and two types of connecting links in Scicos. Basic blocks can have two types of inputs and two types of outputs: regular inputs, activation inputs, regular outputs, and activation outputs. Regular inputs and outputs are interconnected by regular links, and activation inputs and outputs, by activation links (regular input and output ports are placed on the sides of the blocks, activation input ports are on top and activation output ports are on the bottom). Blocks can have an unlimited number of each type of input and output ports.

The function of regular links is fairly clear, but that of activation links is more subtle. An activation link carries an activation signal, which is timing information specifying the times when the blocks receiving the signal should be activated.

9.3.1 Basic Blocks

There are three types of basic blocks in Scicos: Regular basic blocks, zero crossing basic blocks, and synchro basic blocks.

Regular Basic Block

Regular basic blocks (RBB) can have both regular and activation input and output ports. An RBB can have a *continuous state* x and a *discrete state* z. Let the vector function u denote the regular inputs. Then the x state of an RBB, when the block is active, evolves continuously according to

$$\dot{x} = f(t, x, z, u, p, n_e), \tag{9.1}$$

where f is a block-specific function, p is a vector of constant parameters, and n_e is the *activation code*. The activation code is an integer designating the port(s) through which the block is activated. Specifically, if the activating input ports are $i_1, i_2, \ldots, i_n$, then

$$n_e = \sum_{j=1}^{n} 2^{i_j - 1}.$$

The states x and z jump when activated by one or more events at time t_e according to the following equations:

$$
\begin{aligned}
x(t_e) &= g_c(t_e, x(t_e^-), z(t_e^-), u(t_e), p, n_e), & (9.2)\\
z(t_e) &= g_d(t_e, x(t_e^-), z(t_e^-), u(t_e), p, n_e), & (9.3)
\end{aligned}
$$

where g_c and g_d are block-specific functions. Note that z remains constant between any two successive events, so $z(t_e^-)$ can be interpreted as the previous value of the discrete state z.

The regular output of the block is defined by

$$y(t) = h(t, x(t^-), z(t^-), u(t), p, n_e),   \tag{9.4}$$

where h is a block-specific function, when the block is active. The output y remains constant when the block is not active.

RBBs can also generate activation signals on their activation output ports. These activation signals can be only of event type and can be scheduled only when the block is activated by an input event. If the block is activated by an event at time t_e, the time of each output event is generated according to

$$t_{evo} = k(t_e, z(t_e), u(t_e), p, n_e)   \tag{9.5}$$

for a block specific function k and where t_{evo} is a vector of time, each entry of which corresponds to one activation output port. If an element is less than t_e, it simply means the absence of an output activation signal on the corresponding activation output port. The choice $t_{evo} = t$ should be avoided; the resulting causality structure is ambiguous. In particular, $t_{evo} = t$ does not mean that the output event is synchronized with the input event; synchronization of two events is a very restrictive condition (see Synchro blocks on page 311). Two events can have the same time but not be synchronized.

Event generations can also be prescheduled. Prescheduling of events can be done by setting the "initial firing" variables of blocks with event output ports. Initial firing is a vector with as many elements as the block's output event ports. Initial firing can be considered as the initial condition for t_{evo}. By setting the ith entry of the initial firing vector to t_i, an output event is scheduled on the ith output event port of the block at time t_i if $t_i \geq 0$; no output event is scheduled if $t_i < 0$. The scheduled event is then fired when time reaches t_i.

Only one output event can be scheduled on each output event port. Initially and in the course of the simulation, this means that by the time a new event is to be scheduled, the old one must have been fired. This is natural because the register that contains the firing schedule of a block should be considered as part of its state, having dimension equal to the number of the block's output event ports. An interpretation is that as long as the previously scheduled event has not been fired, the corresponding output port is considered busy, and it cannot accept a new event scheduling. If the simulator encounters such a conflict, it stops and returns the error message *event conflict*.

Most of the elementary blocks in Scicos are RBBs; the following are a few examples.

- Static blocks: A static block is one where the regular outputs are static functions of the regular inputs. For example, the block that realizes $y = \sin(u)$ is a static block. Static blocks have no output activation ports, and they have no states. Clearly, these blocks are special cases of RBBs.

  The **Non linear** palette contains a number of static blocks. Note that these blocks have no activation input ports; they inherit their activation times from their regular inputs (see Section 9.3.2).

- Discrete-time state-space systems: A discrete-time system

$$
\begin{aligned}
\xi(k+1) &= m(\xi(k), u(k+1)), \\
y(k+1) &= n(\xi(k), u(k+1))
\end{aligned}
$$

  can be implemented as an RBB if the block receives, on its activation input port, event signals on a regular basis. In this case z is used to store ξ, and the block has no activation output port.

- Clocks: A clock is a generator of event signals on a periodic basis. An RBB cannot act as a clock. The reason is that except for a possible prescheduled initial output event, RBBs must receive an event signal on one of their activation input ports to be able to generate an output event. The way to generate a clock in Scicos is by using an "event delay block." An event delay block is an RBB with one activation input port and one activation output port. When activated by an event, the event delay block schedules an event on its activation output port, i.e., after a period of time, it generates an event on its activation output port. By feeding back the activation output to the activation input (see Figure 9.18), a clock is constructed. To start the process, an output event is prescheduled on the activation output port of the event delay block. This is done by setting the block's initial firing vector to 0 (or any $t \geq 0$ if the clock is to start operating at time t).

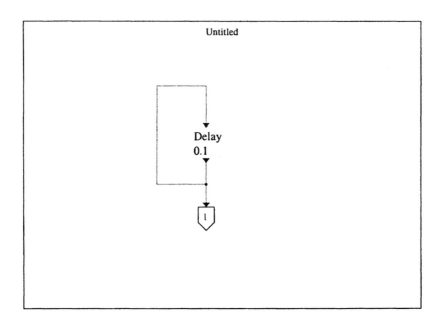

Figure 9.18: *Constructing an event clock using feedback on a delay block*

This way of defining clocks may seem complicated. However, it provides considerable flexibility. For example, systems with multiple asynchronous clocks driving various system components are very easy to model this way, and so is modeling clocks with variable frequencies.

Zero Crossing Basic Block

A zero crossing basic block (ZBB) has a (possibly vector) regular input and activation outputs but no regular outputs or activation inputs. A ZBB can generate event outputs only if at least one of the regular inputs crosses zero (changes sign). In such a case, the generation of the event, and its timing, can depend on the combination of the inputs that have crossed zero and the signs of the inputs just before the crossing occurs. The event generated by zero crossing must be generated with a delay with respect to the time when the actual crossing occurs; but this delay can be arbitrarily short, and in fact, it can safely be set to zero for most practical purposes.

The simplest example of a zero crossing basic block is the **zcross** block in the **Threshold** palette. This block generates an event if the input crosses 0. Other examples are **+ to -** and **- to +**, which generate an output event when the input crosses zero with a negative, respectively positive, slope. The most general form of this block is realized by the block **GENERAL** in the

Threshold palette.

Inputs of ZBBs should not remain zero during simulation; this situation is ambiguous. However, these inputs can start off at zero. Similarly, the input of a zero crossing basic block should not jump across zero. If it does, the crossing may or may not be detected.

ZBBs cannot be modeled as RBBs because in these blocks no output event can be generated unless an input event has arrived beforehand.

Synchro Basic Block

An RBB can only generate activation signals of event type, and moreover, these events are never synchronized (simultaneous) with its input activation signal. Synchro basic blocks (SBB) are the only blocks that generate output activation signals that are synchronized with their input activation signals. These blocks have a unique activation input port (possibly inherited, see Section 9.3.2), a unique (possibly vector) regular input, no state, no parameters, and two or more activation output ports. Depending on the value of the regular input, the incoming activation signal is routed to one of the activation output ports. This means in particular that at every activation, the SBB generates a t_{evo} with all of its entries strictly less than t (current time) except one, which is exactly t.

An example of such a block is the **event select** block in **Branching** palette. The other is the **If-then-else** block in the same palette. These blocks are used for routing activation signals and undersampling event signals. See Figure 9.28 for an example of the use of the **If-then-else** block.

Despite obvious similarities, split and addition operations on activation signals are not SBBs. The split operation corresponds to a duplication of signal, and addition of two activation signals is an activation signal obtained by taking the union of their activation times. Strictly speaking, these operations do not correspond to blocks in the Scicos formalism.

9.3.2 Inheritance and Time Dependence

It would be rather tedious to explicitly include all the activation signals in a Scicos diagram, which is why activation signals can be inherited. If a block has no activation input port, it inherits its activation signal from its regular inputs. In particular, in a first pass, the compiler adds as many activation input ports to such a block as the block has regular input ports and connects them to the activation signals of the blocks generating the regular inputs.

Blocks that are active permanently can be declared as such, and the compiler adds an activation input port (numbered 0) to the block. These

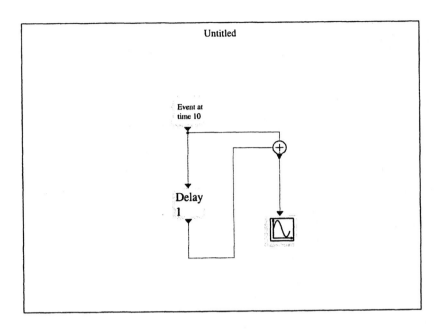

Figure 9.19: *Activation links: split and addition*

blocks are called "time-dependent." Note that time-dependent blocks do not inherit.

9.3.3   Synchronization

Synchronization is an important issue in Scicos. Even if two event signals have the same time, they are not necessarily synchronized, meaning one is fired just before or just after the other but not "at the same time." The only way two event signals can be synchronized is if they can be traced back, through activation links, activation additions, activation splits, and SBBs alone, to a common origin (a single output activation port). See Figure 9.19 for an example of the usage of activation links, additions and splits. This means, in particular, that a block can never have two "synchronized" output activation ports; super blocks, however, can have synchronized output activation ports; see, for example, the **2-freq clock** in the **Event** palette. This block is illustrated in Figure 9.20. The events on the second activation output port of this super block are clearly synchronized with a subset of events on its first output activation port. In fact, the events on the first output activation port of the super block are the union of the output events on both output activation ports of the **M.freq.clock**. This block generates $n - 1$ times an event on its first output activation port, then one event on its second output activation port, always with the same delay d with respect to the input event, and then starts over. This means that the second output activation port acts like a clock having a frequency n times smaller than the input frequency of the block, which is also the first activation output of the super block. Specifically, the frequency of the train of events generated on the first output activation port is $1/d$ and on the second is n/d.

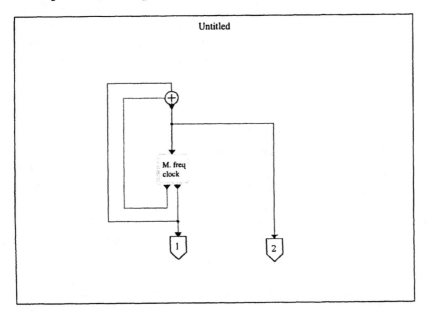

Figure 9.20: *Super block defining a 2-frequency clock*

9.4 Block Construction

In addition to the blocks available in Scicos's palettes, users can use custom blocks. Super blocks allow block functionality to be defined by graphically interconnecting existing blocks, and new blocks can be constructed by compiling super blocks. But the standard way of constructing a new block is by defining a pair of functions: one for handling the user interface (*Interfacing* function) and the other for specifying its dynamic behavior (*Computational* function). The first function is always written in Scilab, but the second can be in Scilab, C, or Fortran. C and Fortran routines dynamically linked or permanently interfaced with Scilab can be used and give the best results as far as simulation performance is concerned. The **scifunc** and **GENERIC** blocks in the **others** palettes are generic *Interfacing* functions, very useful for rapid prototyping and testing user-developed *Computational* functions in the early phases of development.

9.4.1 Super Block

Not all blocks in Scicos's palettes are basic blocks; some are super blocks. A super block is obtained by interconnecting basic blocks and other super blocks. The simplest example of such a block is the **CLOCK**, which is obtained from one regular basic block and two activation links and one output activation port. As far as the user is concerned, in most cases there is no real distinction between basic blocks and super blocks.

To construct a super block, the user can copy the **Super Block** block from the **Others** palette into the Scicos window and open it. This will open up a new Scicos window in which the super block can be defined. The construction of the super block model is done as usual: by copying and connecting blocks from various palettes. Input and output ports of the super block should be specified by input and output block ports available in the **Inputs/Outputs** palette. Super blocks can be used within super blocks without any limit on the number or the depth. Super blocks can also be created using an existing diagram with the **Region to Super Block** button.

If the super block is of interest in other constructions, it can be saved in various formats.

If the super block is used only in a particular construction, then it need not be saved. Selecting the **Exit** button in the **Diagram** menu of the super block window closes the super block window and reactivates the main Scicos window (or another super block window if the super block was being defined within another super block). Saving the block diagram saves automatically the content of all super blocks used inside of it.

At compilation, all the super blocks in the Scicos model are expanded, and a diagram including only basic blocks is generated. This phase is completely transparent to the user.

9.4.2 Scifunc Block

The **Scifunc** block allows the use of the Scilab language for defining Scicos blocks. It provides a generic *Interfacing* function and a generator of a *Computational* function. The *Computational* function is generated interactively, the user needing only to define block parameters such as the number and sizes of inputs and outputs, initial states, type of the block, the initial firing vector, and Scilab expressions for various functions defining the dynamic behavior of the block. This is done interactively by clicking on the **Scifunc** block, once copied in the Scicos window. Besides its performance (the generated function being a Scilab function), the main disadvantage of **scifunc** is that the dialogue for updating block parameters cannot be customized.

9.4.3 GENERIC Block

The **GENERIC** block also provides a generic *Interfacing* function, but the *Computational* function needs to be defined separately, either as a Scilab function, or a Fortran or C function. Compared to the **scifunc** block, the **GENERIC** block requires more information on the properties of the corresponding *Computational* function. Besides the name of the function, the user should specify information such as the type, and whether or not the block contains a direct feedthrough term or whether it is time-dependent.

Another important difference with **scifunc** is that the *Computational* function of **scifunc** is part of the data structure of the diagram, and thus it is saved along with the diagram. But in the case of the **GENERIC** block, only the name of the function figures in the data structure of the diagram. This means that the functions realizing *Computational* functions of **GENERIC** blocks of a Scicos diagram must be saved separately, and loaded or dynamically linked before simulation. One way to make such a diagram self-contained, if the *Computational* functions of all **GENERIC** blocks are Scilab functions, is to define these functions in the "context" of the diagram; we will say more on that later.

9.4.4 Fortran Block and c Blocks

Fortran and C blocks can be used to define on the fly *Computational* functions written in Fortran and C for static blocks (no internal states). The sources of these functions are saved in the data structure of the diagram; they are compiled and dynamically linked automatically when needed. This works only if a proper compiler and linker are available in the system and are accessible by Scilab.

9.4.5 *Interfacing* Function

For defining a fully customizable basic block, the user must define a Scilab function for handling the user interface. This function, referred to as the *Interfacing* function, not only determines the geometry, color, number of ports, and their sizes, icon, etc, but also the initial states, parameters, and the dialogue for updating them.

The *Interfacing* function is used by the Scicos editor only to initialize, draw, connect the block, and to modify its parameters. What the interfacing function should do and should return depends on an input flag **job**. The syntax is as follows:

Syntax

```
[x,y,typ]=block(job,arg1,arg2)
```

Parameters

- job=='plot': the function draws the block.

    - **arg1** is the data structure of the block.
    - **arg2** is not used.
    - **x,y,typ** are not used.

    In general, we can use the **standard_draw** function, which draws a rectangular block, and the input and output ports. It also handles the size, icon, and color aspects of the block.

- job=='getinputs': the function returns position and type of input ports (regular or activation).

    - **arg1** is the data structure of the block.
    - **arg2** is not used.
    - **x** is the vector of x coordinates of input ports.
    - **y** is the vector of y coordinates of input ports.
    - **typ** is the vector of input ports types (1 for regular and 2 for activation).

    In general, we can use the **standard_input** function.

- job=='getoutputs': returns position and type of output ports (regular and activation).

    - **arg1** is the data structure of the block.
    - **arg2** is not used.
    - **x** is the vector of x coordinates of output ports.
    - **y** is the vector of y coordinates of output ports.
    - **typ** is the vector of output ports types .

In general, we can use the `standard_output` function.

- job=='getorigin': returns coordinates of the lower left point of the rectangle containing the block's silhouette.

    - **arg1** is the data structure of the block.
    - **arg2** is not used.
    - **x** is the x coordinate of the lower left point of the block.
    - **y** is the y coordinate of the lower left point of the block.
    - **typ** is not used.

In general, we can use the `standard_origin` function.

- job=='set': opens up a dialogue for block parameter acquisition (if any).

    - **arg1** is the data structure of the block.
    - **arg2** is not used.
    - **x** is the new data structure of the block.
    - **y** is not used.
    - **typ** is not used.

- job=='define': initialization of block's data structure (name of corresponding *Computational* function, type, number and sizes of inputs and outputs, etc).

    - **arg1, arg2** are not used.
    - **x** is the data structure of the block.
    - **y** is not used.
    - **typ** is not used.

Block Data-Structure Definition

Each Scicos block is defined by a Scilab data structure as follows:

```
list('Block',graphics,model,unused,GUI_function)
```

where **GUI_function** is a string containing the name of the corresponding *Interfacing* function, and **graphics** is the structure containing the graphical data:

```
graphics=..
  list([xo,yo],[l,h],orient,dlg,pin,pout,pcin,pcout,gr_i)
```

- xo: x coordinate of block origin

- yo: y coordinate of block origin

- l: block's width

- h: block's height

- **orient**: Boolean, specifies whether block is flipped or not (regular inputs are on the left or right).

- **dlg**: vector of character strings, contains block's symbolic parameters.

- pin: vector, `pin(i)` is the number of the link connected to the ith regular input port, or 0 if this port is not connected.

- pout: vector, `pout(i)` is the number of the link connected to the ith regular output port, or 0 if this port is not connected.

- pcin: vector, `pcin(i)` is the number of the link connected to the ith activation input port, or 0 if this port is not connected.

- pcout: vector, `pcout(i)` is the number of the link connected to the ith activation output port, or 0 if this port is not connected.

- gr_i: character string vector, Scilab instructions used to draw the icon.

The data structure containing simulation information is **model**:

```
model=list(eqns,#input,#output,#clk_input,#clk_output,..
       state,dstate,rpar,ipar,typ,firing,deps,label,unused)
```

- **eqns**: list containing two elements. First element is a string containing the name of the *Computational* function (fortran, C, or Scilab function). Second element is an integer specifying the type of the *Computational* function. The type of a *Computational* function specifies essentially its calling sequence; more on that later.

- **#input**: vector of size equal to the number of block's regular input ports. Each entry specifies the size of the corresponding input port. A negative integer stands for "to be determined by the compiler." Specifying the same negative integer on more than one input or output port tells the compiler that these ports have equal sizes.

- **#output**: vector of size equal to the number of block's regular output ports. Each entry specifies the size of the corresponding output port. Specifying the same negative integer on more than one input or output port tells the compiler that these ports have equal sizes.

- **#clk_input**: vector of size equal to the number of activation input ports. All entries must be equal to 1. Scicos does not support vectorized activation links.

- **#clk_output**: vector of size equal to the number of activation output ports. All entries must be equal to 1. Scicos does not support vectorized activation links.

- **state**: column vector of initial continuous state.

- **dstate**: column vector of initial discrete state.

- **rpar**: column vector of real parameters passed on to the corresponding *Computational* function.

- **ipar**: column vector of integer parameters passed on to the corresponding *Computational* function.

- **typ**: string. Basic block type: `'z'` if ZBB, `'1'` if SBB and anything else, except `'m'` and `'s'`, for RBB.

- **firing**: column vector of initial firing times of size equal to the number of activation output ports of the block. It includes preprogrammed event firing times (< 0 if no firing).

- **deps**: [`udep timedep`]

  - **udep**: Boolean. True if system has direct feedthrough, i.e., at least one of the outputs depends explicitly on one of the inputs.

  - **timedep**: Boolean. True if block is time-dependent.

- **label**: character string, used as block identifier. This field may be set by the `label` button in `Block` menu.

Examples

Example of a static block   The block **ABSBLK** that realizes the absolute value function in the **Non linear** palette has a simple *Interfacing* function because there are no parameters to be set in this case.

```
function [x,y,typ]=ABSBLK_f(job,arg1,arg2)
//Absolute value block
x=[];y=[];typ=[];
select job
case 'plot' then standard_draw(arg1)
case 'getinputs' then [x,y,typ]=standard_inputs(arg1)
case 'getoutputs' then [x,y,typ]=standard_outputs(arg1)
case 'getorigin' then [x,y]=standard_origin(arg1)
case 'set' then x=arg1
case 'define' then
  in=-1;out=in;                    ← ports have equal undefined dimension
  model=list(list('absblk',1),in,out,[],[],[],[],..
                      [],[],'c',[],[%t %f],' ',list())
  gr_i='xstringb(orig(1),orig(2),''Abs'',sz(1),sz(2),..
                                          ''fill'')'
  x=standard_define([2 2],model,' ',gr_i)
end
```

Example of a static block with parameter   The **LOGBLK** block *Interfacing* function is somewhat more complicated because the basis of the log function is a parameter to be set interactively.

```
function [x,y,typ]=LOGBLK_f(job,arg1,arg2)
x=[];y=[];typ=[];
select job
case 'plot' then standard_draw(arg1)
case 'getinputs' then [x,y,typ]=standard_inputs(arg1)
case 'getoutputs' then [x,y,typ]=standard_outputs(arg1)
case 'getorigin' then [x,y]=standard_origin(arg1)
case 'set' then
  x=arg1;
  dlg=x(2)(4)                            ← symbolic parameters
  while %t do                            ← open dialogue window
    [ok,a,dlg]=getvalue('Set log block parameters',..
                        'Basis (>1)',list('vec',1),dlg)
    if ~ok then break,end
    if a<=1 then                  ← check user answers consistency
      x_message('Basis must be larger than 1')
    else
      if ok then                  ← update block's data structure
        x(2)(4)=dlg                      ← parameter expressions
        x(3)(8)=a                          ← parameter value
        break
      end
    end
```

```
      end
case 'define' then
  a=%e
  model=list('logblk',0,-1,-1,[],[],[],[],a,[],..
                                  'c',[],[%t %f],' ',list())
  dlg='%e'                                    ← symbolic parameters
  gr_i=['xstringb(orig(1),orig(2),''Log'',sz(1),sz(2),..
                                      ''fill'');']
  sz=[2 2]                                    ← initial block size
  x=standard_define(sz,model,dlg,gr_i)
end
```

Example of a complex RBB The following is the *Interfacing* function associated with the **Jump (A,B,C,D)** block in the **Linear** palette. This block realizes a continuous-time linear state-space system with the possibility of jumps in the state. The number of inputs to this block is two. The first input vector is the standard input of the system; the second is of size equal to the size of the continuous state x. When an event arrives at the unique activation input port of this block, the state of the system jumps to the value of the second input of the block. This system is defined by four matrices, A, B, C, and D, which are stored in **rpar**, and an initial condition x_0.

```
function [x,y,typ]=TCLSS_f(job,arg1,arg2)
x=[];y=[];typ=[]
select job
case 'plot' then standard_draw(arg1)
case 'getinputs' then [x,y,typ]=standard_inputs(arg1)
case 'getoutputs' then [x,y,typ]=standard_outputs(arg1)
case 'getorigin' then [x,y]=standard_origin(arg1)
case 'set' then
  x=arg1
  dlg=x(2)(4)                                 ← symbolic expressions
  while %t do
   [ok,A,B,C,D,x0,dlg]=getvalue('Set system parameters',..
     ['A matrix';'B matrix';'C matrix';..
      'D matrix';'Initial state'],..
     list('mat',[-1,-1],'mat',['size(x1,2)','-1'],..
          'mat',['-1','size(x1,2)'],'mat',[-1 -1],..
          'vec','size(x1,2)'),dlg)
   if ~ok then break,end
   // Check user answers consistency
   o=size(C,1);if out==0 then o=[],end
   in=size(B,2);if in==0 then in=[],end
   [ms,ns]=size(A)
   if ms<>ns then
```

```
        x_message('A matrix must be square')
    else                                   ← update block input and output
      [model,graphics,ok]=check_io(x(3),x(2),[in;ms],o,1,[])
      if ok then                           ← update block's data structure
        x(2)=graphics
        x(3)=model
        x(2)(4)=dlg;                                ← symbolic parameters
        x(3)(8)=[A(:);B(:);C(:);D(:)];
        x(3)(6)=x0(:)                               ← set new initial state
        if D<>[] then                       ← update input dependency
          if norm(D,1)<>0 then
            x(3)(12)=[%t %t];
          else
            x(3)(12)=[%f %t];
          end
        else
          x(3)(12)=[%f %t];
        end
        break
      end
    end
  end
case 'define' then
  x0=0;A=0;B=1;C=1;D=0                       ← initial values of parameters
  in=1;nx=size(x0,'*');o=1;
  model=list(list('tcslti',1),[in;nx],o,1,[],x0,..
         [],[A;B;C;D],[],'c',[],[%f %t],' ',list())
  // symbolic parameters
  dlg=[strcat(sci2exp(A));strcat(sci2exp(B));strcat(..
       sci2exp(C));strcat(sci2exp(D));strcat(sci2exp(x0))]
  gr_i=['txt=[''Jump'';''(A,B,C,D)''];';
    'xstringb(orig(1),orig(2),txt,sz(1),sz(2),''fill'')']
  sz=[3 2]                                   ← initial block size
  x=standard_define(sz,model,dlg,gr_i)
end
```

The only difficult part of defining an *Interfacing* function is the **'set'** case, which handles user dialogue and determines system parameters. For example, in the case of **TCLSS_f**, the interfacing function should determine whether or not the block has a direct feedthrough term. The **define** case, on the other hand, is used only once, when the block is first placed in a palette.

9.4.6 *Computational* Function

The *Computational* function should evaluate outputs, new states, continuous state derivative, and the output events timing vector, depending on the type of the block and the way it is called by the simulator.

Behavior

Different tasks that need to be performed for the simulator by the *Computational* function are the following:

- **Initialization** The *Computational* function is called once right at the beginning for initialization. At this point, the continuous and discrete states and the outputs of the block can be initialized (if necessary; note that they are already initialized by the interfacing function). Other tasks can also be performed at this occasion; for example, blocks that read or write data from and to files can open the corresponding files on disk, the **scope** block can initialize the graphics window, memory allocation can be done. The inputs of the blocks are not usable at this point

- **Reinitialization** This is another occasion to initialize the states and the outputs of the block. But this time, the inputs are available. A block can be called a number of times for reinitialization.

- **Outputs update** The simulator is requesting the value of the outputs, which means that the *Computational* function should compute them as a function of the states (9.4), the inputs, and the activation code.

- **States update** Upon arrival of one or more events, the simulator calls the *Computational* function to update the states x and z of the block according to (9.2) and (9.3). The states are updated in place (this avoids useless and time-consuming copies, in particular when part or all of x or z are not to be changed).

- **State derivative computation** During the simulation, the solver calls the *Computational* function (very often) for the value of $\dot{x}$. This means that the function f in (9.1) must be evaluated.

- **Output events timing** If a block has output activation ports, then upon the arrival of an event, the simulator calls the *Computational* function inquiring about the timing of the outgoing events. The *Computational* function should return t_{evo} as defined in (9.5).

- **Ending** Once the simulation is done, the simulator calls the *Computational* function once. This is useful for closing files that have been opened by the block at the beginning or during the simulation, to flush buffered data, to free allocated memory, etc.

The simulator specifies which task should be performed through a flag (see Table 9.1).

Flag	Task
0	State derivative computation
1	Outputs update
2	States update
3	Output events timing
4	Initialization
5	Ending
6	Reinitialization

Table 9.1: *Tasks of* Computational *function and their corresponding flags*

Computational Function Types

There exist various types of *Computational* functions supported by Scicos (see Table 9.2). The *Computational* function type is given by the second field of **eqns** (see Section 9.4.5).

Function type	Scilab	Fortran	C	Comments
0	yes	yes	yes	Fixed calling sequence
1	no	yes	yes	Varying calling sequence
2	no	no	yes	Fixed calling sequence
3	yes	no	no	Inputs/outputs are Scilab lists

Table 9.2: *Different types of* Computational *functions. Type 0 is obsolete.*

Computational **function type 0**   For blocks of this type, the simulator constructs a unique input vector by stacking up all the input vectors, and if expects the outputs to be stacked up in a unique vector as well. This type is supported for backward compatibility but should be avoided, since it is not efficient.

The calling sequence is similar to that of *Computational* functions of type 1 having one regular input and one regular output.

Computational **function type 1**   The simplest way of explaining this type is by considering a special case: For a block with two input vectors and three output vectors, the *Computational* function has the following synopsis:

Fortran case

```
      subroutine myfun(flag,nevprt,t,xd,x,nx,z,nz,tvec,
     &    ntvec,rpar,nrpar,ipar,nipar,u1,nu1,u2,nu2,
     &    y1,ny1,y2,ny2,y3,ny3)
c
      double precision t,xd(*),x(*),z(*),tvec(*),rpar(*)
      double precision u1(*),u2(*),y1(*),y2(*),y3(*)
      integer flag,nevprt,nx,nz,ntvec,nrpar,ipar(*)
      integer nipar,nu1,nu2,ny1,ny2,ny3
```

See Table 9.3 for the definitions of the arguments.

C case   Type 1 *Computational* functions can also be written the same way in the C language. Note that arguments must be passed as pointers; see the example below.

The best way to learn how to write these functions is to examine the routines in the Scilab directory **SCIDIR/routines/scicos**. There you will find the *Computational* functions of all Scicos blocks available in various palettes, most of which are Fortran type 0 and 1.

Example   The following is the *Computational* function associated with the **Abs** block; it is of type 1 (also 0, since the input and output are unique). The associated *Interfacing* function was defined on page 319.

```
      subroutine absblk(flag,nevprt,t,xd,x,nx,z,nz,
     &    tvec,ntvec,rpar,nrpar,ipar,nipar,u,nu,y,ny)
c     Scicos block simulator
c     returns Absolute value of the input vector
c
      double precision t,xd(*),x(*),z(*),tvec(*),rpar(*)
      double precision u(*),y(*)
      integer flag,nevprt,nx,nz,ntvec,nrpar,ipar(*)
      integer nipar,nu,ny
c
      do 15 i=1,nu
```

I/O	Args.	Description
I	flag	0, 1, 2, 3, 4, 5 or 6, (see Table 9.1)
I	nevprt	activation code
I	t	time
O	xdot	derivative of the continuous state
I/O	x	continuous state
I	nx	size of x
I/O	z	discrete state
I	nz	size of z
O	tvec	times of output events (for flag=3)
I	ntvec	number of activation output ports
I	rpar	parameter
I	nrpar	size of rpar
I	ipar	parameter
I	nipar	size of ipar
I	ui	ith input (regular), i=1,2, ...
I	nui	size of ith input
O	yj	jth output (regular), j=1,2, ...
I	nyj	size of jth output

Table 9.3: *Arguments of* Computational *functions of type 1.* I: *input,* O: *output.*

```
        y(i)=abs(u(i))                    ← vectorized input/output
15      continue
        end
```

This example is particularly simple, since **absblk** is only called with **flag** equal to 1, 4, and 6. That is because this block has no state and no output event port.

The C version of this block would be

```
#include "<SCIDIR>/routines/machine.h"
void absblk(flag,nevprt,t,xd,x,nx,z,nz,
        tvec,ntvec,rpar,nrpar,ipar,nipar,u,nu,y,ny)
/*
Returns Absolute value of the input vector
*/
double  *t,xd[],x[],z[],tvec[],rpar[],u[],y[] ;
integer *flag,*nevprt,*nx,*nz,*ntvec,*nrpar,ipar[],*nipar;
```

```
integer *nu,*ny ;
{
int i ;
for (i==0;i<*nu;i++)
  y[i]=abs(u[i]) ;
}
```

Example   The following is the *Computational* function associated with the
Jump **(A,B,C,D)** block. This function is clearly of type 1. See page 321 for
the *Interfacing* function.

```
      subroutine tcslti(flag,nevprt,t,xd,x,nx,z,nz,tvec,
     &   ntvec,rpar,nrpar,ipar,nipar,u1,nu1,u2,nu2,y,ny)
c     continuous state space linear system with jump
c     rpar(1:nx*nx)=A
c     rpar(nx*nx+1:nx*nx+nx*nu)=B
c     rpar(nx*nx+nx*nu+1:nx*nx+nx*nu+nx*ny)=C
c     rpar(nx*nx+nx*nu+nx*ny+1:nx*nx+nx*nu+nx*ny+ny*nu)=D
c!
c
      double precision t,xd(*),x(*),z(*),tvec(*),rpar(*)
      double precision u1(*),u2(*),y(*)
      integer flag,nevprt,nx,nz,ntvec,nrpar,ipar(*)
      integer nipar,nu1,nu2,ny
c
      la=1
      lb=nx*nx+la
c
      if(flag.eq.1) then
c     y=c*x+d*u1
         ld=lc+nx*ny
         lc=lb+nx*nu1
         call dmmul(rpar(lc),ny,x,nx,y,ny,ny,nx,1)
         call dmmul1(rpar(ld),ny,u1,nu1,y,ny,ny,nu1,1)
      elseif(flag.eq.2.and.nevprt.eq.1) then
c     x+=u2
         call dcopy(nx,u2,1,x,1)
      elseif(flag.eq.0) then
c     xd=a*x+b*u1
         call dmmul(rpar(la),nx,x,nx,xd,nx,nx,nx,1)
         call dmmul1(rpar(lb),nx,u1,nu1,xd,nx,nx,nu1,1)
      endif
      end
```

The functions **dmmul** and **dmmul1** perform matrix multiplication.

Computational **function type 2**   This *Computational* function type is specific to programming in C. The synopsis is

```
#include "<SCIDIR>/routines/machine.h"
void selector(flag,nevprt,t,xd,x,nx,z,nz,tvec,ntvec,
    rpar,nrpar,ipar,nipar,inptr,insz,nin,outptr,outsz,nout)

integer *flag,*nevprt,*nx,*nz,*ntvec,*nrpar;
integer ipar[],*nipar,insz[],*nin,outsz[],*nout;

double x[],xd[],z[],tvec[],rpar[];
double *inptr[],*outptr[],*t;
```

See Table 9.4 for a description of the arguments.

I/O	Args.	description
I	*flag	0, 1, 2, 3, 4, 5, or 6, (see Table 9.1)
I	*nevprt	activation code
I	*t	time
O	xd	derivative of the continuous state (**flag**= 0)
I/O	x	continuous state
I	*nx	size of x
I/O	z	discrete state
I	*nz	size of z
O	tvec	times of output events (**flag**=3)
I	*ntvec	number of activation output ports
I	rpar	parameter
I	*nrpar	size of rpar
I	ipar	parameter
I	*nipar	size of ipar
I	inptr	inptr[i] is pointer to beginning of ith input
I	insz	insz[i] is the size of the ith input
I	*nin	number of input ports
I	outptr	outptr[j] is pointer to beginning of jth output
I	outsz	outsz[j] is the size of the jth output
I	*nout	number of output ports

Table 9.4: *Arguments of* Computational *functions of type 2. I: input, O: output.*

Example The following is the *Computational* function associated with the **Selector** block. It is assumed here that at most one input activation port of this block is active at any time.

```
#include "<SCIDIR>/routines/machine.h"
void selector(flag,nevprt,t,xd,x,nx,z,nz,tvec,ntvec,
  rpar,nrpar,ipar,nipar,inptr,insz,nin,outptr,outsz,nout)

integer *flag,*nevprt,*nx,*nz,*ntvec,*nrpar;
integer ipar[],*nipar,insz[],*nin,outsz[],*nout;

double x[],xd[],z[],tvec[],rpar[];
double *inptr[],*outptr[],*t;
{
    int k;
    double *y;
    double *u;
    int nev,ic;
    ic=z[0];
    if ((*flag)==2) {
        ic=-1;
        nev=*nevprt;
        while (nev>=1) {                    ← decode activation code
            ic=ic+1;
            nev=nev/2;
        }
        z[0]=ic;                            ← activated input activation port
    }
    else {
    /* copy selected input on the output */
        if (*nin>1) {
            y=(double *)outptr[0];
            u=(double *)inptr[ic];
            for (k=0;k<outsz[0];k++)
                *(y++)=*(u++);
        }
        else {
            y=(double *)outptr[ic];
            u=(double *)inptr[0];
            for (k=0;k<outsz[0];k++)
                *(y++)=*(u++);
        }
    }
}
```

Computational **function type 3**   This *Computational* function type is specific to programming in Scilab. The calling sequence is as follows:

```
[y,x,z,tvec,xd]=test(flag,nevprt,t,x,z,rpar,ipar,u)
```

See table 9.5 for a description of the arguments.

I/O	Args.	description
I	**flag**	0, 1, 2, 3, 4, 5, or 6 (see Table 9.1)
I	**nevprt**	activation code (scalar)
I	**t**	time (scalar)
I	**x**	continuous state (vector)
I	**z**	discrete state (vector)
I	**rpar**	parameter (any type of scilabtt variable)
I	**ipar**	parameter (vector)
I	**u**	**u(i)** is the vector of **i** th regular input (list)
O	**y**	**y(j)** is the vector of **j** th regular output (list)
O	**x**	new x if **flag**=2, 4, 5 or 6
O	**z**	new z if **flag**=2, 4, 5 or 6
O	**xd**	derivative of x if **flag**= 0 (vector), [ ] otherwise
O	**tvec**	times of output events if **flag**=3 (vector), [ ] otherwise

Table 9.5: *Arguments of* Computational *functions of type 3. I: input, O: output.*

There is no predefined computational function written in Scilab code (for efficiency, all blocks are written in C and Fortran). The examples below are just toy codes.

Example   The following is the *Computational* function associated with a block that displays in a Scilab window, every time it receives an event, the number of events it has received up to the current time and the values of its two inputs.

```
function [y,x,z,tvec,xd]=test(flag,nevprt,t,x,z,rpar,ipar,u)
y=list();tvec=[];xd=[]
if flag==4 then
    z=0
elseif flag==2 then
    z=z+1
    write(%io(2),'Number of calls:'+string(z))
```

```
    [u1,u2]=u(1:2)
    write(%io(2),'first input');disp(u1)
    write(%io(2),'second input');disp(u2)
end
```

Example The advantage of coding inputs and outputs as lists is that the number of inputs and outputs need not be specified explicitly. In this example, the output is the elementwise product of all the input vectors, regardless of the number of inputs.

```
function [y,x,z,tvec,xd]=..
    elemprod(flag,nevprt,t,x,z,rpar,ipar,u)
tvec=[];xd=[]
y=u(1)
for i=2:length(u)
    y=y.*u(i)
end
y=list(y)
```

9.5 Example

To define a generic diagram in Scicos, symbolic parameters should be defined in the "context." For a generic system-observer diagram, for example, we may have a "context" as in Figure 9.21, where **A**, **B**, and **C** are system parameters, **dt** is the sampling period, **x0** is the initial condition of the system, and **K** the observer gain matrix.

The corresponding Scicos diagram is given in Figure 9.22; the observed system is the super block in Figure 9.23, and the hybrid observer is the super block in Figure 9.24.

The **A** and **B** matrices are used both in the definition of the linear system and the **Jump linear system** used in the two super blocks; the difference is that in the case of the system, the initial condition **x0** is used (Figure 9.25), whereas in the case of the observer, the initial condition is set to zero. The **C** and **K** matrices are used in the **Gain** blocks. In particular, the observer gain is given in Figure 9.26. The **dt** parameter is used in the **Clock**. The result of the simulation (which gives the observation error of all the components of the state) is given in Figure 9.27.

Note that in this diagram not only the system matrices, but also their dimensions, are arbitrary. You can, for example, change the **C** matrix so that the number of outputs of the system changes.

This example is part of Scicos demos. You can experiment by changing the context (don't forget to use **Eval** for your modifications to be effective).

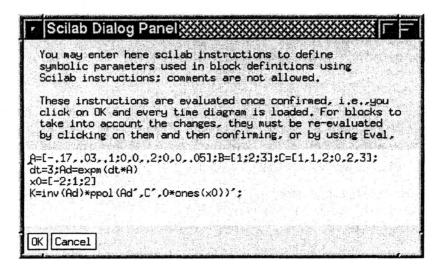

Figure 9.21: *Context of the diagram*

9.6   Palettes

A number of useful blocks can be found in Scicos's palettes. New palettes can also be defined by the user.

9.6.1   Existing Palettes

Scicos comes with a number of predefined blocks organized in seven palettes:

- **Inputs/Outputs** palette contains blocks for reading from and writing to data files, scopes, signal generators, etc.

- **Linear** palette contains blocks realizing linear operators including linear dynamical systems, both discrete and continuous.

- **Non linear** palette contains blocks realizing nonlinear operators.

- **Threshold** palette contains zero crossing blocks.

- **Events** palette contains blocks generating events.

- **Others** palette contains **Scifunc, GENERIC, Super Block**, and C and Fortran blocks.

- **Branching** palette contains **Synchro** blocks, selectors, and multiplexers.

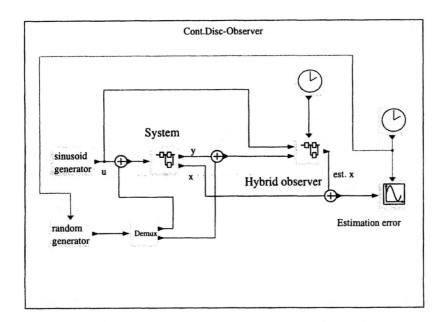

Figure 9.22: *Linear system with a hybrid observer*

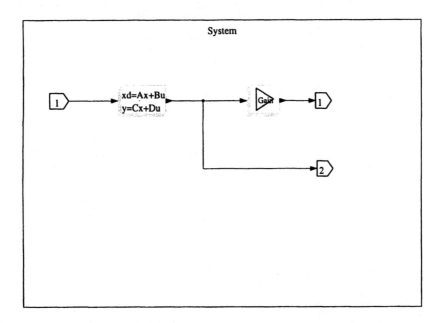

Figure 9.23: *Model of the system. The two outputs are* **y** *and* **x**.

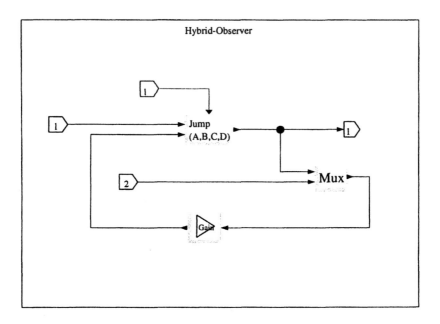

Figure 9.24: *Model of the hybrid observer. The two inputs are* **u** *and* **y**.

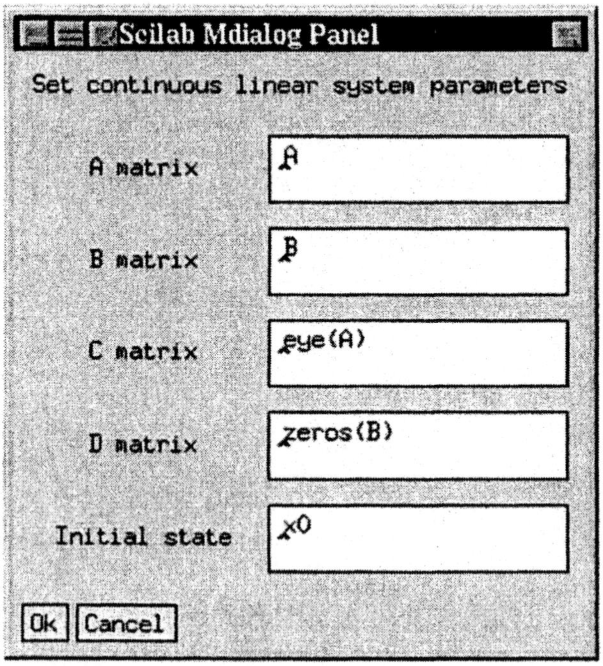

Figure 9.25: *Linear system dialogue box*

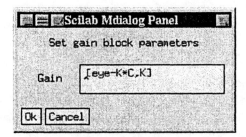

Figure 9.26: **Gain** *block dialogue box*

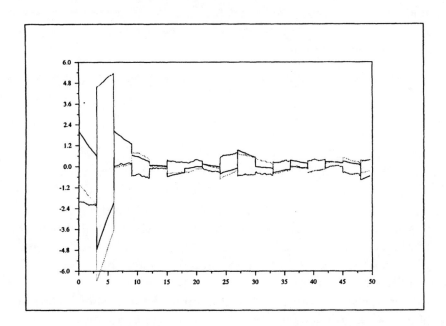

Figure 9.27: *Simulation result*

9.6.2 Constructing New Palettes

The **Add new Block** button of the **Edit** menu can be used to add a user-defined block to a Scicos diagram. In particular, when you select this button, you are asked to enter the name of the new block's *Interfacing* function. If the function is not already defined in the Scilab environment, Scicos asks the user to give the name of the file in which the function is defined. You can then place the block in the diagram.

User-defined blocks can also be placed in palettes. It is very easy to construct and edit new palettes in Scilab. User-defined blocks and (customized) existing blocks of general interest should be placed in palettes for future use.

To place a block in a palette, load the palette as a Scicos diagram and then use the **Add new Block** button of the **Edit** menu; in fact, you can do all the usual editing and customization operations associated with regular Scicos diagrams. You can then save the palette. If you choose the button **Save as Palette** of the **Diagram** menu, you also have the option of having Scicos update your **.scilab** file so that the palette is added to the list of available palettes the next time you run Scilab and Scicos. If the palette you are saving is not new and is already recognized by Scicos, you can simply save it as a regular diagram.

Figure 9.28 shows a super block that realizes the Saint Venant function:

$$
f_{sv}(p) = \begin{cases} \left(\dfrac{2g}{g-1}\right)^{\frac{1}{2}} p^{\frac{1}{g}} \left(1 - p^{\frac{g-1}{g}}\right)^{\frac{1}{2}}, & \text{if } p > \left(\dfrac{2}{g+1}\right)^{\frac{g}{g-1}}, \\[2ex] g^{\frac{1}{2}} \left(\dfrac{2}{g+1}\right)^{\frac{g+1}{2(g-1)}}, & \text{otherwise.} \end{cases}
$$

This function is used in fluid dynamics for computing the mass flow of a compressible fluid through a nozzle, and it would be natural to have it in a palette as a general-purpose block. Here, $0 \le p \le 1$ is the relative pressure and $g > 1$ is a (fluid-dependent) constant. The value of g is defined in the context of the super block.

To put this super block in a palette, first, the user should save it using the **Save Interf. Func.** button in the **Diagram** menu of the super block. This operation creates the *Interfacing* function associated with the super block. Then, the user should load the palette in which he wants to place the super block, as a regular diagram, and use the **Add new Block** button to include the super block in the palette. The palette can then be saved as usual.

This block can now be copied in any Scicos diagram from the palette. Note that if the *Interfacing* function of the block is saved as a block (as opposed to a super block), the value of g in the context of the super block becomes the default value of g in the new block. To change the value of g, g has to be defined in the context of the diagram in which the new block is placed.

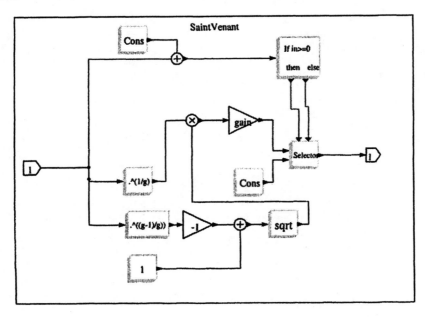

Figure 9.28: *Super block realizing the Saint Venant function. The parameter g is defined in the context of the super block.*

Chapter 10

Symbolic/Numeric Environment

10.1 Introduction

In solving engineering problems including modeling, identification, simulation, and control, the numerical computations usually come at the end of a symbolic solution process. This includes algebraic transformations, derivatives, and integrations before obtaining equations that are to be numerically solved. For instance, for system simulation we often need to compute a Jacobian matrix, and for control problems we often need to compute gradients.

A typical example, which is illustrated later in this chapter, arises in the equations of motion of mechanical systems with constraints. A choice method for deriving these equations is the Lagrange formulation, which leads to the Euler equations. To solve this type of problem we need to obtain the rigid matrix from the Euler equations, and this is obtained through a set of symbolic computations consisting of various manipulations and derivative computations.

Scilab can perform a few symbolic calculations: For instance, the function **derivat** computes the derivative of a rational function:

```
-->x=poly(0,'x');                        ← definition of polynomial

-->derivat((1+x)/(1-x^2))
 ans  =
        1
    ---------
            2
    1 - 2x + x
```

The calculations are done by simple operations on the coefficients of the input polynomials. The computations make use only of floating point arithmetic. In contrast, computer algebra systems (Macsyma, Maple, Mathematica, Mupad) carry out these operations using exact (integer, rational) arithmetic.

It is usually not efficient to make symbolic computations in a numerical system such as Scilab. Computer algebra systems can themselves perform numerical computations. However, as symbolic objects become bigger, the associated numerical computations become less efficient and more time-consuming. This is particularly true for matrix computations.

As an example, we can see the time needed to compute the singular values of a random 100×100 numerical matrix, both in the computer algebra system Maple and in Scilab. First we compute the singular values in Maple (**writemat** is a Maple procedure we have written that saves the random matrix in a file, which will subsequently be loaded into Scilab):

```
> with(linalg):
Warning, new definition for norm
Warning, new definition for trace
> m:=evalm(randmatrix(100,100)/99.0):
> read('writemat.mpl'):writemat(m,'mat100.data'):
> st:=time():v:=evalf(Svd(m)):time()-st;
                        57.261
> v[1],v[100];
                11.46254872, .01816095144
```

Then using Scilab we compute the singular values of the saved matrix:

```
-->m=read('mat100.data',100,100);

-->timer();v=svd(m);timer()
 ans  =
    0.24999

-->[v(1)  v(100)]
 ans  =
!   11.462537      0.0181609 !
```

We can see that Scilab made the computations more than 200 times faster than Maple. This is not particular to the Maple system and is a problem with all computer algebra systems where computations are made with nonhardware floating point numbers. The conclusion is that computer algebra systems such as Maple are inefficient for numerical computations. Nevertheless, with respect to the above-cited engineering problems, *it is clear that it would be useful to link symbolic and numerical environments.* There exist at least three user models for such a link:

Model 1 We use a computer algebra system like Maple, and we call numerical libraries or numerical systems like Scilab to do the numerical computations.

This model does not exist yet in most computer algebra systems. It could be very useful because it seems at first glance more complicated to build a computer algebra system (with high-level data structures) than simply to call fast numerical libraries. But existing computer algebra systems have not been designed for numerical computations, and adding them to an existing system is not easy.

Model 2 We use a numerical system like Scilab and we call a computer algebra system like Maple to do the symbolic computations.

Engineers are interested mainly in numerical computations and prefer powerful numerical systems. Thus adding symbolic computations to such a system seems at the present time the most useful approach. This is the model that exists in the symbolic Maple toolbox for use with Matlab.

Model 3 We use a user interface that allows the use of both a computer algebra system like Maple and a numerical system like Scilab using a common syntax.

This could be the best model, but this requires a user-friendly syntax that is easily adopted by the users of both systems. Many efforts have been made to create such an interface for several mathematical systems. For example, the Open Math project [21] is a first step toward a common mathematical language and interface.

In Scilab we use a weak version of **Model 2**. We use both Scilab and Maple, and we exchange data between them. But the final step, that of numerical computation, will be made in Scilab. A typical engineering problem consists of 5 steps, which are illustrated in Figure 10.1.

1. Modeling the problem. This can be done both in Scilab and in Maple. However, for symbolic computations we usually must begin in Maple, where equation transformations can be easily performed. For instance, for simulation and/or optimization, we need to compute derivatives, Jacobians, or gradients. These are computed in Maple.

2. The result of step 1 is usually a matrix (e.g., the Jacobian) or a vector (e.g., the gradient). Once computed, they need to be transferred to

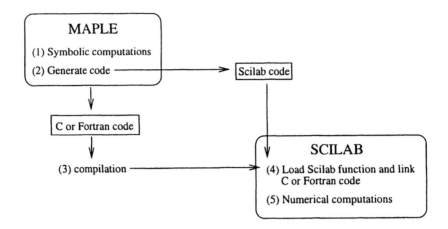

Figure 10.1: *User model of the interface between Scilab and Maple*

Scilab. This is accomplished by having Maple automatically generate Scilab, C, or Fortran code to be subsequently used by Scilab.

3. We compile the Fortran or C code generated in step 2.

4. We link into Scilab the Fortran code compiled in step 3, and we load into Scilab the Scilab code generated in step 2.

5. We solve the simulation or control problem by using standard Scilab functions (such as **ode** or **optim**), where inputs are the functions loaded in step 4.

For the procedure just described, a few tools that already exist in Scilab and Maple are needed:

- To generate C or Fortran code from Maple.

- To generate Scilab code from Maple.

- To link C and Fortran object code into Scilab (see Section 5.6).

We are going to describe the first two tools in the following sections, and in the last section we will show complete examples.

10.2   Generating Optimized Fortran Code with Maple

In every computer algebra system, and in Maple in particular, there are functions that allow the translation of expressions into C or Fortran code. It is also possible to optimize the generated code by grouping common subexpressions into variables.

In the sequel, we will mainly use examples of Fortran code generation. We can generate C code in the same way.

The following is a simple Maple example:

```
> e:=expand((1+sin(x))^10);
                     2            3            4              5
e := 1 + 10 sin(x) + 45 sin(x)  + 120 sin(x)  + 210 sin(x)  + 252 sin(x)

              6            7            8            9          10
   + 210 sin(x)  + 120 sin(x)  + 45 sin(x)  + 10 sin(x)  + sin(x)

> fortran(['e'=e]);
      e = 1+10*sin(x)+45*sin(x)**2+120*sin(x)**3+210*sin(x)**4+252*sin(x
     #)**5+210*sin(x)**6+120*sin(x)**7+45*sin(x)**8+10*sin(x)**9+sin(x)*
     #*10
> fortran(['e'=e],optimized);
      t1 = sin(x)
      t2 = t1**2
      t3 = t2*t1
      t4 = t2**2
      t8 = t4**2
      e = 1+10*t1+45*t2+120*t3+210*t4+252*t4*t1+210*t4*t2+120*t4*t3+45*t
     #8+10*t8*t1+t8*t2
```

We can also translate whole Maple procedures into C or Fortran. For example:

```
> f := proc(x)
>     if abs(x) < 1 then 1+tan(x^2)/6;
>     else 1-tan(x^2)/6; fi;
> end:
> fortran(f,mode=double,precision=double);
      doubleprecision function f(x)
      doubleprecision x

        if (dabs(x) .lt. 1.D0) then
          f = 1.D0+dtan(x**2)/6.D0
        else
          f = 1.D0-dtan(x**2)/6.D0
        endif
      return
      end
```

These tools are very useful, but for most problems we are either in the case of Model 2 described above or in the case where we only want to make a few symbolic computations followed by the generation of Fortran code to numerically solve our simulation or optimization problem. Normally, we do not need to symbolically solve the problem, and we do not want to use the simple **fortran** Maple function: Indeed, we do not want to generate lines of Fortran code from Maple expressions, put them in a file, and put in all the standard Fortran declarations. We need a tool for generating complete Fortran code. The Macrofort package from Maple's share library can accomplish this, as can the MacroC package C code.

MacroC and Macrofort have a very similar syntax, so users may use both with a minimum amount of work and time. The MacroC and Macrofort "languages" are made up of two kinds of instructions: single instructions and macro instructions. Both have the same syntax: a Maple list whose first element is a keyword describing the statement to be generated. The optional other elements are the relevant arguments. Single instructions ordinarily describe a unique statement, and macros describe a set (usually a block) of statements. Keywords are drawn from the Fortran or C instruction name (when it exists, or something close when it does not exist) with respectively an **f** or a **c** at its end for single statements or an **m** at its end for a macro statement.

We are not going to describe here both Macrofort and MacroC in detail (see [30, 18, 31]). Instead, we show a few examples of Macrofort (MacroC works the same way). With Macrofort, the user no longer has to deal with numerical labeling of **DO** loops and can also engage in structured Fortran programming. For instance, after loading Macrofort into Maple (Release V.4) with

```
with(share):readshare(macrofor,'numerics'):
```

or into Maple (Release V.5) with

```
with(share):with(macrofor):
```

we can generate a Fortran matrix with

```
> m:=matrix([[sin(x),sin(x)^2+cos(x)],
          [sin(x)+cos(x)^2,cos(x)]]);
                [                                    2          ]
                [     sin(x)            sin(x)  + cos(x)]
          m := [                                         ]
                [                2                        ]
                [sin(x) + cos(x)            cos(x)        ]

> genfor([matrixm,mat,m]);
        mat(2,2) = cos(x)
        mat(1,1) = sin(x)
        mat(1,2) = sin(x)**2+cos(x)
        mat(2,1) = sin(x)+cos(x)**2
```

and we can optimize the code with

```
> optimized:=true:                          ← to optimize code
> genfor([matrixm,mat,m]);
        t1 = sin(x)
        t2 = t1**2
        t3 = cos(x)
        t5 = t3**2
       mat(1,1) = t1
       mat(1,2) = t2+t3
       mat(2,1) = t1+t5
       mat(2,2) = t3
```

The next example shows how to generate a complete while loop (which does not exist in standard Fortran 77):

```
> l:=[whilem,abs(a)>eps,[equalf,a,big],[equalf,a,a/2.0],1000]:

> genfor(l);
c
c      WHILE   (eps<abs(a)) DO <WHILE_LIST> (1)
c
c      WHILE LOOP INITIALIZATION
           maxwhile1 = 1000
           nwhile1 = 0
         a = big
c
c      WHILE LOOP BEGINNING
 1000 continue
c
c      WHILE LOOP TERMINATION TESTS
      if (eps.lt.abs(a)) then
        if (nwhile1.le.maxwhile1) then
c
c        NEW LOOP ITERATION
           nwhile1 = nwhile1+1
c
c        <WHILE_LIST>
           a = 0.5E0*a
        goto 1000
        else
c
c        WHILE LOOP TERMINATION :
c        BYPASSING THE MAXIMUM ITERATION NUMBER
           write(6,2000)
 2000      format(' maxwhile1 ')
        endif
```

```
c
c      NORMAL WHILE LOOP TERMINATION
       endif
c      WHILE LOOP END (1)
```

Comments are generated automatically (this is optional).

Next we generate a complete double-precision Fortran subroutine:

```
> precision:=double:                        ← specifies double precision
> l:=[[declarem,doubleprecision,[a('*'),b('*')]],
>      [dom,i,1,n,[equalf,a(i),b(i)*sqrt(2)]]]:
> genfor([subroutinem,foo,[a,b,n],l]);
c
c      SUBROUTINE foo
c
       subroutine foo(a,b,n)
         doubleprecision a(*),b(*)
c
         do 1000, i=1,n
           a(i) = b(i)*sqrt(2.D0)
 1000    continue
c
       end
```

Finally, we show a more complete example, where a main Fortran program is generated. The program computes the minimum of a multivariable function by using a gradient algorithm. The formulation of the problem is

$$\min_x f(x), \qquad x \in \mathbb{R}^n, \qquad f(x) \in \mathbb{R}.$$

If we call ∇f the gradient of f with respect to x, a very simple gradient algorithm for finding the optimum of the problem is to start from an initial point x_0 and to iterate on x using $x = x - \rho \nabla f$ until convergence. Here ρ is a given positive real number.

What follows is the Maple procedure using Macrofort that generates a main Fortran program to solve the above problem:

```
gengrad:=proc(func,x,n,x0,ro,epsilon)
global optimize, output, precision;
local eps,f,fold,g,gr,i,j,k,ma;

g:=linalg[grad](func,[x[j]$j=1..n]);       ← compute the gradient of f

ma:=[seq([equalf,x[k],x0[k]],k=1..2),      ← initialization of x
      [equalf,eps,epsilon],                       ← value of ε
      [equalf,f,func],                      ← initial value of f
```

```
        [equalf,fold,f*(1-10*eps)],
        [whilem,abs(f-fold)>eps,[],                    ← convergence test
         [[equalf,fold,f],
          [matrixm,gr,g],                    ← gradient for current value of x
          [dom,i,1,n,[equalf,x[i],x[i]-ro*gr[i]]], ← iteration on x
          [equalf,f,func]],                          ← actualization of f
         1000],
        [writem,output,['5(e14.7,2x)'],[f,x]]];       ← output result
ma:=[programm,'grad',                              ← Fortran declarations
        [[declarem,'implicit doubleprecision',['(a-h,o-z)']],
         [declarem,dimension,[gr(n),x(n)]],op(ma)]];
writeto('grad.f');                                 ← open file for writing
init_genfor();
optimized:=true;                               ← we optimize the Fortran code
precision:=double;                            ← everything is double-precision
genfor(ma);
writeto(terminal);
end;
```

We can test this program on the function $x_1^2 + x_2^2 - 5$. Choosing $(5, 5)$ as a starting point of the algorithm, 0.5 as the value of ρ, and 0.00001 for ϵ,

```
fct:=x[1]^2+x[2]^2-5;
```

```
gengrad(fct,x,2,[5,5],0.5,0.00001);
```

The resulting generated Fortran program is

```
c
c       MAIN PROGRAM grad
c
        program grad
          implicit doubleprecision (a-h,o-z)
          dimension gr(2),x(2)
          x(1) = 5
          x(2) = 5
          eps = 0.1D-4
          t2 = x(1)**2
          t4 = x(2)**2
          f = t2+t4-5
          fold = f*(1-10*eps)
c
c       WHILE  (eps<abs((f+(-1*fold)))) DO <WHILE_LIST> (1)
c
c       WHILE LOOP INITIALIZATION
              maxwhile1 = 1000
              nwhile1 = 0
```

```
c
c          WHILE LOOP BEGINNING
 1000     continue
c
c          WHILE LOOP TERMINATION TESTS
          if (eps.lt.abs(f-fold)) then
            if (nwhile1.le.maxwhile1) then
c
c              NEW LOOP ITERATION
               nwhile1 = nwhile1+1
c
c              <WHILE_LIST>
               fold = f
                gr(1) = 2*x(1)
                gr(2) = 2*x(2)
c
               do 1001, i=1,2
                 x(i) = x(i)-0.5D0*gr(i)
 1001          continue
c
               t2 = x(1)**2
               t4 = x(2)**2
               f = t2+t4-5
            goto 1000
            else
c
c              WHILE LOOP TERMINATION :
c              BYPASSING THE MAXIMUM ITERATION NUMBER
               write(6,2000)
 2000          format(' maxwhile1 ')
            endif
c
c          NORMAL WHILE LOOP TERMINATION
          endif
c          WHILE LOOP END (1)
          write(6,2001) f
 2001     format('value of function: ',e14.7)
          write(6,2002) x
 2002     format('value of argument: ',5(e14.7,2x))
        end
```

Compiling and executing the above program yields

```
make grad
f77    grad.f   -o grad
grad
value of function: -0.5000000E+01
```

```
value of argument:   0.0000000E+00   0.0000000E+00
```

This is a simple example, but it can also be used for much more complicated functions to be minimized. For complicated minimization problems the Maple function described above can generate hundreds of lines of Fortran code. This would be onerous to do by hand, would give rise to many mistakes, and would require substantial debugging efforts. For a more "realistic" application, it would be, of course, simpler to generate just the cost/gradient routine and to use the high-level Scilab function **optim** (see Section 8.4.2) to perform the optimization.

10.3   Maple to Scilab Interface

Since we have Scilab, it is desirable to make use of its computational algorithms. So, instead of writing everything in Fortran as we did in the preceding section, we are going to use Maple to obtain symbolic results that can then be used as input to Scilab solvers.

At the end of step 2 in Figure 10.1, we usually have computed algebraic expressions, vectors, and/or matrices that we want to use in Scilab. The way we do this is with Scilab functions that correspond to Maple objects. For instance, suppose in Maple we have the following matrix:

```
> m:=matrix([[x[1,1]+x[2,2],k*(x[1,2]+x[2,1])],
             [10*k,1],[x[2,3],x[1,3]]]);
```

$$
\begin{bmatrix}
x_{1,1} + x_{2,2} & k\,(x_{1,2} + x_{2,1}) \\
10\,k & 1 \\
x_{2,3} & x_{1,3}
\end{bmatrix},
$$

```
> x:=[[1,2,3],[4,5,6]]:k:=2:
> map(eval,m);
```

$$
\begin{bmatrix}
6 & 12 \\
20 & 1 \\
6 & 3
\end{bmatrix}.
$$

We want to have in Scilab the same matrix available as a Scilab function, which can be called as **mat(k,x)**, where **k** is a scalar, **x** is a 2×3 matrix, and the result is the numerical 3×2 matrix

```
-->mat(2,[1,2,3;4,5,6])
 ans  =
!   6.       12.  !
!   20.       1.  !
!   6.        3.  !
```

At the present time, there are two ways to do this:

1. To automatically generate a Scilab coded function.

2. To automatically generate a Fortran subroutine or C function that can be compiled and linked to Scilab as a Scilab function.

The difference between the two previous approaches is that the generation of Fortran and C code is much slower than the generation of Scilab code, since Maple optimization is not used for Scilab code generation. Also, the compilation of Fortran or C code can take a long time for big files, but their execution time is much faster (on the order of 10 to 100 times) than that of Scilab code. Moreover, Fortran is faster than C.

A good strategy is the following:

- If you want to compute a matrix in Scilab just a few times, generate a Scilab code function.

- If you want to compute the matrix many times (e.g., for some iterative algorithm) and you have a Fortran compiler, generate Fortran code, otherwise generate C code.

We have created a Maple procedure for doing this job easily; it is freely distributed with Scilab.

This Maple procedure is

maple2scilab(fname,expr,parameters,code,[dirname]):

fname is the name of the Scilab function that will be generated.

expr is the name of the Maple matrix, vector, or algebraic expression.

parameters is a Maple list containing the parameters of the arguments of the Scilab function.

code is the character **f** for the generation of Fortran code, **c** for the generation of C code, and **s** to generate a Scilab coded function.

dirname is an optional argument. It is the pathname of the directory where the code of the generated Scilab function and Fortran routines will reside. When this argument is not given, the default is the current directory.

The generation of Fortran and C code is accomplished using respectively the functions Macrofort and MacroC (see Section 10.2). The generated code is optimized using the Maple optimizer (common subexpressions in matrices are shared by introduction of new variables).

As an example of the above discussion we solve a hot rolling mill problem. This problem leads to a nonlinear system of three equations:

$$
\begin{aligned}
f_1(F, h_2, \Phi) &= h_2 - S - \frac{F + a_2(1 - e^{a_3 F})}{a_1}, \\
f_2(F, h_2, \Phi) &= F + \frac{R\xi T_1}{h_2} \\
&\quad - lkR\left(\tfrac{1}{2}\pi\sqrt{\tfrac{h_2}{R}}\arctan\sqrt{r} - \tfrac{\pi\xi}{4} - \ln\left(\tfrac{h_N}{h_2}\right) + \tfrac{1}{2}\ln\left(\tfrac{h_1}{h_2}\right)\right), \\
f_3(F, h_2, \Phi) &= \arctan\left(\Phi\sqrt{\tfrac{R}{h_2}}\right) \\
&\quad - \tfrac{1}{2}\sqrt{\tfrac{h_2}{R}}\left(\tfrac{\pi}{4}\ln\left(\tfrac{h_2}{h_1}\right) + \sqrt{\tfrac{R}{h_2}}\arctan\sqrt{r} - \tfrac{T_1}{klh_1} + \tfrac{T_2}{klh_2}\right),
\end{aligned}
$$

with

$$
r = \frac{h_1 - h_2}{h_2}, \qquad \xi = \sqrt{\frac{h_1 - h_2}{R}}, \qquad h_N = h_2 + R\Phi^2,
$$

and where $a_1, a_2, a_3, l, k, R, T_1, T_2, h_1$, and S are given.

One way to solve this problem, i.e., to compute F, h_2, and Φ, is to compute the Jacobian matrix of the system in order to use a Newton-like method to solve the system. This can easily be done in Maple.

We will generate both the function that will compute values of the system, `fct(x)`, and the function that will compute values of the Jacobian, `fjac(x)`, where the vector **x** corresponds to the vector of variables $[f, h_2, \Phi]$. Then we will use the `fsolve` function (see Section 8.4.3) to find the solution. Each of the three methods (Fortran, C, and pure Scilab) will be used, and we will then compare them.

In Maple, we use the following code to define the system of equations and the generation of the Jacobian:

```
> data:=a1=610,a2=648,a3=-.00247,l=1250,k=1.4*10^(-2),
>    gr=360,t1=12,t2=35,h1=24,s=12:
> exp1:=h2-s-(f+a2*(1-exp(a3*f)))/a1:
> exp2:=f-l*k*gr*(Pi*sqrt(h2/gr)*arctan(sqrt(r))/2-Pi*csi/4
>        -log(hn/h2)+log(h1/h2)/2)+gr*csi*t1/h2:
> exp3:=arctan(phi*sqrt(gr/h2))-sqrt(h2/gr)*(Pi*log(h2/h1)/4
```

```
>          +sqrt(gr/h2)*arctan(sqrt(r))-t1/k/l/h1+t2/k/l/h2)/2:
> r:=(h1-h2)/h2:
> csi:=sqrt((h1-h2)/gr):
> hn:=h2+gr*phi^2:
>
> lf:=evalf(subs(data,f=x[1],h2=x[2],phi=x[3],
>          vector([exp1,exp2,exp3]))):
> j:=linalg[jacobian](lf,[x[1],x[2],x[3]]):   ← compute Jacobian
>
> x:=vector(3):
> read('/usr/local/lib/scilab-2.3/maple/maple2scilab.mpl'):
> optimized:=true:
>
> maple2scilab(fct_f,lf,[x],'f');      ← generate function via Fortran

Fortran file created (to be compiled): fct_f.f
Scilab file created: fct_f.sci

Usage in Scilab: link('fct_f.o','fct_f');
                 getf('fct_f.sci');
                 out=fct_f(x)
> maple2scilab(fjac_f,j,[x],'f');       ← generate Jacobian via Fortran

Fortran file created (to be compiled): fjac_f.f
Scilab file created: fjac_f.sci

Usage in Scilab: link('fjac_f.o','fjac_f');
                 getf('fjac_f.sci');
                 out=fjac_f(x)

> maple2scilab(fct_c,lf,[x],'c');            ← generate function via C

C file created (to be compiled): fct_c.c
Scilab file created: fct_c.sci

Usage in Scilab: link('fct_c.o','fct_c');
                 getf('fct_c.sci');
                 out=fct_c(x)
> maple2scilab(fjac_c,j,[x],'c');            ← generate Jacobian via C

C file created (to be compiled): fjac_c.c
Scilab file created: fjac_c.sci

Usage in Scilab: link('fjac_c.o','fjac_c');
                 getf('fjac_c.sci');
                 out=fjac_c(x)
```

```
> maple2scilab(fct_s,lf,[x],'s');          ← generate function via Scilab

Scilab file created: fct_s.sci

Usage in Scilab: getf('fct_s.sci');
                 out=fct_s(x)
> maple2scilab(fjac_s,j,[x],'s');          ← generate Jacobian via Scilab

Scilab file created: fjac_s.sci

Usage in Scilab: getf('fjac_s.sci');
                 out=fjac_s(x)
```

Once the Fortran and C files are compiled, everything is loaded into Scilab:

```
-->link('fct_f.o','fct_f');getf('fct_f.sci');
linking files fct_f.o  to create a shared executable
shared archive loaded
Linking fct_f (in fact fct_f_)
Link done

-->link('fjac_f.o','fjac_f');getf('fjac_f.sci');
linking files fjac_f.o  to create a shared executable
shared archive loaded
Linking fjac_f (in fact fjac_f_)
Link done

-->link('fct_c.o','fct_c');getf('fct_c.sci');
linking files fct_c.o  to create a shared executable
shared archive loaded
Linking fct_c (in fact fct_c_)
Link done

-->link('fjac_c.o','fjac_c');getf('fjac_c.sci');
linking files fjac_c.o  to create a shared executable
shared archive loaded
Linking fjac_c (in fact fjac_c_)
Link done

-->getf('fct_s.sci');getf('fjac_s.sci');
```

Finally, the problem is solved 100 times for each method in order to compare the running times:

```
-->timer();for i=1:100, fsolve([1000,10,0],fct_f,fjac_f); end;
-->timer()
 ans   =
     1.716598

-->timer();for i=1:100, fsolve([1000,10,0],fct_c,fjac_c); end;
-->timer()
 ans   =
     1.74993

-->timer();for i=1:100, fsolve([1000,10,0],fct_s,fjac_s); end;
-->timer()
 ans   =
     11.24955
```

As we can see, the fastest result is obtained using the generated Fortran code.

10.4   First Example: Simulation of a Rolling Wheel

We will show in this section an example where it is very useful to use both symbolic and numerical computations. This is an example of a mechanical system with dynamic and geometric constraints. The derivation of Euler equations from the Lagrange equation is easy to do but involves a great deal of tedious symbolic computations.

The problem consists in simulating the dynamics of a purely rolling (no sliding) wheel. The wheel is modeled using a Lagrangian formulation. The Lagrangian (i.e., the kinetic minus the potential energy) for the wheel is

$$\mathcal{L} = \frac{1}{2}m\left(\dot{x}^2 + \dot{y}^2 + \dot{z}^2\right) + \frac{1}{2}J_1\left(\sin(\phi)^2\dot{\theta}^2 + \dot{\phi}^2\right) + \frac{1}{2}J_3\left(\dot{\psi} + \cos(\phi)\dot{\theta}\right)^2$$
$$-m\,g\,z,$$

where most variables are described in Figure 10.2. Here J_1 and J_2 are the moments of inertia for the wheel, m is the mass of the wheel, and θ, ϕ, ψ are the wheel's Euler angles.

The Euler equations are used to obtain the wheel dynamics from the expression of the Lagrangian:

$$\frac{d}{dt}\left(\frac{\partial\mathcal{L}}{\partial\dot{q}}\right) - \frac{\partial\mathcal{L}}{\partial q} = -A(q)\lambda, \tag{10.1}$$

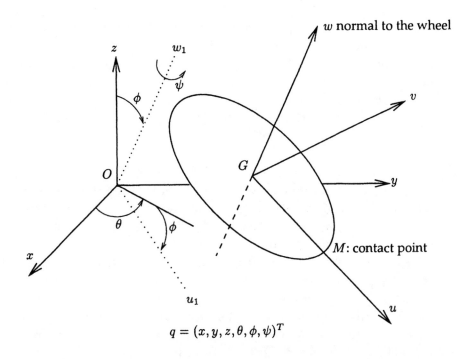

$$q = (x, y, z, \theta, \phi, \psi)^T$$

Figure 10.2: *Wheel description*

where q is the vector of generalized coordinates (see Figure 10.2) and λ is a Lagrange multiplier associated to the set of geometric and dynamic constraints. We will come back to $A(q)$ in a moment. To obtain the dynamic equations of motion for the wheel, we must evaluate the Euler equations. This consists in differentiating the Lagrangian, which can be done in Maple.

The system of Euler equations can be rewritten as follows:

$$M(q)\ddot{q} + C(q)\dot{q}^2 + R(q, \dot{q}) + A(q)\lambda = 0, \tag{10.2}$$

where $M(q)$ and $C(q)$ are 6×6 matrices, $\dot{q}^2$ is the componentwise square of $\dot{q}$, $(\dot{q}_1^2, \ldots, \dot{q}_6^2)^T$, and $R(q, \dot{q})$ is a vector. The manipulations necessary to transform (10.1) into (10.2) can be made in Maple. Once (10.2) is obtained, it is possible to evaluate q by solving this differential system. However, there are also constraints that need to be satisfied. For this problem there are both geometric and dynamic constraints.

The geometric constraint for the problem expresses the fact that the wheel is in contact with the floor. This equation, where the speed is not involved, is

$$z = r \sin(\phi). \tag{10.3}$$

The dynamic constraint express the nonrolling conditions on the floor. There are three equations where the speed is involved: $\vec{V_G} + \vec{\Omega} \wedge \overrightarrow{GM} = 0$, where $\vec{V_G}$ is the speed of the center of inertia (G) of the wheel, $\vec{\Omega}$ is the rotation vector, M is the contact point of the wheel with the floor, and $\wedge$ denotes the cross product of two vectors.

These constraints are computed using Maple and are put into the following form:

$$A(q)^T \dot{q} = 0, \tag{10.4}$$

where $A(q)^T$ is a 3×6 matrix. One can check that the time derivative of the geometric constraint (10.3) is contained in equation (10.4). So, we will not try to use this geometric constraint to eliminate the coordinate z in the dynamics of the wheel. This will be done at the end.

We wish to solve system (10.2) subject to the constraint in (10.4). This constraint expresses the fact that $\dot{q}$ belongs to the kernel of $A(q)^T$: This can be rewritten explicitly as $\dot{q} = S(q)\eta$, where η is a 3-dimensional time-dependent vector and $S(q)$ is a 6×3 matrix that spans the kernel of $A(q)^T$.

Using Maple, we can compute $S(q)$ and eliminate λ from the Euler equation (10.2) by left multiplication with $S(q)^T$, and we simplify the Euler equation by using the fact that $S(q)^T A(q) = 0$, where $S(q)$ has the following form:

$$S(q) = \begin{pmatrix} r\cos(\phi)\sin(\theta) & r\sin(\phi)\cos(\theta) & r\sin(\theta) \\ -r\cos(\phi)\cos(\theta) & r\sin(\phi)\sin(\theta) & -r\cos(\theta) \\ 0 & r\cos(\phi) & 0 \\ 1 & 0 & 0 \\ 0 & 1 & 0 \\ 0 & 0 & 1 \end{pmatrix}.$$

We can remark (since a submatrix of matrix S is the identity matrix) that $\eta(t) = (\dot{\theta}, \dot{\phi}, \dot{\psi})^T$. So, denoting by μ be the vector $(\theta, \phi, \psi)^T$, equation (10.2) left multiplied by $S(q)^T$ becomes the differential system

$$M'(\mu)\ddot{\mu} + R'(\mu, \dot{\mu}) = 0, \tag{10.5}$$

where the nonzero elements of the matrix $M'(\mu)$ are

$$
\begin{aligned}
M'(\mu)_{1,1} &= \tfrac{1}{2}\left(J_1 + r^2 m\cos(2\mu_2) + r^2 m + J_3\cos(2\mu_2) + J_3 \right. \\
&\quad \left. - J_1\cos(2\mu_2)\right), \\
M'(\mu)_{1,3} &= \cos(\mu_2)(r^2 m + J_3), \\
M'(\mu)_{2,2} &= J_1 + r^2 m, \\
M'(\mu)_{1,3} &= \cos(\mu_2)(r^2 m + J_3), \\
M'(\mu)_{3,3} &= r^2 m + J_3,
\end{aligned}
$$

and $R'(\mu, \dot{\mu})$ is defined by

$$
\begin{aligned}
R'(\mu, \dot{\mu})_1 &= -\dot{\mu}_2 \left(r^2 \, m \, \dot{\mu}_1 \, \sin(2\,\mu_2) + J_3 \, \sin(\mu_2) \, \dot{\mu}_3 - J_1 \, \dot{\mu}_1 \, \sin(2\,\mu_2) \right. \\
&\quad \left. + \, J_3 \, \dot{\mu}_1 \, \sin(2\,\mu_2) \right), \\
R'(\mu, \dot{\mu})_2 &= \tfrac{1}{2} \, r^2 \, m \, \dot{\mu}_1^2 \, \sin(2\,\mu_2) - \tfrac{1}{2} \, J_1 \, \dot{\mu}_1^2 \, \sin(2\,\mu_2) + r^2 \, \sin(\mu_2) \, m \, \dot{\mu}_1 \, \dot{\mu}_3 \\
&\quad + r \, \cos(\mu_2) \, m \, g + J_3 \, \sin(\mu_2) \, \dot{\mu}_1 \, \dot{\mu}_3 + \tfrac{1}{2} J_3 \, \dot{\mu}_1^2 \, \sin(2\,\mu_2), \\
R'(\mu, \dot{\mu})_3 &= -\sin(\mu_2) \, \dot{\mu}_1 \, \dot{\mu}_2 \left(2\,r^2 \, m + J_3 \right).
\end{aligned}
$$

Similarly, the constraint equation $\dot{q} = S(q)\eta$ can be rewritten as $\dot{q} = S(\mu)\dot{\mu}$.

Writing $\xi = (x, y, z)^T$, the first three equations of the constraints give $\dot{\xi} = S'(\mu)\dot{\mu}$.

Finally, setting $\chi = (\chi_1, \chi_2, \chi_3)^T$ where $\chi_1 = \mu$, $\chi_2 = \dot{\mu}$, and $\chi_3 = \xi$, we obtain the differential system

$$
\begin{aligned}
\dot{\chi}_1 &= \chi_2, \\
M'(\chi_1)\dot{\chi}_2 &= -R'(\chi_1, \chi_2), \\
\dot{\chi}_3 &= S'(\chi_1)\chi_2.
\end{aligned}
\tag{10.6}
$$

Note that $z = \chi_3(3)$ appears only in equation (10.6) as $\dot{z} = r\cos(\phi)\dot{\phi}$ (which is the time derivative of the geometric constraint (10.3)). So, z can be removed from the state variables of the preceding system. After numerical integration, $z(t)$ will be calculated as $r \sin \phi(t)$.

To numerically solve this system, we will use a numerical integrator in Scilab. To accomplish this we use the **maple2scilab** procedure to generate three Scilab functions corresponding to the computation of the three matrices M', R', and S'. For numerical efficiency, we generate Fortran code.

Mprim, **Rprim**, and **Sprim** are the Maple matrices corresponding respectively to M', R', and S', and **mprim**, **rprim**, and **sprim** will be the names of the generated Fortran subroutines and Scilab functions:

```
optimized:=true:
J:=vector(3);
g:=9.81;
maple2scilab('mprim',Mprim,[chi,J,'r','m'],'f');
maple2scilab('rprim',Rprim,[chi,J,'r','m'],'f');
maple2scilab('sprim',Sprim,[chi,J,'r','m'],'f');
```

The inputs to the Scilab functions are the vector **chi** corresponding to the unknown ξ, and the parameters **J**, **r**, and **m**. We now link the Fortran subroutines into Scilab as Scilab functions (as shown in Section 10.3) and create a Scilab function that computes $\dot{\chi}$ as a function of χ and t:

```
function [chidot]=wheel(t,chi)
```

```
mp=mprim(chi,J,r,m)
rp=rprim(chi,J,r,m)
sp=sprim(chi,J,r,m)
chidot=0*chi
chidot(1:3)=chi(4:6)
chidot(4:6)=-mp\rp
chidot(7:8)=sp(1:2)
```

All that remains to do is to integrate the above function by passing it as an argument to **ode**, which is the ordinary differential equation solver of Scilab (see Section 8.2):

```
r=1;
m=1;
J=[0.3,0.4,1.0];
tmin=0.0;tmax=15;nn=300;
times=(0:(nn-1));
times=tmax*times/(nn-1) +tmin*((nn-1)*ones(times)-times);
```
                                                          initial conditions:
```
x0=[0;                                                        ← θ
%pi/2+0.1;                                                    ← φ
0;                                                            ← ψ
5.0;                                                          ← θ̇
0.0;                                                          ← φ̇
4.0;                                                          ← ψ̇
0;                                                            ← x
0];                                                           ← y
```

```
x=ode(x0,tmin,times,wheel);                              ← simulation
```

After the simulation, a conversion to Cartesian coordinates is performed, and the result can be used to perform a Scilab animation of the simulation of the wheel (as shown in the Scilab demo panel). Figure 10.3 is a snapshot of this animation.

10.5 Second Example: Control of an n-Link Pendulum

This problem consists in first simulating the behavior of a simple n-link pendulum with one end point constrained to move only along a horizontal line with specified dynamics; this is the same type of problem as in the preceding section. In the second part of the problem we linearize the equations of the n-link pendulum around the unstable equilibrium point (i.e., the pendulum extended vertically from its constrained end point), and we find a feedback control that stabilizes the unconstrained end point. In this work,

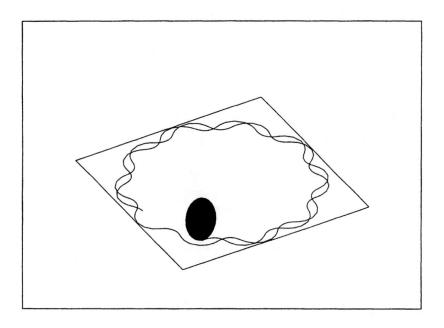

Figure 10.3: *Snapshot of wheel animation*

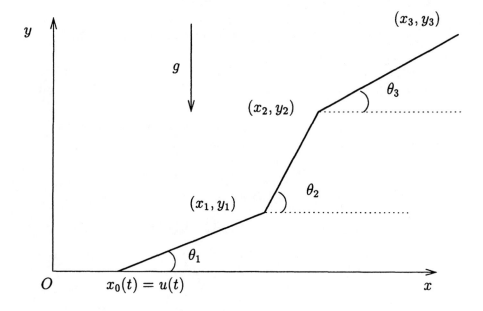

Figure 10.4: *3-link pendulum description*

Maple is used for deriving the dynamic equations of the pendulum and for linearizing them. Scilab is used for the numerical simulation and feedback control gain computation. The modeling is valid for any number of links, but we will use 3 links in this example to simplify the equations.

10.5.1   Simulation of the n-Link Pendulum

As in the previous section, the pendulum is modeled using a Lagrangian formulation:

$$L(q, \dot{q}) = \sum_{i=1}^{n} E_i^c - E_i^p,$$

where E_i^c and E_i^p are respectively the kinetic energy and the potential energy of link i:

$$
\begin{aligned}
E_i^c &= \frac{m_i}{2}\left((\dot{x}_{i-1} - \frac{l_i}{2}\sin(\theta_i)\dot{\theta}_i)^2 + (\dot{y}_{i-1} + \frac{l_i}{2}\cos(\theta_i)\dot{\theta}_i)^2 \right) + \frac{J_i}{2}\dot{\theta}_i^2, \\
E_i^p &= m_i g \left(y_{i-1} + \frac{l_i}{2}\sin(\theta_i) \right),
\end{aligned}
$$

where m_i, l_i, and $J_i = m_i l_i^2 / 12$ are respectively the mass, length, and moment of inertia of link i.

The generalized coordinates q are equal to the set $(x_i, y_i, \theta_i)_{i=1,\ldots,n}$. The point (x_0, y_0) is the end point of the pendulum with the given dynamics: $\ddot{x}_0(t) = \Gamma(t), y_0(t) = 0$. For the simulation, $\Gamma(t)$ is specified as a horizontal acceleration of the end point of the pendulum. Note that x_0 and $\dot{x}_0$ are not considered to be generalized coordinates.

With these generalized coordinates, the pendulum is subject to the following constraints (which do not depend on speed):

$$
\begin{aligned}
x_i - x_{i-1} &= l_i \cos(\theta_i), \\
y_i - y_{i-1} &= l_i \sin(\theta_i),
\end{aligned}
\qquad i = 1, \ldots, n. \tag{10.7}
$$

As usual, we use the Euler equation already described in equation (10.1),

$$\frac{d}{dt}\left(\frac{\partial \mathcal{L}}{\partial \dot{q}} \right) - \frac{\partial \mathcal{L}}{\partial q} = -A(q)\lambda, \tag{10.8}$$

to obtain the equations of the pendulum, where the Lagrange multiplier λ is associated with the constraints (10.7) (in fact, the time derivative of constraints (10.7)).

The time derivatives of the constraint equations give

$$\dot{x}_i - \dot{x}_{i-1} + l_i \sin(\theta_i)\dot{\theta}_i = 0, \qquad i = 1, \ldots, n,$$
$$\dot{y}_i - \dot{y}_{i-1} - l_i \cos(\theta_i)\dot{\theta}_i = 0,$$

which is of the form $A(q)^T \dot{q} = 0$. The derivation of the equations is the same as for the wheel example in the previous section. With the help of Maple, we compute the kernel of $A(q)^T$, which gives us the equation $\dot{q} = S(q)\eta$, where η is a 3 dimensional time-dependent vector. From the particular structure of $S(q)$, we can see that in fact, $\eta(t) = (\dot{\theta}_1, \dot{\theta}_2, \dot{\theta}_3)^T$.

We can see that rewriting the Euler equation, taking into account the constraints, leads to an equation that depends only on θ. So, if μ denotes the vector $(\theta_1, \theta_2, \theta_3)^T$, the Euler equation (10.8) can be rewritten as

$$M'(\mu)\ddot{\mu} + C'(\mu)\dot{\mu}^2 + R'(\mu, \Gamma) = 0,$$

where $\dot{\mu}^2$ stands for the vector $(\dot{\mu}_1^2, \dot{\mu}_2^2, \dot{\mu}_3^2)^T$, and with

$$
\begin{aligned}
M'(\mu)_{1,1} &= l_1{}^2 m_3 + J_1 + l_1{}^2 m_2 + \tfrac{1}{4} m_1 l_1{}^2, \\
M'(\mu)_{1,2} &= \tfrac{1}{2} l_1 l_2 \cos(\theta_1 - \theta_2)(m_2 + 2m_3), \\
M'(\mu)_{1,3} &= \tfrac{1}{2} l_1 m_3 l_3 \cos(\theta_1 - \theta_3), \\
M'(\mu)_{2,1} &= \tfrac{1}{2} l_1 l_2 \cos(\theta_1 - \theta_2)(m_2 + 2m_3), \\
M'(\mu)_{2,2} &= \tfrac{1}{4} m_2 l_2{}^2 + l_2{}^2 m_3 + J_2, \\
M'(\mu)_{2,3} &= \tfrac{1}{2} l_2 m_3 l_3 \cos(\theta_2 - \theta_3), \\
M'(\mu)_{3,1} &= \tfrac{1}{2} l_1 m_3 l_3 \cos(\theta_1 - \theta_3), \\
M'(\mu)_{3,2} &= \tfrac{1}{2} l_2 m_3 l_3 \cos(\theta_2 - \theta_3), \\
M'(\mu)_{3,3} &= \tfrac{1}{4} m_3 l_3{}^2 + J_3,
\end{aligned}
$$

$$
\begin{aligned}
C'(\mu)_{1,1} &= 0, \\
C'(\mu)_{1,2} &= \tfrac{1}{2} l_1 l_2 \sin(\theta_1 - \theta_2)(m_2 + 2m_3), \\
C'(\mu)_{1,3} &= \tfrac{1}{2} l_1 m_3 l_3 \sin(\theta_1 - \theta_3), \\
C'(\mu)_{2,1} &= -\tfrac{1}{2} l_1 l_2 \sin(\theta_1 - \theta_2)(m_2 + 2m_3), \\
C'(\mu)_{2,2} &= 0, \\
C'(\mu)_{2,3} &= \tfrac{1}{2} l_2 m_3 l_3 \sin(\theta_2 - \theta_3), \\
C'(\mu)_{3,1} &= -\tfrac{1}{2} l_1 m_3 l_3 \sin(\theta_1 - \theta_3), \\
C'(\mu)_{3,2} &= -\tfrac{1}{2} l_2 m_3 l_3 \sin(\theta_2 - \theta_3), \\
C'(\mu)_{3,3} &= 0,
\end{aligned}
$$

and

$$R'(\mu) = \frac{1}{2} \left(\begin{array}{c} l_1\,(2\,m_3 + 2\,m_2 + m_1)\,(g\,\cos(\theta_1) - \Gamma\,\sin(\theta_1)) \\ l_2\,(2\,m_3 + m_2)\,(g\,\cos(\theta_2) - \Gamma\,\sin(\theta_2)) \\ m_3\,l_3\,(-\sin(\theta_3)\,\Gamma + g\,\cos(\theta_3)) \end{array} \right).$$

We introduce the vector $\chi = (\chi_1, \chi_2, \chi_3, \chi_4)^T$, where $\chi_1 = \mu$, $\chi_2 = \dot{\mu}$, $\chi_3 = x_0$, and $\chi_4 = \dot{x}_0$. The preceding matrices depend only on χ_1 and Γ, the given horizontal acceleration of the end point. Thus, the differential system to be solved is

$$
\begin{aligned}
\dot{\chi}_1 &= \chi_2, \\
M'(\chi_1)\dot{\chi}_2 &= -R'(\chi_1, \Gamma) - C'(\chi_1)\chi_2^{\,2}, \\
\dot{\chi}_3 &= \chi_4, \\
\dot{\chi}_4 &= \Gamma.
\end{aligned}
\tag{10.9}
$$

We compute $\chi(t)$ by solving the ODE (10.9), and then we obtain $x_i(t)$ and $y_i(t)$ using equation (10.7).

Just as for the problem of the wheel, we use Scilab to solve the preceding ODE. For that, we use the **maple2scilab** procedure to generate three Scilab functions corresponding to the computation of the three matrices M', R', and C'. For numerical efficiency we generate Fortran code.

Mprim, **Rprim**, and **Cprim** are the Maple matrices corresponding respectively to M', R', and C', and **mprim**, **rprim**, and **cprim** will be the names of the generated Fortran subroutines and Scilab functions:

```
optimized:=true:
J:=vector(3);
g:=9.81;
maple2scilab('mprim',Mprim,[chi,'J','l','m'],'f');
maple2scilab('rprim',Rprim,[chi,'J','l','m','Gamma'],'f');
maple2scilab('cprim',Cprim,[chi,'J','l','m'],'f');
```

The inputs of these Scilab functions are the vector **chi** corresponding to the unknown χ, and the parameters **J**, **l**, and **m**. The input **Gamma** is a real number that represents the horizontal acceleration of the end point. This is followed by linking the Fortran subroutines into Scilab as Scilab functions as shown in 10.3 and by creating a Scilab function that computes $\dot{\chi}$ as a function of χ and t:

```
function [chidot]=npend(t,chi)
chidot=0*chi
chidot(2*n+1)=chi(2*n+2)
chidot(2*n+2)=Gamma(t)
```

```
mp=mprim(chi,J,l,m)
rp=rprim(chi,J,l,m,Gamma(t))
cp=cprim(chi,J,l,m)
chidot(1:n)=chi(n+1:2*n)
x=cp*(chi(n+1:2*n).*chi(n+1:2*n))+rp
chidot(n+1:2*n)=-mp\x
```

This function is suitable for **n** links in the pendulum. **Gamma(t)** is a given function that is seen by **npend** as a global variable. It is then used as an input argument to the Scilab ODE solver:

```
n=3;
l=2*ones(1,n);
m=ones(1,n);
J=0.3*ones(1,n);
tmin=0.0;tmax=10;nn=300;
times=linspace(tmin,tmax,nn);
deff('[y]=Gamma(t)','y=-k^2*sin(k*t)')    ← x₀(t) = sin(kt)
                                            initial conditions:
k=3
y0=[-%pi/2*ones(n,1);                       ← θ
    zeros(n,1);                             ← θ̇
    0;                                      ← x₀(0)
    k];                                     ← ẋ₀(0)
yt=ode(y0,0,times,npend);                   ← simulation
```

where the arrows denote, in order: $x_0(t) = \sin(kt)$; initial conditions: $\leftarrow \theta$; $\leftarrow \dot{\theta}$; $\leftarrow x_0(0)$; $\leftarrow \dot{x}_0(0)$; $\leftarrow$ simulation.

With the above tools we are now able to simulate the movement of the 3-link pendulum. Figure 10.5 is a snapshot of such a simulation showing the somewhat chaotic trajectory of the pendulum's free end point.

10.5.2 Control of the n-Link Pendulum

The purpose of this section is to compute a control that stabilizes the n-link pendulum near the unstable equilibrium point using linear control techniques. For that, we need to linearize equation (10.9) to obtain a controlled linear system for which the tools of Chapter 6 are used. We will study the nonlinear system with a linear feedback control.

Maple is used to linearize the system. Equation (10.9) is linearized around the points $\chi_1 = (\pi/2, \pi/2, \pi/2)^T$, $\chi_2 = 0$, $\chi_3 = 0$, $\chi_4 = 0$, which corresponds to the unstable equilibrium point of the pendulum with $\Gamma = 0$. This yields a linear system with the linearized variables $\delta\mu$ and $\delta\Gamma$:

$$M_1 \dot{\widehat{\delta\mu}} + R_1 \delta\mu + R_2 \delta\Gamma = 0.$$

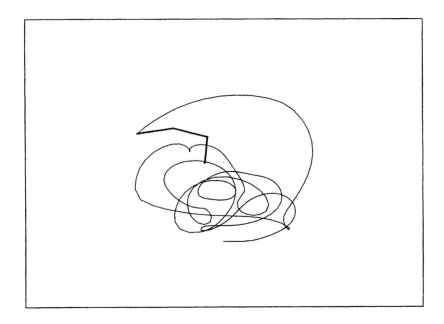

Figure 10.5: *Snapshot of 3-link pendulum animation*

As was done for the simulation, we introduce a new variable χ, and from the preceding equation we obtain the linear system

$$\dot{\widehat{\delta\chi}} = \left(\begin{array}{cc} \begin{array}{c|c} 0 & I \\ \hline -M_1^{-1}R_1 & 0 \end{array} & 0 \\ 0 & \begin{array}{c|c} 0 & 1 \\ \hline 0 & 0 \end{array} \end{array} \right) \delta\chi + \left(\begin{array}{c} 0 \\ -M_1^{-1}R_2 \\ 0 \\ 1 \end{array} \right) \delta\Gamma.$$

Again, the M_1, R_1, and R_2 matrices are transferred from Maple to Scilab:

```
maple2scilab('m1',M1,['J','l','m'],'f');
maple2scilab('r1',R1,['J','l','m'],'f');
maple2scilab('r2',R2,['J','l','m'],'f');
```

and in Scilab we compute the (A, B) pair of the linear system using the functions **m1**, **r1**, and **r2**, which were automatically generated by Maple:

```
mp=m1(J,l,m)
rp=r2(J,l,m);
cp=r1(J,l,m);
A=[0*mp,eye(mp); inv(mp)*(-cp),0*mp];
B=[0*rp;inv(mp)*(-rp)];
```

```
A=[A,0*ones(2*n,2);0*ones(2,2*n),[0,1;0,0]];
B=[B;0;1];
```

Now all the tools described in Chapter 6 can be used to compute a stabilizing feedback for the linearized pendulum system. Here, to simplify, we suppose that the whole state of the system is observed. Otherwise, it would be necessary to construct an observer.

```
-->contr(A,B)                              ← testing controllability
  ans  =
     8.
```

```
-->K=ppol(A,B,-[4*ones(1,2*n),1,1]);   ← computing a feedback gain
                                          (real closed-loop poles)
```

The closed-loop nonlinear system can now be described by the following function:

```
function [chidot]=cnpend(t,chi)
u=-K*(chi-chi_eq)                          ← feedback control
chidot=0*chi
chidot(2*n+1)=chi(2*n+2)
chidot(2*n+2)=u
mp=mprim(chi,J,1,m)
rp=rprim(chi,J,1,m,u)
cp=cprim(chi,J,1,m)
chidot(1:n)=chi(n+1:2*n)
chidot(n+1:2*n)=-mp\(cp*(chi(n+1:2*n).*chi(n+1:2*n))+rp)
```

Here `chi_eq` is the unstable equilibrium point: It is a global variable, which must be set in the Scilab environment.

The closed-loop system converges toward an equilibrium point only if the initial point of the system (initial position of the pendulum) is close to the unstable equilibrium point. For instance, we give an example of control for the 3-link pendulum with initial state as follows:
`chi_0=chi_eq+[%pi/20;0.01;0.01;0.01;0.01;0.01;0;0]`.
We solve the corresponding ODE as follows:

```
yt=ode(chi_0,0,times,cnpend);
```

Figure 10.6 displays the time evolution of the control (horizontal acceleration of the end point), together with the controlled end point and the angles of the pendulum.

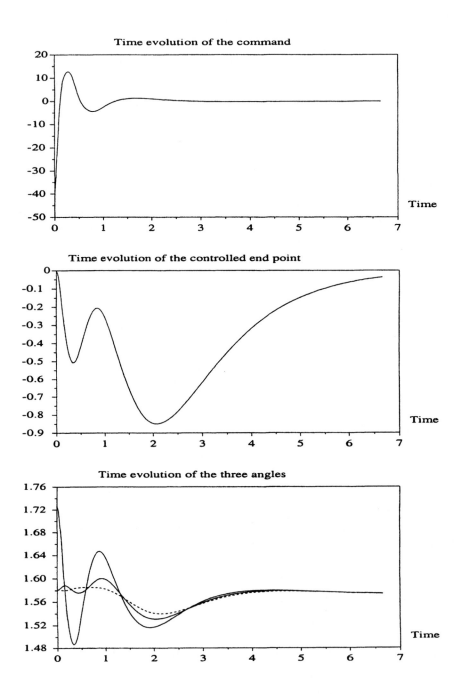

Figure 10.6: *Time evolution of the controlled pendulum*

Chapter 11

Graph and Network Toolbox: Metanet

Graphs are involved in many areas of computer science and automatic control. They are a powerful tool for modeling a number of mathematical problems, mainly in combinatorics. The graph data structure is also used in computer science for representing data structures and programs. Graphs are preferential tools for modeling complex structures and relations between objects; in particular, they can represent the links between information in a data base. Their applications are numerous in various domains, mainly in system modeling for economics or automatic control. In particular, graphs are useful for utility networks such as those for electricity, water transportation, and road networks. The types of problems that we find in these domains are simulation (pipe network problem) and optimization (capacity problems, minimum linear cost flow problems, and minimum quadratic cost flow problems).

Metanet is a Scilab toolbox for graph and network flow computations. It comes with a collection of Scilab functions and with a special graphics window for displaying and modifying graphs. It has been especially designed to handle large graphs (with thousands of edges) and to facilitate the rapid construction of programs for use on graphs.

Programming with graphs To work efficiently on graphs it is necessary to have data structures for manipulating them and for handling them as a single abstract object in a program. We must also have functions that can easily access the graph object and modify it. In working on graphs, it is important to be able to visualize them. It is possible to find graph visualization programs (such as GraphEd [34] or CabriGraph [7]), with libraries of Fortran, C, or C++ programs for graph computations. However, in such packages creation of new programs or the simple operation of putting colors and labels on a set of arcs or nodes requires the construction of C, C++, or Fortran code. Thus, the overall process is not user friendly. Few systems offer facilities to add new functionalities easily.

In Metanet the graph list is a new abstract Scilab object that represents a graph and that can be handled like other Scilab objects such as matrices and linear systems. The graph list can be passed as an argument to functions and be returned by them. Consequently, Scilab's interactive interpreter and high-level language can be used to *program with graphs and networks* for computing and/or testing algorithms.

In this chapter we describe the main functionalities of Metanet, illustrating the types of problems it solves. After we define the notion of graph, in the section that follows the emphasis of the discussion is on the types of objects Metanet can handle, with the corresponding various representations. Then we show how to create and load graphs in Metanet and how to generate graphs and networks automatically. Then after presenting the functions of the Metanet toolbox, at the end of the chapter, we will show two examples using Metanet.

11.1   What Is a Graph?

We define here the notion of graph and discuss which types of graphs can be handled by Metanet.

A graph $\mathcal{G} = (\mathcal{N}, \mathcal{A})$ (see Figure 11.1) is defined by:

- A nonempty set of n points $\mathcal{N}$, which are the nodes of the graph.

- A nonempty set of m arcs, or edges, $\mathcal{A}$, which is a set of pairs (i, j) with $i \in \mathcal{N}$ and $j \in \mathcal{N}$. An element a of $\mathcal{A}$ is called an arc if the pair (i, j) is directed, i.e., a is a directed link from tail node i to head node j. An element a of $\mathcal{A}$ is called an edge if the pair (i, j) is undirected, i.e., a is an undirected link between node i and node j. Note that an arc is the same as a directed edge.

In fact, with this definition we can have directed or undirected multigraphs (there can be more than one edge between two nodes) with loops (edge from a node to itself) allowed. We suppose also that the graph has at least one node and one edge. These are the types of graphs that can be handled by Metanet.

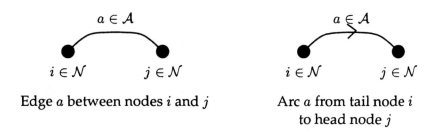

Edge a between nodes i and j Arc a from tail node i
to head node j

Figure 11.1: *Definition of edges and arcs*

11.2 Representation of Graphs

There are many ways to represent graphs. We have to choose a representation that is easy to use in Scilab and that is minimal with respect to the
size of the corresponding data structure. Indeed, if we want to handle large
graphs with thousands edges, the representation must be as compact as
possible and use the least amount of memory possible.

We present here the various ways to represent graphs in Metanet, showing the advantage of each one and how to use it.

11.2.1 Standard Tail/Head Representation

We describe here the default representation of graphs.

A minimal way to represent a graph is to give the number of nodes,
n, the list of the tail nodes, and the list of the head nodes. Each node has
a number, and each edge has a number. The numbers of nodes are consecutive, and the numbers of edges are consecutive. In Metanet, these lists of
tails and heads are represented by row vectors. So, if we call **tail** and **head**
these row vectors, the edge number **a** goes from node number **tail(a)**
to node number **head(a)**. The number of nodes is needed, since isolated
nodes (those unlinked to any other node) may exist. The size of the vectors
tail and **head** is equal to the number of edges, m, of the graph. This is the
standard representation of graphs in Metanet, and this representation uses
$2m$ double-precision floating-point numbers.

The above representation is very general and yields both directed and
undirected multigraphs (graphs with multiple arcs between two nodes)
possibly with loops and isolated nodes. This corresponds to the definition
given in Section 11.1.

The standard function to create graphs is **make_graph**. For instance,
we can create a small directed graph with a loop and an isolated node (see

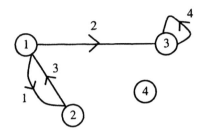

Figure 11.2: *Small directed graph*

Figure 11.2) by using node number = 4, tail = [1,1,2,3], head = [2,3,1,3], or in Scilab

```
g=make_graph('foo',1,4,[1 1 2 3],[2 3 1 3]);
```

The second argument, 1, means that the graph is directed.

The data structure that contains this standard representation of graphs in Metanet is a typed list (see Section 2.4.1) and is referred to as a *graph list*. As usual, the first element of a typed list is a vector of strings that defines its type, **'graph'**, and all the access functions to the other elements. The graph list has 33 elements (not counting the first one, defining the type). Only the first five elements must have a value in the list; all the others can be given the empty vector **[]** as a value, and then a default is used. These five required elements are:

name name of the graph (a string)

directed flag equal to 1 if the graph is directed or equal to 0 if the graph is undirected

node_number number of nodes

tail row vector of the tail node numbers

head row vector of the head node numbers

A graph must have at least one arc, so that **tail** and **head** are not empty. The other elements of the graph list are additional information on edges and nodes such as graphical data (coordinates, colors ...) and attributes (node demands, edge costs, edge capacities, ...).

You can access and/or modify the elements of a graph list by using the name of the elements. For instance, to get the value of the **node_number** element of graph **g**, you have only to type **g('node_number')**. To change this value to **m**, for instance, you have only to type **g('node_number')=m;**.

Remark In this chapter we will speak about edge and node numbers. These are in fact internal numbers. The internal numbers are automatically generated in a consecutive fashion when graphs are loaded or created. *Only internal node and edge numbers are used for computations.* This convention is for simplification. Each node and each edge has a *name*, which can be any string. Node names must be unique. There is a one-to-one correspondence between node names and internal node numbers. If no node name or no edge name is given, they are taken as the string of the corresponding internal number. Moreover, labels can be given to nodes and edges: They can be any string and are used for displaying information on nodes and edges in the Metanet window.

The functions to create graph lists are described in Section 11.3.

11.2.2 Other Representations

Other graph representations are better suited for some algorithms. For instance, we often need to obtain the list of edges linked to a given node or the list of nodes linked to a given node in order to go quickly through the graph. The tail/head representation is an inefficient representation for these types of operations. So, other representations, different from the standard tail/head representation, can be obtained in Metanet by means of functions to convert from one representation into another. They are described below.

Adjacency Lists

Representation by an adjacency list is often used by algorithms. If n is the number of nodes and m is the number of arcs of the graph, the adjacency list representation consists of three row vectors, the pointer array **lp** (size = $n + 1$), the node array **ls** (size = m), and the arc array **la** (size = m). If the graph is undirected, each edge corresponds to two arcs. This representation uses $n + 2m + 1$ double-precision floating-point numbers.

With this type of representation, it is easy to determine the successors of a node (see Figure 11.3):

- node number **i** has **lp(i+1)-lp(i)** successors

- the successors of node number **i** are nodes with numbers from **ls(lp(i))** to **ls(lp(i+1)-1)**

- the outgoing arcs from node number **i** are arcs with numbers from **la(lp(i))** to **la(lp(i+1)-1)**.

Usually, a program uses **lp**, **ls**, and **la** in the following loop:

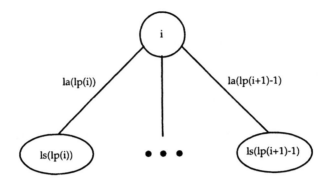

Figure 11.3: *Adjacency list representation of graphs*

for i=1 : <number of nodes>
        for j=lp(i) : lp(i+1)-1
                <work on arc la(j)>
                <work on node ls(j)>
        end
end

See the example of Section 11.3.3 to see how adjacency lists are used to make a depth-first search in a graph.

As an example, the adjacency list representation of the graph (vectors **lp, ls**, and **la**) of Figure 11.2 is given below:

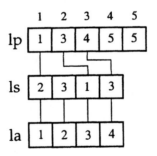

The function used to compute the adjacency list representation of a graph is **adj_lists**. We can easily check the adjacency list representation of the previous graph:

```
-->g=make_graph('foo',1,4,[1 1 2 3],[2 3 1 3]);

-->[lp,la,ls]=adj_lists(g)
 ls  =
```

```
!   2.     3.     1.     3. !
  la  =
!   1.     2.     3.     4. !
  lp  =
!   1.     3.     4.     5.     5. !
```

Node–Arc Incidence Matrix

For a directed graph, if n is the number of nodes and if m is the number of arcs of the graph, the node–arc incidence matrix of the graph is the $n \times m$ matrix A such that

- if $A(i, j) = +1$, then node i is the tail of arc j,

- if $A(i, j) = -1$, then node i is the head of arc i.

This matrix is often simply called the "incidence matrix."

If the graph is undirected and m is the number of edges, the node–arc incidence matrix A is also an $n \times m$ matrix such that

- if $A(i, j) = 1$, then node i is an end of edge j.

Note that with this type of representation it is impossible to have loops.

The problem of this representation is that it leads to very big matrices. For a graph with n nodes and m arcs, the node–arc incidence matrix uses $n \times m$ double-precision floating-point numbers compared with $n + 2m + 1$ for the adjacency list representation. So, this matrix is represented in Scilab as a sparse matrix (see Section 2.2.2).

`graph_2_mat` and **mat_2_graph** are the functions used to compute the node–arc incidence matrix of a graph and to return to a graph from a node–arc incidence matrix. For instance, you can find below the node–arc incidence matrix of the graph corresponding to Figure 11.2 with loop arc number 4 deleted and the matrix for the same undirected graph:

```
-->g=make_graph('foo',1,4,[1 1 2],[2 3 1]);
```

```
-->a=graph_2_mat(g);                          ← a is sparse
```

```
--> full(a)                    ← use full to have a beautiful display
  ans  =
!    1.     1.   - 1. !
! - 1.     0.     1. !
!    0.   - 1.     0. !
!    0.     0.     0. !
```

```
-->g('directed')=0; a=graph_2_mat(g); full(a)
 ans   =
!   1.      1.      1. !
!   1.      0.      1. !
!   0.      1.      0. !
!   0.      0.      0. !
```

One advantage of this representation is that it gives a correspondence between a graph and a matrix. For instance, you can apply transformations to the node–arc incidence matrix and see the result on the graph, or conversely, you can see the effect of a graph algorithm on the node–arc incidence matrix. Another advantage of the node–arc incidence matrix is that it appears naturally in flow problems (see Section 11.5.6). If it is not used very often for flow computations, it can be used to see the links between linear cost flow problems and linear programming (see Section 11.5.6), at least for educational purposes.

Node–Node Incidence Matrix

Another interesting representation is the $n \times n$ node–node incidence matrix A of the graph, where $A(i,j) = 1$ if there is one arc from node i to node j. This is another way to link graphs and matrices, but only 1 to 1 graphs (no more than one arc from one node to another) can be represented. This matrix is also known as the "adjacency matrix."

The same functions used to compute the node–arc incidence matrix (see above) of a graph are used to compute the node–node incidence matrix: **graph_2_mat** and **mat_2_graph**. To specify that we are working with the node–node incidence matrix, the flag **'node-node'** must be given as the last argument of these functions.

The node–node incidence matrix uses n^2 double-precision floating-point numbers, and in the same way as for arc–arc incidence matrix, it leads to very big matrices. So, the node–node incidence matrix is also represented as a Scilab sparse matrix.

For instance, you can find below the node–node incidence matrix of the graph corresponding to Figure 11.2 and the matrix for the same undirected graph:

```
-->g=make_graph('foo',1,4,[1 1 2 3],[2 3 1 3]);
```

```
-->a=graph_2_mat(g,'node-node');                    ← a is sparse
```

```
-->full(a)                          ← use full  to have a beautiful display
 ans   =
```

```
!   0.    1.    1.    0. !
!   1.    0.    0.    0. !
!   0.    0.    1.    0. !
!   0.    0.    0.    0. !

-->g('directed')=0; a=graph_2_mat(g,'node-node'); full(a)
 ans  =
!   0.    1.    1.    0. !
!   1.    0.    0.    0. !
!   1.    0.    1.    0. !
!   0.    0.    0.    0. !
```

As an example, we have applied the **bandwr** Metanet function (see Section 11.5.8) to the node–node incidence matrix of a graph. This function reduces the bandwidth of a matrix by renumbering the elements of the matrix:

```
g=load_graph('toto');

show_graph(g);

a=graph_2_mat(g,'node-node');

iperm=bandwr(tril(a)'+eye);                    ← get an upper diagonal matrix

g('node_name')=string(iperm);

show_graph(g,'new');
```

After loading the graph **'toto'** shown in a Metanet window (see Section 11.3.3) in Figure 11.4, we have computed the node–node incidence matrix **a**. This matrix is symmetric, with zeros on the main diagonal. To use the **bandwr** function, we need to transform this matrix into an upper triangular matrix with a full main diagonal. Then **bandwr** gives the **iperm** permutation, which changes the numbering of the elements of the matrix. We have to give this permutation only for the new numbers of the nodes and display the new graph in another Metanet window, shown in Figure 11.5. We can see now in this new graph that linked nodes have numbers closer than before.

Chained Lists

The chained list representation is used by some algorithms. If n is the number of nodes of the graph and m is the number of arcs of the graph, this

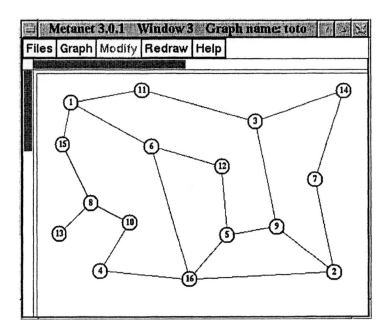

Figure 11.4: *Graph* `toto`

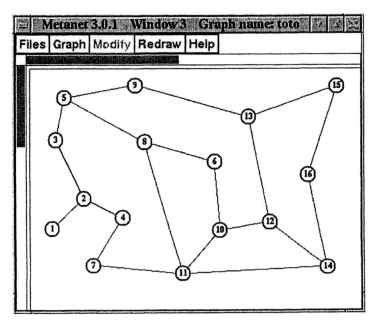

Figure 11.5: *Graph* `toto` *after bandwidth reduction of its node–node incidence matrix*

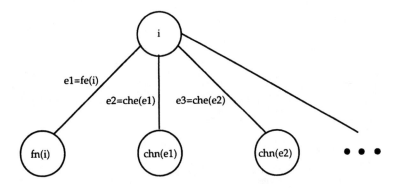

Figure 11.6: *Chained list representation of graphs*

representation uses four vectors, **fe** (size = n), **che** (size = m), **fn** (size = n), and **chn** (size = m), which are described below:

- **e1=fe(i)** is the number of the first edge starting from node **i**.

- **e2=che(e1)** is the number of the second edge starting from node **i**, **e3=che(e2)** is the number of the third edge starting from node **i**, and so on until the value is 0.

- **fn(i)** is the number of the first node reached from node i.

- **chn(i)** is the number of the node reached by edge **che(i)**.

All this can be more clearly seen in Figure 11.6. The chained list representation uses $2(n + m)$ double-precision floating-point numbers.

You can use the **chain_struct** function to obtain the chained list representation of a graph from the adjacency lists representation (see above).

Usually, a program uses **fe, fn, che,** and **chn** in the following loop:

```
for  i=1 : <number of nodes>
     j=fe(i)
     k=fn(i)
     while (j > 0)
             <work on arc j>
             <work on node k>
             k=chn(j)
             j=che(j)
     end
end
```

11.2.3 Graphs and Sparse Matrices

We have seen that it is possible to obtain the sparse node–node incidence matrix from a 1-to-1 graph and conversely. In fact, the link is closer. The way to enter sparse matrices in Scilab is by using vectors like the tail and head vectors of a graph. For instance, to obtain the node–node incidence matrix of a 1-to-1 directed graph, it is sufficient to execute

```
n=g('node_number');
m=arc_number(g);
ta=g('tail'); he=g('head');
a=sparse(ij=[ta' he'],ones(m,1),[n n])
```

11.3 Creating and Loading Graphs

Metanet comes with a graphical window for displaying and editing graphs, called the Metanet window. There are two ways for handling graphs in Metanet: by using the Metanet window or directly via Scilab functions. The Metanet window is another program, which runs independently of Scilab and which is launched automatically when needed. This section shows how to use Scilab to create and load graphs and how to use Metanet to display them.

11.3.1 Creating Graphs

The standard function for making a graph list is **make_graph**. The first argument is the name of the graph; the second argument is a flag, which can be 1 (directed graph) or 0 (undirected graph); the third argument is the number of nodes of the graph, and the last two arguments are the tail and head vectors of the graph.

We have already seen that the graph named **'foo'** in Figure 11.2 can be created by the command

```
g=make_graph('foo',1,4,[1 1 2 3],[2 3 1 3]);
```

As an example, the simplest graph we can create in Metanet is

```
g=make_graph('min',1,1,[1],[1]);
```

It is directed, has one node and one loop arc on this node, and can be seen in Figure 11.7.

The following graph, shown in Figure 11.8, is the same as the graph in Figure 11.2, but it is undirected:

Figure 11.7: *Smallest directed graph*

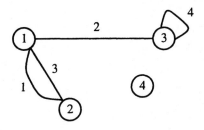

Figure 11.8: *Small undirected graph*

```
g=make_graph('ufoo',0,4,[1 1 2 3],[2 3 1 3]);
```

You can also give 0 as the third argument of **make_graph** (number of nodes). This means that **make_graph** will itself compute from its last arguments (the tail and head vectors) the number of nodes of the graph. Consequently, this graph has no isolated nodes, and the node numbers are taken from the numbers in the tail and head vectors. For instance, if you enter

```
g=make_graph('foo1',1,0,[1 1 2 3],[2 3 1 3]);
```

then the graph (shown in Figure 11.9) has three nodes with numbers 1, 2, and 3, no isolated nodes, and four arcs. Note the difference with the graph of Figure 11.2.

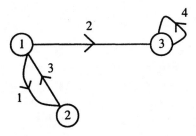

Figure 11.9: *Directed graph*

The other elements of the graph list can be entered by using the names of the elements. For instance, to give the graph `'foo'` coordinates for the nodes, you can enter

```
g=make_graph('foo',1,4,[1 1 2 3],[2 3 1 3]);
g('node_x')=[42 108 176 162];
g('node_y')=[36 134  36  93];
```

Another simple example: If you want to transform the directed graph **g** into an undirected graph, you have only to execute `g('directed')=0;`.

11.3.2 Loading and Saving Graphs

Graphs are saved in ASCII files, called *graph files*. A graph file has the extension `.graph`. It is possible to create a graph file with a text editor and to load it into Scilab, but it is a very cumbersome job. Programs are available to generate graphs (see Section 11.4).

To load a graph into Scilab, use the `load_graph` function. Its argument is the absolute or relative pathname of the graph file; if the `.graph` extension is missing, it is assumed. The command `load_graph` returns the corresponding graph list. For instance, to load the graph **foo**, which is in the current directory, and put the corresponding graph list in the Scilab variable **g**, execute `g=load_graph('foo');` or `g=load_graph('foo.graph');`. To load the graph **mesh100** given in the Scilab distribution, execute `g=load_graph(SCI+'/demos/metanet/mesh100.graph');`.

To save a graph, use the `save_graph` function. Its first argument is the graph list, and its second argument is the name or the pathname of the graph file; if the `.graph` extension is missing, it is assumed. If the path is the name of a directory, the name of the graph is used as the name of the file. For instance, the following command saves the graph **g** into the graph file **foo.graph**: `save_graph(g,'foo.graph')`.

11.3.3 Using the Metanet Window

Metanet windows can be used to visualize graphs. They are a powerful tool for creating and modifying graphs. You can have as many Metanet windows open simultaneously as you want. Each Metanet window is a separate program: The communication between Scilab and the Metanet windows is managed with Scilab's communication toolbox, called GeCI [51].

An important point is that *there is no link between the graph displayed in the Metanet window and the graphs loaded into Scilab.* So, when you have created or modified a graph in the Metanet window, you have to save it as a graph file (see Section 11.3.2) and load it again in Scilab. Conversely, when you have modified a graph in Scilab, you have to display it again in the Metanet window by using the **save_graph** function (see Section 11.3.2). The philosophy is that computations are made only in Scilab, and the Metanet window is used only to display, create, or modify graphs. So, you can use the Metanet toolbox without using a Metanet window.

Another way to see a graph is to plot it in a Scilab graphical window with the **plot_graph** function, but there is no possibility, at the present time, to modify the displayed graph using this method.

The standard way of using the Metanet window is from Scilab. Indeed, the Metanet window is opened only when needed as a new process. Since many Metanet windows can be opened at the same time, each Metanet window is labeled by an integer beginning from 1. One of these windows is the *current Metanet window*. The current Metanet window is where graphs are displayed by default.

The **metanet** function opens a new Metanet window and returns its number. A path can be given as an optional argument: It is the directory where graph files are searched; by default, graph files are searched in the working directory. The **metanet** function is mainly used when we want to create or modify a new graph.

The way Metanet windows are used to create and modify graphs is not explained here. Metanet windows are self documented and function with pull-down menus and the mouse in a user-friendly way. The Metanet window allows you to edit and/or create graphs. From the Metanet window you can also load and save graphs.

We describe below the functions used in conjunction with the Metanet window.

Displaying a Graph

The first thing we would like to do is to see the graph we are working with. This can be accomplished by using the **show_graph** function:

```
show_graph(g)
```

It displays the graph **g** in the current Metanet window. If there is no current Metanet window, a new Metanet window is created, and it becomes the current Metanet window. If there is already a graph displayed in the current Metanet window, the new graph is displayed instead. The number of the current Metanet window, where the graph is displayed, is returned by **show_graph**.

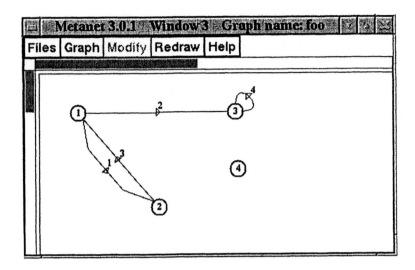

Figure 11.10: *Directed graph in Metanet window*

Two optional arguments can be given to **show_graph** after the graph list argument. If an optional argument is equal to the string **'new'**, a new Metanet window is created. If an optional argument is a positive number, it is the value of the scale factor for drawing the graph. For instance, **show_graph(g,'new',2)** displays the graph **g** in a new Metanet window with the scale factor equal to 2.

The following Scilab session creates the graph **foo** of Figure 11.2 and displays it in a Metanet window, shown in Figure 11.10. Note that we have given coordinates to nodes in order to have a nice layout of the graph.

```
-->g=make_graph('foo',1,4,[1 1 2 3],[2 3 1 3]);

-->g('node_x')=[40 145 245 248];                ←   x coordinates

-->g('node_y')=[40 160 39 112];                 ←   y coordinates

-->show_graph(g)
  ans  =
     1.
```

Showing Arcs and Nodes

In Metanet, a set of arcs in a graph is nothing but a row vector of edge numbers, and a set of nodes is a row vector of node numbers. So, a path is

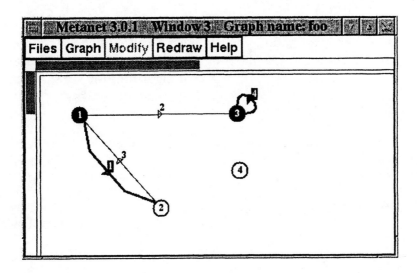

Figure 11.11: *Highlighted arcs and nodes in Metanet window*

also represented by a row vector of edge numbers. A very useful thing to do is to distinguish a set of nodes and/or a set of arcs in the displayed graph. This is done by highlighting nodes and/or arcs using the **show_arcs** and **show_nodes** functions.

The arguments of the **show_arcs** and **show_nodes** functions are respectively a row vector of arc numbers (or edge numbers if the graph is undirected) and a row vector of node numbers. These sets of arcs and nodes are highlighted in the current Metanet window. Note that the corresponding graph must be displayed in this window; otherwise, the numbers might not correspond to appropriate arc or node numbers (see below for changing the current Metanet window). By default, using one of these functions switches off any previous highlighting. If you want to keep previous highlighting, use the optional argument **'sup'**.

For instance, after displaying the graph **foo** shown in Figure 11.10, the following session highlights 2 arcs and 2 nodes:

```
show_arcs([1 4]); show_nodes([1 3],'sup');
```

The result is shown in Figure 11.11.

Note that another way to distinguish between arcs and nodes in a displayed graph is to give them colors. For that you have to use the elements **edge_color** and **node_color** of the graph list. But you have to modify the graph list of the graph and use **show_graph** again to display the graph with new colors.

By using the **show_arcs** and **show_nodes** functions, or by changing the colors of edges and nodes, you can even do animations in the Metanet window, for instance, to show the progress of an algorithm. As an example, we are going to write a depth-first search algorithm for a graph and follow the progress of the search by highlighting all the nodes reached by the algorithm. This can be done by the following code in Scilab:

```
g=load_graph('tree'); show_graph(g);   ← loading and displaying graph
n=g('node_number');
[lp,la,ls]=adj_lists(g);                ← compute adjacency lists
pw=zeros(1,n);
for i=1:n, nn(i)=lp(i+1)-lp(i); end;    ← nn(i) = number of
                                           nodes following i
pw(1)=1;
i=1; show_nodes(i);                     ← highlight root node
while %T
  while nn(i)<>0                        ← depth-first search loop
    ll=lp(i)+nn(i)-1;
    j=ls(ll);
    nn(i)=nn(i)-1;
    if pw(j)==0 then
      pw(j)=i;
      i=j; show_nodes(i);              ← highlight node
    end
  end
  if i==1 then break, end
  i=pw(i); show_nodes(i);              ← highlight node
end
```

The algorithm is the classical Tarjan algorithm in a nonrecursive form. We use here the adjacency lists (see Section 11.2.2) because it is the best way to obtain the list of nodes linked to a given node. Every time we reach a node i, we use the **show_nodes(i)** command to highlight it. Here **'tree'** is the name of a tree graph on which we try the algorithm.

Managing Metanet Windows

The **netwindow** function is used to change the current Metanet window. For instance, **netwindow(2)** chooses Metanet window number 2 as the current one.

The **netwindows** function returns a list. Its first element is the row vector of all the Metanet window numbers, and the second element is the number of the current Metanet window. This number is 0 if no current Metanet window exists.

In the following example, there are two Metanet windows with numbers 1 and 3, and window number 3, is the current Metanet window:

```
-->netwindows()
 ans =
       ans(1)

!    1.    3. !

       ans(2)

   3.
```

11.4 Generating Graphs and Networks

In working with graphs and particularly with networks, it is very useful to generate them automatically. This section shows how to generate such graphs in Metanet.

The function **gen_net** can be used in Metanet to generate graphs. It uses a triangulation method to generate a planar connected graph. But it also uses user information to give edges reasonable values for costs and maximal capacities in order to have a working network on which to test algorithms.

As an example, we generate a network with 10 nodes. To do this we must give **gen_net** a 12-vector. The vector entries have the following significance and values:

1 Number for initialization of random generation: 6
   each integer number gives a different graph
2 Number of nodes: 10
3 Number of sources: 1
4 Number of sinks: 1
5 Minimum cost: 0
6 Maximum cost: 100
7 Input supply: 100
8 Output supply: 100
9 Minimum capacity: 10
10 Maximum capacity: 50
11 Percentage of edges with costs: 100
12 Percentage of edges with capacities: 100

The preceding values correspond to the following calling sequence:

```
-->v=[6,10,1,1,0,100,100,100,10,50,100,100];

-->g=gen_net('net',1,v);                              ← 1 means directed

-->arc_number(g)
 ans  =
    22.

-->g('edge_max_cap')
 ans  =
        column 1 to 7

!   49.    42.    19.    24.    21.    20.    40. !

        column  8 to 14

!   32.    31.    44.    41.    33.    29.    13. !

        column 15 to 21

!   49.    26.    38.    45.    41.    41.    20. !

        column 22

!   36. !

-->g('edge_cost')
 ans  =
        column 1 to 8

!   10.    2.    33.    94.    79.    59.    48.    70. !

        column  9 to 15

!   14.    37.    93.    79.    59.    48.    70. !

        column 16 to 22

!   14.    37.    93.    79.    59.    48.    9. !
```

The generated network has 22 arcs, and the values of maximum capacities and costs are as indicated above. The graph is shown in Figure 11.12.

If **gen_net** is called without arguments (i.e., **gen_net();**), a window appears and prompts the user to enter the appropriate values interactively.

Special nodes are distinguished as source nodes and sink nodes. Source nodes are drawn with a large arrow going into the node, and sink nodes

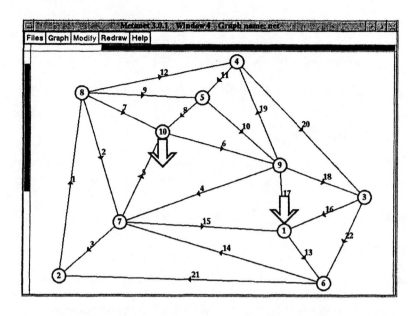

Figure 11.12: *Generated network*

are drawn with a large arrow going out of the node. Source nodes and sink nodes correspond respectively to entry and output nodes of the flow of the network (see Section 11.5.6). Such nodes exist mainly for convenience and are drawn differently in the Metanet window. Source nodes have the smallest numbers, and sink nodes have the biggest numbers. In the preceding generated graph, there is one source (number 1) and one sink (number 10).

In the following, the generated network of Figure 11.12 will be used for demonstrating the use of some flow functions (see Section 11.5.6).

11.5 Graph and Network Computations

Most functions of the Metanet toolbox are used to make computations on graphs and networks. We are not going to describe each function. However, we do give an overview of the most important functions in the following sections.

11.5.1 Getting Information About Graphs

The following functions are used to get simple values not included in the graph list:

- **arc_number**: returns the number of arcs of a graph

- **edge_number**: returns the number of edges of a graph

- **node_number**: returns the number of nodes of a graph

11.5.2 Paths and Nodes

The following functions make a link between a set of nodes and a path of arcs. They also compute information about how the nodes are linked.

- **find_path**: finds a path between two nodes

- **neighbors**: gets the nodes connected to a node

- **nodes_degrees**: computes the degrees of the nodes of a graph

- **nodes_2_path**: gets the path of arcs from a set of nodes

- **path_2_nodes**: gets the set of nodes included in a path of arcs

- **predecessors**: gets the tail nodes of incoming arcs of a node

- **successors**: gets the head nodes of outgoing arcs of a node

11.5.3 Modifying Graphs

The following functions modify existing graphs. The same operations can be performed directly in the Metanet window, but these functions are useful for programming.

- **add_edge**: adds an edge or an arc between two nodes

- **add_node**: adds an isolated node to a graph

- **contract_edge**: contracts edges between two nodes

- **delete_arcs**: deletes all the arcs or edges between a set of nodes

- **delete_nodes**: deletes nodes

- **split_edge**: splits an edge by inserting a node in it

- **supernode**: replaces a group of nodes with a single node

11.5.4 Creating New Graphs From Old Ones

The following functions make standard computations on one or two graphs to obtain another graph.

- **arc_graph**: computes the graph with nodes corresponding to the arcs of another graph

- **graph_complement**: computes the complement of a graph

- **graph_power**: computes the kth power of a directed 1-graph

- **graph_simp**: converts a graph to a simple undirected graph

- **graph_sum**: gets the sum of two graphs

- **graph_union**: gets the union of two graphs

- **line_graph**: computes the graph with nodes corresponding to the edges of another graph

- **subgraph**: gets the subgraph of a graph

11.5.5 Graph Problem Solving

The following functions execute the main standard algorithms for graph problems (see, for instance, [32]). All these algorithms are implemented in Fortran code and are suited for handling large graphs, i.e., with thousands of arcs and nodes.

articul: finds one or more articulation points in a graph

best_match: computes the maximum matching in a graph

circuit: finds a circuit or the rank function in a directed graph

con_nodes: gets the set of nodes of a connected component

connex: computes the connected components

cycle_basis: computes the basis of cycles of a simple undirected graph

girth: computes the Girth number of a directed graph

graph_center: computes the center of a graph

graph_diameter: computes the diameter of a graph

hamilton: computes a Hamiltonian circuit of a graph

is_connex: checks the connectivity of a graph

max_clique: computes the maximum clique of a graph

min_weight_tree: computes a minimum-weight spanning tree in a graph

perfect_match: finds min-cost perfect matching in a graph

shortest_path: computes the shortest path between two nodes

strong_con_nodes: gets the set of nodes of a strong connected component

strong_connex: computes the strong connected components

trans_closure: computes the transitive closure of a graph

11.5.6   Network Flows

There are Scilab functions that make computations on utility networks flows. The problems they solve are very important optimization problems in many engineering domains, so we will describe all of them with examples.

Capacities and Flows

A utility network is a graph with arc and node attributes that satisfy certain properties.

We consider a connected graph $\mathcal{G} = (\mathcal{N}, \mathcal{A})$, where $\mathcal{N}$ and $\mathcal{A}$ are respectively the sets of nodes and arcs (see Section 11.1). There are n nodes and m arcs. Two numbers are associated with each arc (i, j) (from node i to node j): $l_{i,j}$ and $u_{i,j}$, which are respectively the lower and upper capacities of the arc. Usually, they are positive numbers, and lower capacities are often equal to zero.

Metanet can compute the maximum-capacity path from node i to node j. Between all possible paths from node i to node j, it is the path composed by the set of arcs A for which $\min u_{i,j}, (i, j) \in A$ is maximum. Only maximum capacities are taken into account and are supposed to be positive.

The **max_cap_path** function is used to compute this path. We have applied it to the network 'net' generated in Section 11.4 to find a path with maximum capacity from source node number 1 to sink node number 10 (see Figure 11.13):

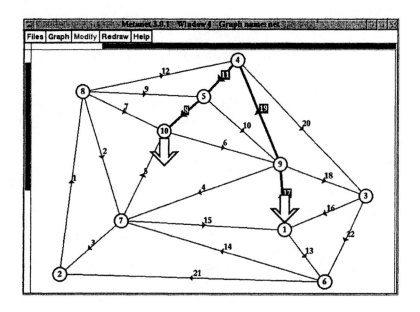

Figure 11.13: *Maximum-capacity path from source node to sink node*

```
-->[p,cap]=max_cap_path(1,10,g)
 cap  =
    32.
 p  =
!  17.    19.    11.    8. !

-->show_arcs(p)
```

The **max_cap_path** function returns the path that we have highlighted in Figure 11.13 and the value of the capacity of the path.

We can check the values of the maximum capacities of the arcs of the path (which must be greater than or equal to the value 32 computed above):

```
-->g('edge_max_cap')(p)
 ans  =
!  38.    41.    41.    32. !
```

A real number b_i is associated with each node i, which can be positive (supply, source node), negative (demand, sink node), or zero (pure transshipment node). Note that the element **'node_demand'** of the graph list is the opposite of b (supply = − demand).

If we suppose that the graph is directed, a flow on $\mathcal{G}$ is a vector x having a value on each arc and satisfying Kirchhoff's first law:

$$\sum_{j:(i,j)\in\mathcal{A}} x_{i,j} - \sum_{j:(j,i)\in\mathcal{A}} x_{j,i} = b_i \qquad \forall i \in \mathcal{N}.$$

If A is the $n \times m$ node–arc incidence matrix of the graph (see Section 11.2.2), if x denotes the m column vector of flows (indexed by arc numbers), and if b denotes the n column vector of elements b_i, then Kirchhoff's first law is just the matrix equality $Ax = b$.

Metanet can compute the maximum flow between two nodes i and j: Between all possible paths between i and j there is a path for which the flow on this path is maximum. This is defined to be the maximum flow. All the capacities and flows are supposed to be integers (i.e., if they are not integers, they are rounded).

The **max_flow** function is used to compute the path of the maximum flow. We have applied it to the network **'net'** generated in Section 11.4 to compute the maximum flow from source node number 1 to sink node number 10:

```
-->[v,phi,flag]=max_flow(1,10,g)
 flag  =
    1.
 phi  =
          column 1 to 8

!  16.     0.     0.     8.    21.     0.    16.    30. !

          column  9 to 16

!   0.     0.    30.     0.    29.    13.     0.     0. !

          column 17 to 22

!  38.     0.    30.     0.    16.     0. !
  v   =
    67.

-->p=find(phi<>0)
 p  =
          column 1 to 8

!   1.     4.     5.     7.     8.    11.    13.    14. !

          column  9 to 11
```

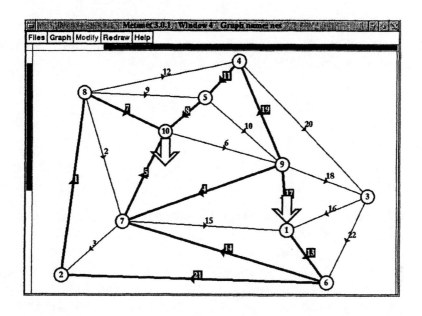

Figure 11.14: *Maximum flow from source node to sink node*

```
!   17.    19.     21.  !

-->g('edge_label')=string(phi);

-->g('edge_label')(find(phi==0))=" ";

-->show_graph(g);show_arcs(p)
```

The **max_flow** function returns a flag equal to 1 (which means that the problem is feasible), the vector of the flow on the arcs, and the value of the maximum flow. We have found the path **p** where the flow is not equal to zero, and we have highlighted it. Moreover, we have given these arcs the value of the flow as labels. All this is displayed in the Metanet window shown in Figure 11.14.

We can check that the maximum-capacity path shown in Figure 11.13 (arc numbers 17, 19, 11, and 8) is used by the maximum flow path **p** computed above.

Linear Cost Network Flow Problem

If we associate a cost $c_{i,j}$ with each arc, we can solve minimum cost flow problems.

The general minimum linear cost network flow problem consists in finding a flow that minimizes the linear cost on the arcs and that satisfies the capacity constraints. We call it problem $(\mathcal{LNF})$:

$$(\mathcal{LNF}) \quad \begin{cases} \min \sum_{(i,j)\in\mathcal{A}} c_{i,j} x_{i,j}, \\ \sum_{j:(i,j)\in\mathcal{A}} x_{i,j} - \sum_{j:(j,i)\in\mathcal{A}} x_{j,i} = b_i \quad \forall i \in \mathcal{N}, \\ l_{i,j} \le x_{i,j} \le u_{i,j} \quad \forall (i,j) \in \mathcal{A}. \end{cases}$$

Note that problem $(\mathcal{LNF})$ can have a solution only if $\sum b_i = 0$. When this condition is satisfied, we say that the problem is feasible.

If c denotes the m-dimensional column vector of costs (indexed by arc numbers) and l and u denote respectively the m-dimensional column vectors of lower and upper capacities (indexed by arc numbers), A, x, and b having the same meaning as for the Kirchhoff law above, the minimum linear cost network flow problem can be expressed as the following linear programming problem $(\mathcal{LP})$:

$$(\mathcal{LP}) \quad \begin{cases} \min c^T x & c \in \mathbb{R}^m, \quad x \in \mathbb{R}^m, \\ Ax = b & A \in \mathbb{R}^{n\times m}, \quad b \in \mathbb{R}^n, \\ l \le x \le u & l \in \mathbb{R}^m, \quad u \in \mathbb{R}^m. \end{cases}$$

Metanet can solve variations of the $(\mathcal{LNF})$problem described below. Metanet can solve a problem we refer to as $(\mathcal{LNF}_1)$. This is problem $(\mathcal{LNF})$ where the capacities are nonnegative integer numbers, the costs must be nonnegative real numbers, and the vector b is equal to 0 (i.e., there are only pure transshipment nodes). Metanet uses the out-of-kilter algorithm to solve $(\mathcal{LNF}_1)$. The **min_lcost_flow1** function is used to solve problem $(\mathcal{LNF}_1)$. Note that if the minimum capacities are equal to 0, the problem is trivial and the optimal flow is also equal to 0.

Metanet can solve a problem we refer to as $(\mathcal{LNF}_{1c})$. This is problem $(\mathcal{LNF}_1)$ with a constrained value of the flow from one node to another node. Metanet uses the algorithm of Busacker and Goven to solve $(\mathcal{LNF}_{1c})$. The **min_lcost_cflow** function is used to solve problem $(\mathcal{LNF}_{1c})$.

Metanet can solve a problem we refer to as $(\mathcal{LNF}_2)$. This is problem $(\mathcal{LNF})$ where the minimum capacities are equal to zero and the costs and the demands must be nonnegative integers. Metanet uses a relaxation algorithm due to D. Bertsekas [12] to solve $(\mathcal{LNF}_2)$. To solve problem $(\mathcal{LNF}_2)$ the **min_lcost_flow2** function is used. We can apply it to the **'net'** graph again. We need to give feasible values for the supply vector b. We consider that all nodes but source node 1 and sink node 10 are pure transshipment nodes (demand equal to 0). We have seen above that the maximum flow from source node to sink node is equal to 67, so the demand and supply of these nodes must be less than or equal to this value.

```
-->g('node_demand')=[-67 0 0 0 0 0 0 0 0 67];

-->[c,phi,flag]=min_lcost_flow2(g)
 flag  =
     1.
 phi  =
          column 1 to 8

!   36.    0.    16.    24.    21.    0.    36.    10. !

          column  9 to 16

!   0.    0.    10.    0.    29.    13.    0.    0. !

          column 17 to 22

!   38.    4.    10.    0.    20.    4. !
 c  =
    14060.

-->p=find(phi<>0);

-->g('edge_label')=string(phi);

-->g('edge_label')(find(phi==0))=" ";

-->show_graph(g);show_arcs(p)
```

We can see that the problem is feasible (**flag** is equal to 1), and the value of the cost is 14,060. We do the same as for the maximum flow problem and show the graph with the value of the flow as labels on the arcs in Figure 11.15.

As an educational example and to show the interest of the Metanet toolbox in Scilab, we can try to solve the minimum linear cost problem ($\mathcal{LNF}$) as the linear programming problem ($\mathcal{LP}$) by using the **linpro** function:

```
-->p=g('edge_cost')';

-->C=full(graph_2_mat(g));              ← linpro needs a full matrix
```

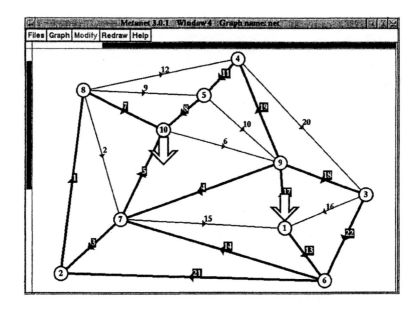

Figure 11.15: *Maximum flow from source node to sink node*

```
-->b=-g('node_demand')';

-->mi=g('node_number');

-->ci=g('edge_min_cap')'; cs=g('edge_max_cap')';

-->[x,lagr,f]=linpro(p,C,b,ci,cs,mi);

-->f
 f  =
    14060.

-->max(abs(phi-x'))
 ans  =
    9.415D-14
```

We can see that the minimum cost has the same value 14,060 and that the difference between the flow vectors is nearly zero (about 10^{-13}). But usually, solving directly the linear programming problem for large problems is less efficient.

Quadratic Cost Network Flow Problem

All the flow problems in the previous section are linear. We can consider a minimum quadratic cost network flow problem. We call it problem $(\mathcal{QNF})$:

$$
(\mathcal{QNF}) \quad
\begin{cases}
\min \sum_{(i,j)\in A} \frac{1}{2} w_{i,j}[(x_{i,j} - \bar{x}_{i,j})^2], \\[2mm]
\sum_{j:(i,j)\in A} x_{i,j} - \sum_{j:(j,i)\in A} x_{j,i} = b_i \quad \forall i \in \mathcal{N}, \\[2mm]
l_{i,j} \le x_{i,j} \le u_{i,j} \quad \forall (i,j) \in \mathcal{A},
\end{cases}
$$

where $w_{i,j}$ and $\bar{x}_{i,j}$ are given.

Metanet can solve a problem we refer to as $(\mathcal{QNF}_1)$. This is problem $(\mathcal{QNF})$ where the vector b is equal to 0, i.e., there are only pure transshipment nodes. Metanet uses an algorithm due to M. Minoux [32] to solve $(\mathcal{QNF}_1)$. The `min_qcost_flow` function is used to solve problem $(\mathcal{QNF}_1)$.

As for the minimum linear cost problem above, the minimum quadratic cost problem can be solved using standard quadratic optimization programs. An example is shown in Section 8.4.1.

11.5.7 The Pipe Network Problem

With the notations introduced above, the pipe network problem $(\mathcal{PN})$ consists in finding the flow x on the arcs and the potential π at the nodes satisfying the following matrix equations:

$$
(\mathcal{PN}) \quad Ax = b, \qquad A^T \pi = -Rx, \qquad \sum_{i=1}^{n} b_i = 0,
$$

where A is the $n \times m$ node–arc incidence matrix, b is the supply at the nodes, and R is the $m \times m$ diagonal matrix of the resistances of the arcs. We have seen that the first equation is the first Kirchhoff law. The second equation is the second Kirchhoff law. The last equation is necessary for the system to be feasible (supply equals demand).

We can transform the equations in the following way. The rank of A is $n - 1$ (indeed, each $(n - 1) \times (n - 1)$ submatrix of A has its determinant equal to 1 or -1). Thus, A', the A matrix with its last row deleted, has full rank. We call b' and π' the corresponding $n-1$ vectors. In fact, the potentials are defined up to an additive constant because only the differences $\pi_i - \pi_j$ appear in the vector $A^T \pi$; so we can choose $\pi_n = 0$. Then we can eliminate the flow from system $(\mathcal{PN})$ which yields the following matrix equation for the potentials:

$$\left(A'R^{-1}A'^{T}\right)\pi' = -b'.$$

It can be shown that the matrix of the preceding system is sparse, symmetric, and nonsingular. This system can be solved using Scilab's sparse matrix solver, which means that Metanet can be used to solve the pipe network problem in a computationally efficient manner. This is another example where a simple function that uses Metanet graph objects can solve a complete network problem. The **pipe_network** function is used to solve the pipe network problem.

11.5.8   Other Computations

The following functions do not perform computations directly on graphs and networks but are strongly related to graph and network problems:

bandwr: does a bandwidth reduction of a sparse matrix by renumbering the elements of the matrix

convex_hull: computes the convex hull of a set of points in the plane

knapsack: solves a 0-1 multiple knapsack problem

mesh2d: computes a triangulation of n points in the plane

qassign: solves a quadratic assignment problem

salesman: solves the traveling salesman problem

11.6   Examples Using Metanet

In this section we will show two examples using Metanet. The first example is the Scilab demo of Metanet and consists in finding an optimal path in a network using the direct functionalities of Metanet. The second example is a complete and realistic study of the optimization of a network of automated cars for transporting people around a town.

11.6.1 Routing in the Paris Metro

We briefly present the simple example of the Metanet demo, as found by clicking on the demo button in the main Scilab panel. This demo is based on the Paris metro. The network is limited to the stations located within the city limits of Paris proper as well as some stations located in the nearby suburbs (stations that constitute the termini of lines).

The underlying graph is directed, composed of 388 nodes and 948 arcs. There are 13 bidirectional metro lines (with some lines ending with a fork). A line is a succession of stations that can be visited without changing trains; this description corresponds to the natural way one uses the metro maps of the city and to indicate the directions between 2 locations.

The graph is constructed using the 283 real Paris metro stations; 220 of these stations belong to only a single line and are thus just simple nodes of the graph. But for each station that permits transfers between lines we have a multinode system: In fact, such a node corresponds to an ensemble of subnodes that correspond to the number of interconnected lines and are connected by special arcs representing the travel on foot needed to change from one line to another. Sixty-three stations involving 86 nodes are in such a situation, most modeling 2 interconnected lines, although one has connections for up to 5 lines (République).

Nineteen extra nodes have been added: Two of them represent special stations, and 17 are artificial nodes defined for specific problems. For example, to the west of Paris 2 stations are connected to a third one by 2 unidirectional lines, which then merge with a bidirectional line: The additional node corresponds to the two-way junction of the bidirectional line.

The travel times for the different arcs are defined by the element **g('edge_length')** of the graph list. These times are not obtained from measurements but are estimated mean values: It is presumed that the travel for an ordinary arc (between two stations) is 1.5 minutes, although some arc times are taken to be 0.75 minute. Furthermore, the time required to change lines is taken to be 4 minutes (this figure could be improved, since the standard deviation for this estimate can be quite high).

Two very similar graphs have been defined: The first one is a directed graph **dsubw** used for computations, and the second one, **udsubw**, is undirected and used for graphical display: These 2 graphs have exactly the same nodes. The Scilab program of the demo begins with the function **x_message**, which provides for a dialogue with the user so that the starting and the ending stations of travel can be defined. The route is then found by simply using the function **shortest_path** applied to graph **dsubw**. Having the path for this graph, we use the simple following trick for the display

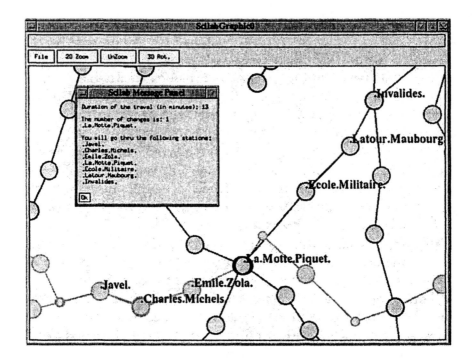

Figure 11.16: *A travel in the Paris metro*

using the **udsubw** graph: With the use of the function **path_2_nodes** we get
the nodes of the path, i.e., the stations corresponding to the route; the node
numbers are the same for the two graphs, and so on the undirected graph
we get the corresponding path with **nodes_2_path**.

There are also additional details required for a clean presentation of the
results: For example, the double occurrences of node names corresponding
to a transfer at interconnected lines must be suppressed, and transfer times
for the final station must not be included in computations.

The complete code of this example can be found in the directory
SCIDIR/demos/metanet of the Scilab distribution.

An example of the results is presented in Figure 11.16.

11.6.2   Praxitele Transportation System

Praxitele ([2], [3], Web site http://www-rocq.inria.fr/praxitele) is a transportation system based on a fleet of electric cars available in self-service
mode in a limited area. These cars are equipped with electronic means for
control and fleet management. The first experiment using Praxitele was

launched at the end of 1977 in the town of Saint Quentin en Yvelines, near Paris. The network consists of 50 electric vehicles and 6 stations. The locations of the stations are:

- 2 stations in the center of town: one at the main bus and railway station and one at the main mall.

- 2 stations in 2 different industrial areas.

- 2 stations for 2 private companies.

Vehicle status is relayed to the management center via a real-time digital communication system.

The transportation system Praxitele can be described as follows: Specific cars, called *praxicars*, are available in self-service parks, called *praxiparks*, distributed throughout the city. A customer, with either a special card or a simple credit card, can take a car in a *praxipark*, use it as long as desired, and then leave it in another *praxipark*.

With the unrestricted evolution of the system some problems will occur: Accumulation of praxicars in some praxiparks will give rise to a shortage of spaces, and others praxiparks will empty, resulting in a shortage of available cars. Service vehicles are used to carry off some cars from full praxiparks and to distribute them to empty praxiparks.

Given the above problem formulation we must solve a logistics problem and a control problem. For the logistics problem one must determine an appropriate number of praxicars, sizes of the different praxiparks, and the number of required service vehicles. We suppose here that the random demand, the location of the praxiparks, and the possible routes are given. The control problem is, then, given the routes and the logistics, to determine optimal routes for the service vehicles and the number of praxicars that need to be transported from some praxiparks to others.

We first define the objective function for the control problem. Then we have to solve a nonlinear flow problem to get a balanced system: We compute the sizes of the praxiparks, the number of service vehicles, and the corresponding routes.

Problem Formulation

We will briefly describe the formulation of the complete problem. You can see in Figure 11.17 all the data and variables of this problem.

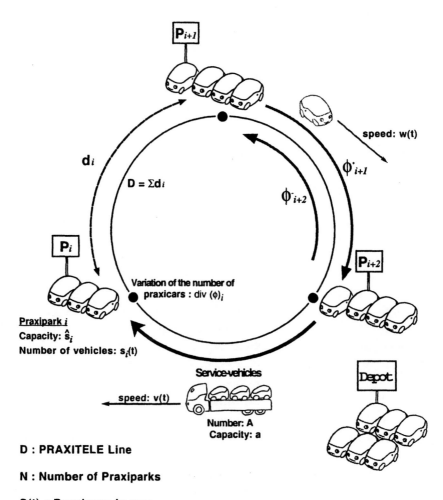

Figure 11.17: *The elements of a Praxitele line*

Modeling   A Praxitele network is defined as follows:

- $(\mathcal{N}, \mathcal{A})$ is the underlying directed graph, where $\mathcal{N}$ is the set of nodes of the network numbered from 1 to N and $\mathcal{A}$ is the set of arcs that are all the possible routes of the network.

- $a \in \mathcal{A}$ is a route between 2 praxiparks i and $i' \in \mathcal{N}$, $a^- = i$ is the departure praxipark, $a^+ = i'$ is the destination praxipark, and $F \overset{\text{def}}{=} |\mathcal{A}|$ is the number of arcs of the graph.

- τ_a is the mean time of the travel for the arc a; the days are divided into periods where the demand can be considered as stationary; on each period, τ_a is supposed to be constant (τ_C is the time for the trip along a cycle C).

- $s : k \in \mathbb{R}^+ \mapsto (s_1(k), \dots, s_N(k)) \in \prod_{i=1}^N [0, \hat{s}_i]$ is the number of cars in the praxiparks at time k, where $\hat{s} \in \mathbb{N}^N$ is the array of the sizes of the praxiparks.

- $[\text{div}(\phi)]_i(k)$ is the net flow out of praxipark i. In fact, we have

$$[\text{div}(\phi)]_i(k) = \sum_{a^+=i} \phi_a - \sum_{a^-=i} \phi_a,$$

where ϕ_a is the flow of travel claimed for the arc a.

- θ_a is the flow of service vehicles on arc a (value given in vehicles/hour) (θ_C is the same flow on a cycle C).

- ψ_a is the flow of praxicars carried by the service vehicles on arc a (praxicars/hour).

- A is the total number of service vehicles used on the network (A_a are those used on arc a, and A_C on a cycle C).

- $p : k \in \mathbb{R}^+ \mapsto (p_1(k), \dots, p_A(k)) \in \prod_{j=1}^A [0, \hat{p}_j]$ is the number of praxicars carried by the service vehicles at time k, where $\hat{p} \in \mathbb{N}^A$ is the array of the capacities of the service vehicles (in our case they are all the same).

- $U_{ji}(k)$ is the number of praxicars to be loaded or unloaded at the praxipark i by the service vehicle j at time k. A necessary condition for $U_{ji}(k)$ to be nonzero is, of course, that the service vehicle j is at praxipark i at time k.

- f_i is the frequency of service vehicles at praxipark i.

- c_v is the cost per time unit for a service vehicle.

- c_p is the cost per time unit for a praxipark space.

Evolution equations of the system   The praxicars are distributed in the praxiparks and in the service vehicles. So, the equations of the Praxitele system are defined as follows:
The number of praxicars in praxipark i at time $k + 1$ is

$$s_i(k + 1) = s_i(k) + [\text{div}(\phi)]_i(k) + \sum_{j=1}^{A} U_{ji}(k). \qquad (11.1)$$

The number of praxicars in service vehicle j at time $k + 1$ is

$$p_j(k + 1) = p_j(k) - \sum_{i=1}^{N} U_{ji}(k). \qquad (11.2)$$

Equations (11.1) and (11.2) are the evolution equations of the system where the state vector is $(s, p)^T$ and the control is U.

Control Problem

The sizes of the praxiparks, the number of service vehicles, and their size (which is unique) are supposed known.

We want now to solve a control problem where the objective function is chosen to minimize the probability of shortage: the shortage of a praxicar for a new customer or the shortage of spaces in a praxipark at the arrival of a praxicar. To this end we choose to maintain the number of praxicars in the praxiparks and in the service vehicles at a specified value.

We implement this strategy by applying proportional regulation using a control $U_{ji}(k)$, which is the control applied to praxipark i by service vehicle j at time k. Thus, we define $U_{ji}(k)$ by

$$U_{ji}(k) = \alpha \left(s_i(k) - \frac{\widehat{s_i}}{2} - p_j(k) + \frac{\widehat{p_j}}{2} \right),$$

where α is chosen in order to ensure the system's stability.

Fleet Size, Praxipark Sizes, and Routes

We now implement the previous control, which leads us to the question of how to solve a financial problem. How should we choose the correct number of service vehicles, of spaces in each praxipark, and routes of service vehicles?

The problem to be solved can be formulated as

$$\min c_v \sum_{a \in \mathcal{F}} A_a + c_p \sum_{i \in \mathcal{N}} \widehat{s}_i, \tag{11.3}$$

subject to the following constraints. The capacity of service vehicle flow

$$0 \leq \psi_a \leq \widehat{p}.\theta_a \ \forall a \in \mathcal{F},$$

the net flow of praxicars at the praxiparks

$$[\operatorname{div}(\psi)]_i = -[\operatorname{div}(\phi)]_i \ \forall i \in \mathcal{N},$$

the net flow of service vehicles,

$$[\operatorname{div}(\theta)]_i = 0 \ \forall i \in \mathcal{N}.$$

The number of service vehicles on arc a can be rewritten using the flow of service vehicles on the arc: $A_a = \tau_a \theta_a$.

The number of spaces needed in praxipark i is given by the frequency f_i of service vehicles at the praxipark i: $\widehat{s}_i = |[\operatorname{div}(\phi)]_i|/f_i$.

Then the problem can be written as

$$\min c_v \sum_{a \in \mathcal{F}} \tau_a \theta_a + c_p \sum_{i \in \mathcal{N}} \frac{|[\operatorname{div}(\phi)]_i|}{f_i}.$$

We now use the decomposition of flow on a cycle basis, decomposition adapted to the case where flows are circuits (i.e., the Kirchhoff law is satisfied for each node). This formulation is reasonable, since the global net flow of service vehicles at praxiparks must be zero.

We choose a cycle basis $\mathcal{C}$ with a given direction for each cycle. The problem is a multicommodity problem. The two commodities are the service vehicles and the praxicars to be carried, and they are connected by the capacity constraint. The flows for the praxicars are not decomposed on the cycle basis.

On the cycle basis $\mathcal{C}$ the flows of service vehicles are

$$\theta_a = \sum_{a \in C; C \in \mathcal{C}} \theta_C . \gamma_C,$$

where $\gamma = (\gamma_C)_{C \in \mathcal{C}}$ is the vector associated to the basis.

We define

$$\chi(a) \stackrel{\text{def}}{=} \{C \in \mathcal{C} : a \in C\}$$

and

$$\chi(i) \stackrel{\text{def}}{=} \{C \in \mathcal{C} : i \in C\},$$

and θ_a becomes

$$\theta_a = \sum_{C \in \chi(a)} \gamma_C \theta_C .$$

We now write the optimization problem defined by equation (11.3) on the cycle basis $\mathcal{C}$. The first sum becomes $\sum_{C \in \mathcal{C}} \tau_C \theta_C$. For the second term, we use the fact that the frequencies f_i of the service vehicles at praxipark i are added and that the frequency at a given praxipark of the service vehicles belonging to a given cycle is equal to the flow on the cycle. So, in the second term, $\hat{s}_i$ becomes $|[\text{div}(\phi)]_i| / \sum_{C \in \chi(i)} \theta_C$.

A constraint on the praxicar flow on one arc has to be added for all the cycles including the arc:

$$0 \leq \psi_a \leq \hat{p} \sum_{C \in \chi(a)} \theta_C \gamma_C, \quad \forall a \in \mathcal{F}, \ \forall C \in \mathcal{C}.$$

The other constraints remain the same.

Equation (11.3) is now

$$\min c_v \sum_{C \in \mathcal{C}} \tau_C \theta_C + c_p \sum_{1 \leq i \leq N} \frac{|[\text{div}(\phi)]_i|}{\theta_i},$$

with $\theta_i = \sum_{C \in \chi(i)} \theta_C$ under the following constraints. The capacity of the flow of service vehicles,

$$0 \leq \psi_a \leq \hat{p} \sum_{C \in \chi(a)} \theta_C \gamma_C, \quad \forall a \in \mathcal{F},$$

the praxicar flow at the praxiparks,

$$[\text{div}(\psi)]_i = -[\text{div}(\phi)]_i, \quad \forall i, 1 \leq i \leq N,$$

and positivity, $\theta_C \geq 0, \ \forall C \in \mathcal{C}.$

The program needed to solve this problem is fairly large. The first step, the construction of the Praxitele network, is accomplished with Metanet. This consists in specifying the network data: location and size of the praxiparks, travel times for the different routes, demand, etc. We compute the cycle basis using the Metanet function `cycle_basis`. Then the optimization problem is solved using Scilab's nonlinear optimization tool, `optim` (see Section 8.4.2). The constraints are handled in the following way: Equality constraints are penalized, and inequality constraints are transformed to equality constraints by introducing gap variables and then are penalized. A static problem is solved for every time period where the demand is supposed to be stationary. The results are plotted with Scilab or visualized with the Metanet graphics window.

This program gives the flow of service vehicles θ to be realized on the cycles and the flow of praxicars ψ to be loaded and unloaded in the praxiparks. The use of ψ for the open-loop control is not implemented because we want a robustness property, and for this reason we have chosen a feedback control where the feedback is the number of praxicars in the praxipark.

In this example, which corresponds to a praxitele network situated in Saint Quentin en Yvelines, we have constructed a simple graph. The problem is that the demand cannot be precisely modeled. So, we model demand as a random process with known mean and with a standard variation equal to 20% of the mean.

The cycle basis is composed of 6 cycles. The results of the sizing problem are given in Figure 11.18. These results show that for the stationary case, nearly all the customers are satisfied. Furthermore, the objective function allows a high number of trips for the praxicars: for 60 time steps of 15 minutes each, each praxicar is used for 19 trips (each praxicar is used for about 5 hours).

Figure 11.19 shows that the regulation of inventory levels gives good results. Indeed, for praxipark 1, we can see that the level never goes negative (the demand is always satisfied), but it is sometimes higher than the maximum capacity of the praxipark. On the other hand, for praxipark 2, we can see that for one time step the level goes negative (indicating a shortage), and for another time step one space is lacking.

Figure 11.20 shows the number of praxicars in service vehicles 1 and 3 corresponding to the preceding regulation.

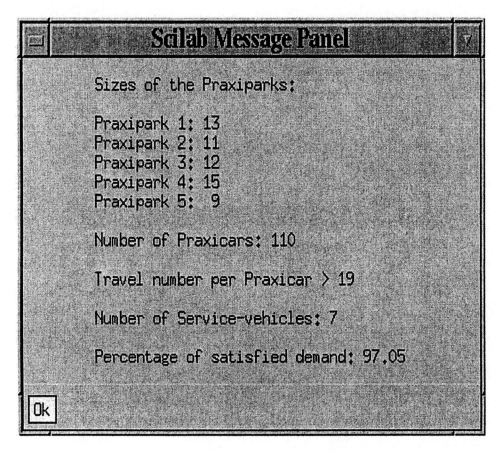

Figure 11.18: *Simulation results of the sizing problem*

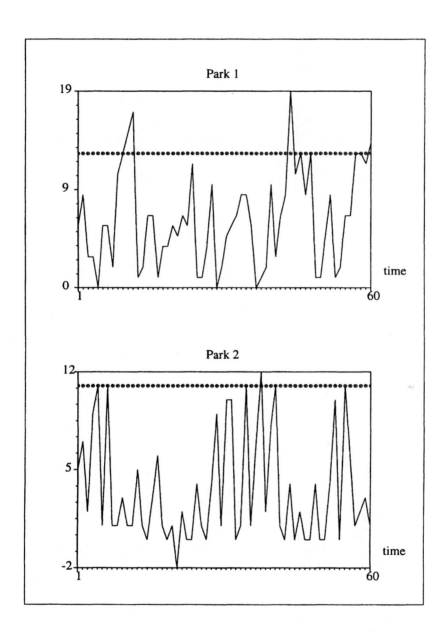

Figure 11.19: *Number of praxicars of praxiparks 1 and 2*

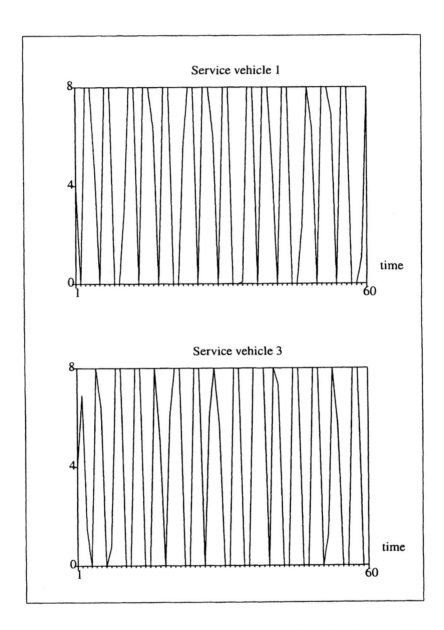

Figure 11.20: *Number of praxicars in service vehicles 1 and 2*

Part III

Applications

Chapter 12

Modal Identification of a Mechanical Structure

In this chapter a Scilab application concerning the identification of the modal characteristics of a mechanical structure under operating conditions is presented. The material covered in this chapter presupposes a knowledge of signal processing and the theory of vibration.

Classical identification techniques are generally applied to models with known inputs and outputs. Some of these techniques identify the underlying model, and others attempt to simulate, by an appropriate excitation (geometric design of the input), every mode shape of the structure. In the case considered here, we suppose that the structure is operating under normal conditions and is excited by external (unknown) phenomena. Examples of such situations are offshore structures subject to swell, cars running on imperfect roads, and wings and structures of aircraft during flight. It is presumed that the excitation coming from the environment is nonstationary.

Modal identification can be followed by change detection, model validation, or optimal sensor location techniques. For modal identification alone the basic approach is described in [28] and [1], which presents vibration monitoring via subspace-based fault detection methods. The bibliographies of these two references give an extensive review of the subject.

12.1 Modeling the System

The behavior of a vibrating structure is approximately linear and can be modeled by the following matrix differential equation:

$$M\ddot{Z}_t + C\dot{Z}_t + KZ_t = E_t, \qquad (12.1)$$

where

- Z_t is the vector of the displacements for the n different degrees of freedom of the structure;

- $M, C,$ and K are $n \times n$ matrices representing the mass, damping, and stiffness of the structure, respectively;

- E_t is the vector of external forces applied to the structure.

The external forces in this problem are not measured but are presumed to be well modeled by nonstationary white noise. Finally, we also have the observation vector

$$Y_t = LZ_t, \qquad (12.2)$$

which has r degrees of freedom, where L is an $r \times n$ matrix representing the presence of sensors. That is to say that $L_{ij} = 1$ if the sensor number i observes the degree of freedom j, and 0 otherwise.

The identification problem considered here is to determine the modal characteristics of the structure, that is, the solution (λ, Φ_λ) of the following equations corresponding to equation (12.1):

$$\begin{aligned}
\det(M\lambda^2 + C\lambda + K) &= 0, \\
(M\lambda^2 + C\lambda + K)\Phi_\lambda &= 0.
\end{aligned}$$

In these preceding equations, M, C, and K are real matrices, and so the eigenfrequencies $\lambda = c + i\omega$ appear in complex conjugate pairs. The ω's are the eigenfrequencies, and the corresponding damping coefficients are

$$\frac{c}{\sqrt{c^2 + \omega^2}} \quad (c < 0).$$

The modal identification problem (with a given observation) is to get the pair $(\Lambda, L\Phi)$, where Λ is the diagonal matrix made by the λ's and $L\Phi$ is the observed part of the modes Φ_λ (the observed modes are the columns of the matrix $L\Phi$).

12.2 Modeling the Excitation

12.2.1 Decomposition of the Unknown Input

In the general case, the excitation E_t can be modeled as the sum of two terms:

- colored noise E_t^b, which represents the excitation due to the environment (wind, swell, irregularities of a road,etc.);

- pure sinusoids E_t^h, which are harmonics of the main excitation (for rotating machinery this can be a consequence of imperfect load balancing).

Using this decomposition, we can write the model (12.1) and (12.2) in the following way:

$$\begin{array}{rcl} M\ddot{Z}_t + C\dot{Z}_t + KZ_t & = & E_t^b + E_t^h, \\ Y_t & = & LZ_t. \end{array} \qquad (12.3)$$

Defining the augmented state $X_t = \begin{bmatrix} Z_t \\ \dot{Z}_t \end{bmatrix}$, system (12.3) can be written as

$$\dot{X}_t = FX_t + \begin{bmatrix} 0 \\ M^{-1}E_t \end{bmatrix}, \qquad (12.4)$$

where

$$F = \begin{bmatrix} 0 & I \\ -M^{-1}K & -M^{-1}C \end{bmatrix}.$$

By linearity, we can separate the contributions of E_t^b and E_t^h with the following additive decompositions:

$$Z_t = Z_t^b + Z_t^h$$

and

$$Y_t = Y_t^b + Y_t^h.$$

Considering now the decomposed augmented states

$$X_t^b = \begin{bmatrix} Z_t^b \\ \dot{Z}_t^b \end{bmatrix} \text{ and } X_t^h = \begin{bmatrix} Z_t^h \\ \dot{Z}_t^h \end{bmatrix},$$

system (12.4) becomes

$$\dot{X}_t^b + \dot{X}_t^h = F(X_t^b + X_t^h) + \left[\begin{array}{c} 0 \\ M^{-1}(E_t^b + E_t^h) \end{array} \right].$$

Summing up, we will study

- the following model, corresponding to the contribution of the non-stationary noise:

$$\begin{array}{rcl} \dot{X}_t^b & = & FX_t^b + \left[\begin{array}{c} 0 \\ M^{-1}E_t^b \end{array} \right], \\ Y_t^b & = & [\, L \;\; 0 \,]\, X_t^b; \end{array} \qquad (12.5)$$

- the following model, corresponding to the contribution of the part of the excitation composed of harmonics:

$$\begin{array}{rcl} \dot{X}_t^h & = & FX_t^h + \left[\begin{array}{c} 0 \\ M^{-1}E_t^h \end{array} \right], \\ Y_t^h & = & [\, L \;\; 0 \,]\, X_t^h. \end{array} \qquad (12.6)$$

12.2.2   Contribution of the Colored Noise

The term E_t^b can be defined by the state representation

$$\begin{array}{rcl} \dot{\xi}_t & = & F_b\xi_t + G_bW_t, \\ E_t^b & = & H_b\xi_t, \end{array} \qquad (12.7)$$

where W_t is a (possibly nonstationary) white noise. Taking together (12.5) and (12.7), we get

$$\begin{array}{rcl} \left[\begin{array}{c} \dot{\xi}_t \\ \dot{X}_t^b \end{array} \right] & = & \left[\begin{array}{cc} F_b & 0 \\ \left[\begin{array}{c} 0 \\ M^{-1}H_b \end{array} \right] & F \end{array} \right] \left[\begin{array}{c} \xi_t \\ X_t^b \end{array} \right] + \left[\begin{array}{c} G_b \\ 0 \end{array} \right] W_t, \\ Y_t^b & = & [\, 0 \;\; [\, L \;\; 0 \,]\,] \left[\begin{array}{c} \xi_t \\ X_t^b \end{array} \right] \end{array} \qquad (12.8)$$

which is nothing but the state representation of the augmented state $\left[\begin{array}{c} \xi_t \\ X_t^b \end{array} \right]$

and where Y_t^b is the observation.

Let $\tilde{F}$ be the state matrix of (12.8):

$$\tilde{F} = \left[\begin{array}{cc} \left[\begin{array}{c} F_b \\ 0 \\ M^{-1}H_b \end{array} \right] & \begin{array}{c} 0 \\ \\ F \end{array} \end{array} \right].$$

The spectral analysis of $\tilde{F}$ gives the poles of the structure but also (unfortunately) the poles of the part of the excitation coming from the colored noise. Generally, it will be easy for a mechanical engineer studying a specific structure to discriminate between the poles of the structure and the poles of the excitation. Some classical and simple examples of known frequencies are the following: (1) the fundamental 50 Hz (or 60 Hz) and its harmonics for the rotating machines of electrical power plants and (2) the frequency used for the command (stepwise command at a given frequency used for space launchers).

Through the observation matrix $\left[\begin{array}{cc} 0 & \left[\begin{array}{cc} L & 0 \end{array} \right] \end{array} \right]$ of (12.8), the observed part of the eigenvectors of $\tilde{F}$ is decomposed into:

- the observed part of the mode shapes of the structure defined by F,

- the observed part of the mode shapes of the part of the excitation composed of the colored noise. The eigenvectors of F_b are obtained filtered by the transfer function of the structure.

Consider now an eigenvector $\tilde{\psi}$ of $\tilde{F}$, where $\tilde{\psi}$ is associated with the eigenvalue μ of F_b. Also, it is presumed that μ is not an eigenvalue of F (we assume that the spectra of F and F_b are distinct). We decompose $\tilde{\psi}$ into $\left[\begin{array}{c} \psi_1 \\ \psi_2 \end{array} \right]$ according to the decomposition of the state $\left[\begin{array}{c} \xi_t \\ X_t^b \end{array} \right]$. Then it can be shown that ψ_1 is the eigenvector of F_b associated with the eigenvalue μ, and ψ_2 is given by

$$\psi_2 = [\mu I - F]^{-1} \left[\begin{array}{c} 0 \\ M^{-1}H_b \end{array} \right] \psi_1.$$

The observed part of $\tilde{\psi}$ is then

$$\left[\begin{array}{cc} 0 & \left[\begin{array}{cc} L & 0 \end{array} \right] \end{array} \right] \tilde{\psi} = \left[\begin{array}{cc} L & 0 \end{array} \right] \psi_2 = \left[\begin{array}{cc} L & 0 \end{array} \right] [\mu I - F]^{-1} \left[\begin{array}{c} 0 \\ M^{-1}H_b \end{array} \right] \psi_1.$$

Applying the formula for the inverse of block matrices, we get

$$[\mu I - F]^{-1} = \left[\begin{array}{cc} \dfrac{1}{\mu}(I - S_\mu^{-1}K) & S_\mu^{-1}M \\ -S_\mu^{-1}K & \mu S_\mu^{-1}M \end{array} \right],$$

with

$$S_\mu = \mu^2 M + \mu C + K,$$

and the observed part of $\tilde\psi$ is finally

$$\begin{bmatrix} 0 & [\, L \quad 0 \,] \end{bmatrix} \tilde\psi = LS_\mu^{-1} H_b \psi_1, \tag{12.9}$$

where $LS_\mu^{-1} H_b$ is the transfer function of a system filtering the modes of the excitation.

Of course, the same result holds in discrete-time state-space representation. With a sampling period δt, vectors $\begin{bmatrix} \xi_t \\ X_t^b \end{bmatrix}$ and Y_t^b become respectively the sampled state and observation, denoted by χ_t^b and $\mathcal{Y}_t^b$. The discrete-state space representation is then

$$\begin{aligned} \chi_{t+\delta t}^b &= \mathcal{F}\chi_t^b + V_{t+\delta t}, \\ \mathcal{Y}_t^b &= \begin{bmatrix} 0 & [\, L \quad 0 \,] \end{bmatrix} \chi_t^b, \end{aligned} \tag{12.10}$$

where the state matrix is defined by

$$\mathcal{F} = e^{\bar F \delta t},$$

and $\{V_t\}$ is a discrete-time white noise with covariance matrix

$$\operatorname{cov}(V_t) \;=\; \int_t^{t+\delta t} e^{\bar F \tau} \tilde Q_\tau e^{\bar F^T \tau} d\tau,$$

where

$$\tilde Q_t \;=\; \begin{bmatrix} G_b Q_t G_b^T & 0 \\ 0 & 0 \end{bmatrix}$$

and

$$Q_t = \operatorname{cov}(W_t).$$

12.2.3 Contribution of the Harmonics

Consider now the system defined by (12.6). We will simply compute the harmonic response. The input E_t^h can be written as the real part:

$$E_t^h = \Re \left[\sum_{k=1}^K E_k e^{i\omega_k t} \right], \tag{12.11}$$

where the ω_k are the harmonic angular frequencies and the E_k are complex vectors, which may be taken to be the "modes" of the excitation representing the geometry of the sinusoidal components of the input.

Taking $f_k = \omega_k/(2\pi)$, we let $X^h(f)$ be the Fourier transform of the complex state X_{ct}^h associated with X_t^h. Using (12.6) and (12.11), we obtain

$$X^h(f) = \frac{1}{2}(2i\pi f I - F)^{-1} \begin{bmatrix} 0 \\ M^{-1} \sum_{k=1}^{K} E_k[2\pi\delta(f - f_k)] \end{bmatrix}.$$

But $\delta(f - f_k) = 0$ if $f \neq f_k$, so we have

$$X^h(f) = \frac{1}{2}\sum_{k=1}^{K}(2i\pi f_k I - F)^{-1} \begin{bmatrix} 0 \\ M^{-1}E_k[2\pi\delta(f - f_k)] \end{bmatrix}.$$

Then the Fourier transform of the observation is

$$Y^h(f) = \frac{1}{2}\sum_{k=1}^{K} \begin{bmatrix} L & 0 \end{bmatrix}(2i\pi f_k I - F)^{-1} \begin{bmatrix} 0 \\ M^{-1}E_k[2\pi\delta(f - f_k)] \end{bmatrix}.$$

This means that the spectra of X_t^h and Y_t^h are composed of tonal lines. By inverse Fourier transforming we get

$$Y^h(f) = \Re \begin{bmatrix} \sum_{k=1}^{K} L S_{2i\pi f_k}^{-1} E_k e^{i\omega_k t} \end{bmatrix}.$$

By this simple manipulation, we see that the so-called modes of the input are filtered by the structure in the same way as the modes of the colored noise (see equation (12.9)).

These preceding results do not change when a discrete-state representation is used. If χ_t^h and $\mathcal{Y}_t^h$ denote, for a sampling period δt, the sampled versions of X_t^h and Y_t^h, their spectra are composed of lines and are obtained by taking the periodic version of the continuous state and observations X_t^h and Y_t^h. The state representation is then written without noise on the state as

$$\begin{aligned} \chi_{t+\delta t}^h &= \mathcal{F}'\chi_t^h, \\ \mathcal{Y}_t^h &= \begin{bmatrix} L & 0 \end{bmatrix}\chi_t^h. \end{aligned} \tag{12.12}$$

The eigenvalues and eigenvectors of $\mathcal{F}'$ are respectively $e^{2i\pi f_k \delta t} = e^{i\omega_k \delta t}$ and $(e^{i\omega_k \delta t}I - F)^{-1} \begin{bmatrix} 0 \\ M^{-1}E_k \end{bmatrix}.$

The observed part of the eigenvectors is $L S_{e^{i\omega_k \delta t}}^{-1} E_k.$

12.2.4   The Final Discrete-State Model

Let us define $\chi_t = \begin{bmatrix} \chi_t^b \\ \chi_t^h \end{bmatrix}$, $\mathcal{Y}_t = \mathcal{Y}_t^b + \mathcal{Y}_t^h$, and $H = \begin{bmatrix} L & 0 \end{bmatrix}$.
Then (12.10) and (12.12) give

$$
\begin{aligned}
\chi_{t+\delta t} &= \begin{bmatrix} \mathcal{F} & 0 \\ 0 & \mathcal{F}' \end{bmatrix} \chi_t + \begin{bmatrix} V_{t+\delta t} \\ 0 \end{bmatrix}, \\
\mathcal{Y}_t &= \begin{bmatrix} 0 & H & H \end{bmatrix} \chi_t.
\end{aligned}
$$

With the following notation,

$$
\mathcal{F}_1 = \begin{bmatrix} \mathcal{F} & 0 \\ 0 & \mathcal{F}' \end{bmatrix}, \quad V_t = \begin{bmatrix} V_t \\ 0 \end{bmatrix}, \quad H_1 = \begin{bmatrix} 0 & H & H \end{bmatrix},
$$

the representation becomes

$$
\begin{aligned}
\chi_{t+\delta t} &= \mathcal{F}_1 \chi_t + V_{t+\delta t}, \\
\mathcal{Y}_t &= H_1 \chi_t,
\end{aligned}
$$

which is a discrete-state representation of (12.3).

The analysis of the eigenstructure of $\mathcal{F}$ gives:

- The eigenfrequencies of the structure and the modal shapes, i.e., the pairs (λ, ϕ_λ) that are solutions of

$$
\begin{aligned}
\det(\lambda^2 M + \lambda C + K) &= 0, \\
(\lambda^2 M + \lambda C + K)\phi_\lambda &= 0.
\end{aligned}
$$

  Note that the observed part is $L\phi_\lambda$.

- The frequencies and the "modes" of the excitation colored noise E_t^b, these "modes" being filtered by $LS_\mu^{-1}H_b$.

  This means that we obtain the pairs $(\mu, LS_\mu^{-1}H_b\psi_\mu)$ that are solutions of

$$
\begin{aligned}
\det(\mu I - F_b) &= 0, \\
(\mu I - F_b)\psi_\mu &= 0.
\end{aligned}
$$

- The frequencies and the "modes" of the periodic excitation E_t^h, these "modes" being filtered in the same manner as in the previous item, that is, $(f_k, LS_{f_k}^{-1}E_k)$.

We have seen that we can consider a very general input, which need not be white or stationary and which can have a deterministic part. With such an input, model (12.1) can be written, at least with an appropriate extended state, as a canonical representation that allows the use of state realization algorithms to obtain the eigenfrequencies and modal shapes of the vibrating structure.

12.3 State-Space Representation and an ARMA Model

Using the augmented state

$$X_t = \begin{bmatrix} Z_t \\ \dot{Z}_t \end{bmatrix}$$

we have seen on page 415 that system (12.1) can be written with the following state-space representation:

$$\begin{aligned} \dot{X}_t &= F X_t + V_t, \\ Y_t &= \begin{bmatrix} L & 0 \end{bmatrix} X_t, \end{aligned}$$

where

$$F = \begin{bmatrix} 0 & I \\ -M^{-1}K & -M^{-1}C \end{bmatrix}$$

and

$$V_t = \begin{bmatrix} 0 \\ -M^{-1}E_t \end{bmatrix}.$$

Now we choose a new state representation to get the canonical representation

$$\begin{aligned} \dot{X}_{e_t} &= F_e X_{e_t} + V_{e_t}, \\ Y_t &= H_e X_{e_t} \end{aligned}$$

such that V_{e_t} is a *white noise*, possibly nonstationary, and where X_{e_t} is the augmented state and F_e the associated state matrix. The eigenstructure of F_e contains the eigenpairs of the triplet (M, C, K), which are the modal shapes of the structure, and possibly the "modes" of the excitation (colored noise, harmonics, etc.).

In discrete time, this state-space representation is

$$\begin{aligned} X_{t+\delta t} &= F X_t + W_t, \\ Y_t &= H X_t, \end{aligned} \tag{12.13}$$

where δt is the sampling period, W_t is a discrete-time white noise (stationary or not), and $F = e^{F_e \delta t}$.

The eigenstructure of the state matrix is, of course, preserved by the sampling operation; more precisely, if (α, V_α) and (β, V_β) denote respectively the eigenpairs of F_e and F, we have the following correspondence:

$$
\begin{aligned}
\beta &= e^{\alpha \delta t}, \\
V_\beta &= V_\alpha.
\end{aligned}
$$

The state-space model being given by (12.13), the observation Y_t can be defined by the equation (ARMA model)

$$
Y_t = \sum_{i=1}^{p} A_i Y_{t-i} + \sum_{j=0}^{p-1} B_j(t) \varepsilon_{t-j},
$$

where:

- ε_t is a white series.

- The B_j are the moving average (MA) coefficients. These parameters define the characteristics of the excitation noise that will be observed later in the model. The characteristics of these coefficients are time varying if E_t is a nonstationary input.

- The A_i are the autoregressive (AR) coefficients, and they are necessarily a solution of the linear system of equations

$$
HF^p = \sum_{i=1}^{p} A_i H F^{p-i}. \tag{12.14}
$$

These coefficients contain the same modal information as the (H, F) pair: the eigenvalues β of F and the part of the eigenvectors V_β of F observed by H, i.e., HV_β. These pairs (β, HV_β) are called the observed eigenstructure. This eigenstructure can be obtained from the algorithm described below.

Right-multiplying (12.14) by V_β, we get

$$
HF^p V_\beta = \sum_{i=1}^{p} A_i H F^{p-i} V_\beta.
$$

But V_β is an eigenvector of F, so we get

$$(\beta^p I_r - \sum_{i=1}^{p} \beta^{p-i} A_i) H V_\beta = 0,$$

where r is the degrees of freedom already defined on page 414.

The polynomial matrix $P(z) = z^p I_r - \sum_{i=1}^{p} z^{p-i} A_i$ is the characteristic polynomial of the following matrix in companion form:

$$F_c = \begin{bmatrix} 0 & I_r & & & \\ & 0 & & & \\ & & & & \\ & & & 0 & I_r \\ A_p & \cdots & & A_2 & A_1 \end{bmatrix}.$$

Therefore, we know that the pair (H_c, F_c), with $H_c = [I_r 0_r \cdots 0_r]$, is a non-minimal state-space representation equivalent to (H, F).

Thus, the problem of identification of the modal structure is equivalent to the identification of the autoregressive coefficients. When the parameters $(A_i)_{i=1,\ldots,p}$ are identified, we can build the companion matrix F_c, and the observed eigenstructure is obtained by diagonalizing this matrix.

12.4 Modal Identification

12.4.1 Instrumental Variable Method

To obtain the eigenstructure of (H, F) described in the previous section the autoregressive coefficients must be determined. This section presents a classical method for the identification of the autoregressive parameters. The basis of the method is the orthogonality (with respect to the covariance) of the residual variables defined by

$$W_t = Y_t - \sum_{i=1}^{p} A_i Y_{t-i},$$

with the past observations being

$$Y_{t-p}, Y_{t-p-1}, Y_{t-p-2}, \ldots$$

This is simply written

$$\mathbb{E}(Z_t W_t^T) = 0, \tag{12.15}$$

where

$$Z_t = \begin{bmatrix} Y_{t-p} \\ \vdots \\ Y_{t-p-(N-1)} \end{bmatrix}$$

is the vector of past variables, called the *instrumental variables* vector (for any $N \geq p$) and where $\mathbb{E}$ is the expectation operator.

Equation (12.15) is a system of N delayed Yule–Walker equations, which can be written

$$\begin{bmatrix} A_p & \cdots & A_1 & -I_r \end{bmatrix} \mathcal{H}_{p+1,N} = 0, \qquad (12.16)$$

where

$$\mathcal{H}_{p+1,N} = \begin{bmatrix} R_0 & R_1 & \cdots & R_p & \cdots & R_{N-1} \\ R_1 & R_2 & & & & \\ \cdot & & & & & \\ \cdot & & & & & \\ \cdot & & & & & \\ R_p & \cdots & & \cdots & R_{N-1} & \cdots & R_{p+N-1} \end{bmatrix} \qquad (12.17)$$

is the theoretical Hankel matrix built from the covariances $R_m = \mathbb{E}(Y_{t+m} Y_t^T)$. In practical applications the theoretical covariances will be replaced by the empirical covariances:

$$R_m(s) = \frac{1}{s} \sum_{t=1}^{s} Y_{t+m} Y_t^T,$$

where $Y_1, \ldots, Y_s$ are the recorded data and s is the length (duration) of the record. Subsequently, (12.16) is solved using these covariances. This solution, in the mean square sense, if $N > p$, is the *instrumental variable identification method*.

The main theoretical result associated with the above method (see [10]) shows that the estimators $\hat{A}_i$ of the parameters A_i are consistent. That is, they converge with probability 1 to the true value when the size of the record goes to infinity. This is true even in the case when the MA part of the ARMA model is nonstationary.

Starting with a Hankel matrix **hk**, the following Scilab code implements the instrumental variable method:

```
// GLOBAL LOOP OF YULE-WALKER INSTRUMENTAL VARIABLE METHOD
for kk   = nmin:pas:nfin,
  npp    = kk*ny;
  hkn    = hk(3:(npp+ny),1:mh);          ← truncated Hankel matrix
  bb     = hk(1:ny,1:mh);
  xx     = hkn'\ bb';                     ← AR coefficients
  hcom   = zeros(npp,npp);
  for i = 1:(npp-ny),                     ← companion matrix
    hcom(i,i+ny) = 1;
  end;
  for j = 1:ny,
    for i = 1:npp,
      hcom(npp-j+1,i) = xx(npp+1-i,j);
    end;
  end;
  chcom = hcom + %i*0;
  [val,vect] = bdiag(chcom);            ← eigenvalues and eigenvectors
  poles = diag(val);                 ← end of the instrumental variable step
  Hhat   = zeros(ny,npp);
  Hhat(1:ny,1:ny) = eye(ny,ny);
  icosub= 0;
  niter = niter+1;
  for i = 1:npp,                                    ← sorting poles
    if ((imag(poles(i)) < 0) | (abs(poles(i)) < 0.2))
      then ri=0.;
      else
        c=log(abs(poles(i)));
        a=abs(atan(imag(poles(i)),real(poles(i))));
        rr=100.*c./sqrt(c.*c+a.*a);
        if (abs(rr) > 0.2) then ri(i)=0.;
        else
          ri=(fech*0.5/%pi)*a;
          icosub=icosub+1;
          freqp(niter,icosub)=ri;
          amort(niter,icosub)=rr;
                                             ← saving eigenvectors
          vecpr(:,icompt1+icosub) = real(Hhat*vect(:,i));
          vecpi(:,icompt1+icosub) = imag(Hhat*vect(:,i));
        end;
      end;
    end;
  icompt(niter)=icosub;
  icompt1=icompt1+icosub;
end;
```

The following is a very simple test example in Scilab:

```
// 4 (2x2) rotation blocks
VV=[0 0; 0 0];
th1=3.16;th2=3.30;th3=3.60;th4=4.30;
X1=[sin(th1) cos(th1); -cos(th1) sin(th1)];
X2=[sin(th2) cos(th2); -cos(th2) sin(th2)];
X3=[sin(th3) cos(th3); -cos(th3) sin(th3)];
X4=[sin(th4) cos(th4); -cos(th4) sin(th4)];
GG=[X1 VV VV VV; VV X2 VV VV; VV VV X3 VV; VV VV VV X4 ];
BA=rand(8,8);MA=(1/BA)*GG*BA;
// time simulation with white noise input
e1=rand(8,9000,'normal');
xx=zeros(8,1);outp=zeros(8,1);
for i=1:9000,
xx=MA*xx+0.01*e1(:,i);
outp=[outp xx];
end;
outp=outp+rand(8,9001,'normal');
```

The theoretical eigenfrequencies (for an a priori nominal sampling frequency of 100 Hz) are

```
 ans  =
!   25.292962     27.521131     32.29578     43.436626  !
```

The transition matrix is built on four (2×2) rotation blocks on the diagonal, and Figure 12.1 gives the spectrum of one component of the output. In this simple example, by examining only the output we easily get the 4 poles by the peaks of the spectrum. Thus, we have the following eigenfrequencies:

```
 ans  =
!   25.280762     27.514648     32.299805     43.432617  !
```

We now construct the Hankel matrix for 2 output components:

```
A=outp(2,501:8692); AA=outp(4,501:8692);
//LAGS
nl=256;
dnl=2*nl;
//SIMPLE CORRELATION AND CROSS-CORRELATION COMPUTING
c1=corr(A,nl);c2=corr(A,AA,nl);
c3=corr(AA,A,nl);c4=corr(AA,nl);
//
cov=zeros(2,dnl);
ii=[1:2:(dnl-1)];ii1=ii+1;
```

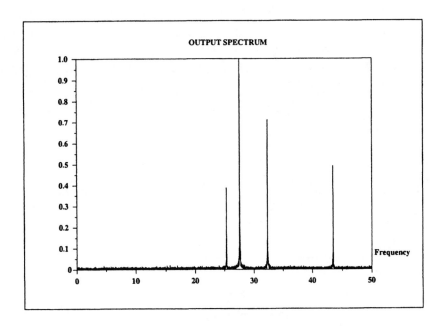

Figure 12.1: *Spectrum of output component 1*

```
//ARRANGING THE CORRELATIONS
cov(1,ii)=c1;cov(1,ii1)=c2;
cov(2,ii)=c3;cov(2,ii1)=c4;
[hk]=hank(128,128,cov);
```

We then apply the previous identification algorithm.

The algorithm is, in fact, an identification procedure for increasing numbers of instrumental variables (in the following case we have chosen the orders from 30 to 64). For each order we get some identified eigenfrequencies and the corresponding eigenvectors. Then a plot, called the stabilization diagram, is constructed by plotting the identified frequencies. This is shown in Figure 12.2, which represents the identified eigenfrequencies (on the y-axis) for the different orders (on the x-axis). This is a common technique for order determination with many identification methods (e.g., instrumental variables, polyreference, CVA).

A frequency is retained as an eigenfrequency for the system if we get an alignment, which means that this frequency has been identified for many different orders; in fact, the values of this frequency are not exactly the same for the different orders, but they are very close. So the selection procedure is the following: For an identified frequency f_0 at any given order we choose a "frequency window" (for example, $[f_0 - 0.1, f_0 + 0.1]$) and we count the number of identified frequencies in this interval for the different orders. The final selected frequencies correspond to the occurrences greater than a given bound. The result is shown in Figure 12.3. In this case the output of the selection operation is

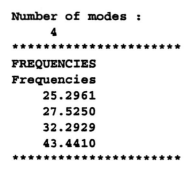

```
Number of modes :
      4
*********************
FREQUENCIES
Frequencies
      25.2961
      27.5250
      32.2929
      43.4410
*********************
```

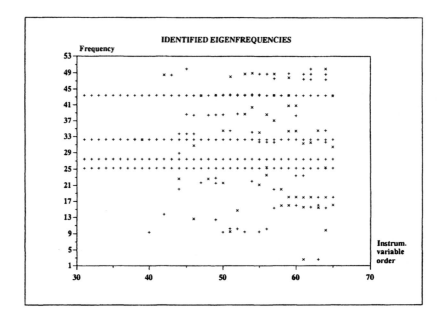

Figure 12.2: *Stabilization diagram*

By this method we get more interesting results for a concrete case: the damping coefficients and the eigenvectors. The eigenvectors may be useful for very close eigenfrequencies and then can be used for the animation of a structure if the finite-element model is available. The comparison of eigenvectors is a very precise criterion: If different values for the frequencies are mixed, we can decide whether 2 values (for 2 different orders) correspond

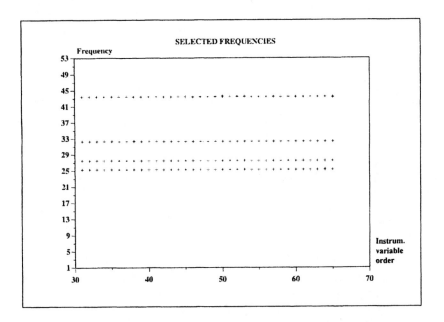

Figure 12.3: *Selection with a window focus*

to the same eigenfrequency by computing the normalized scalar product of the eigenvectors. This value is close to 1 for a sole eigenfrequency. If the observation points are correctly chosen, the eigenvectors give the modal shape of the structure, which can be animated (simulation of the deformation). The precision of the results given by the balanced realization method described in the next section is probably better than for this method. But this method can be easily used for test problems. It is very easy to get the residual by a simple simulation and so to implement hypothesis testing algorithms (for failure detection or model updating).

12.4.2 Balanced Realization Method

With the balanced realization method the eigenstructure of (H, F) can be directly identified with the record Y_t. The method is based on the singular value decomposition (SVD) of the Hankel matrix $\mathcal{H}_{N,N}$ (see page 424):

$$\mathcal{H}_{N,N} = \sum_{k=1}^{N} \sigma_k u_k v_k^T,$$

where

- $\sigma_1 \geq \sigma_2 \geq \cdots \geq \sigma_N$ are the singular values of $\mathcal{H}_{N,N}$;

- the vectors u_k and v_k are the corresponding left and right singular vectors, respectively.

Considering

$$
\begin{aligned}
U(s) &= \left[\begin{array}{cccc} u_1 & u_2 & \cdots & u_p, \end{array}\right] \\
V(s) &= \left[\begin{array}{cccc} v_1 & v_2 & \cdots & v_p \end{array}\right], \\
D(s) &= \mathrm{diag}(\sqrt{\sigma_1}, \sqrt{\sigma_2}, \ldots, \sqrt{\sigma_p}),
\end{aligned}
$$

we can write $\mathcal{H}_{N,N}$ in the following way:

$$
\mathcal{H}_{N,N} = \hat{\mathcal{O}}_N(s)\hat{\mathcal{C}}_N(s) + \sum_{k>p} \sigma_k u_k v_k^T, \tag{12.18}
$$

where $\hat{\mathcal{O}}_N(s) = U(s)D(s)$ and $\hat{\mathcal{C}}_N^T(s) = V(s)D(s)$.

The balanced realization algorithm [28] gives the estimators $\hat{H}(s)$ and $\hat{F}(s)$ for the matrices H and F such that:

- $\hat{H}(s)$ is the first row-block of $\hat{\mathcal{O}}_N(s)$,

- $\hat{F}(s)$ is the mean square solution of

$$
\hat{\mathcal{O}}_{N-1}^{\uparrow}(s) = \hat{\mathcal{O}}_{N-1}(s)\hat{F}(s),
$$

where

- $\hat{\mathcal{O}}_{N-1}(s)$ is the matrix composed by the first $N-1$ row blocks of $\hat{\mathcal{O}}_N(s)$,

- $\hat{\mathcal{O}}_{N-1}^{\uparrow}(s)$ is the matrix composed by the last $N-1$ row blocks of $\hat{\mathcal{O}}_N(s)$,

- and a row block is composed of r rows, r being the number of sensors.

In this way we get a pair of estimators $(\hat{H}(s), \hat{F}(s))$, which is consistent. That is to say, there exists a sequence of matrices $T(s)$ such that $T(s)$ and $T^{-1}(s)$ are uniformly bounded and

$$
\begin{aligned}
T^{-1}(s)\hat{F}(s)T(s) &\longrightarrow F, \\
\hat{H}(s) &\longrightarrow H
\end{aligned}
$$

as s tends to infinity. The modal characteristics are simply the eigenpairs of $\hat{F}(s)$ and $(\hat{\beta}, \hat{V}_\beta)$, observed by $\hat{H}(s)$, i.e., $(\hat{\beta}, \hat{H}(s)\hat{V}_\beta)$. The balanced realization method often gives better results than the instrumental variable method. For practical implementations this method requires an a posteriori reworking by hand, as is also the case with the instrumental variable method. As we have done in the preceding section for the orders of the instrumental variable method, we compare the results for increasing values of p, which is the autoregressive order in (12.16) or the truncation order of $\mathcal{H}_{N,N}$ in (12.18), and we get the *modal signature*, which characterizes the vibrating behavior of the structure, behavior observed by the measurement $\{Y_t\}$.

12.5 Numerical Experiments

The techniques described so far are the preliminary work associated with the Eureka project called Sinopsys (European research and development project on IN OPeration SYStems with 4 academic and 6 industrial partners). The data corresponding to the results presented in this section can be published, so they are not those of industrial cases. A software module developed by the company LMS International[1] and based on the preceding descriptions has been tested on various structures such as cars, bridges, space launchers, and aircraft. The data used in this section are measurements for laboratory mockups: a subframe test structure (of a car) and a plane (GARTEUR structure). These data have been made available courtesy of LMS International and the company SOPEMEA.[2]

For the first example we have 12 acceleration time signals with 4096 samples/channel. The sampling frequency is 410.004 Hz, and the frequency range of interest is the band (0 65 Hz).

12.5.1 Basic Computations

Preprocessing

The first step is the computation of the covariances of the signal in order to build a Hankel matrix. The following results correspond to 256 and 512 lags for the covariance calculations. A Hankel matrix of 2 and 4 records (2 or 4 sensors) is constructed. These computations are accomplished with several lines of Scilab code and are centered on the use of the **fft** command.

The simplest Scilab sequence of commands for 2 records is presented here:

```
//READING THE DATA OF THE FIRST 2 SENSORS
A1=read('sensor1',4096,1);
A2=read('sensor2',4096,1);
```

---

[1]http://www.lmsintl.com
[2]http://www.messier-bugatti.com/comppro/sopemea.htm

```
//LAGS
nl=128;
dnl=2*nl;
//SIMPLE CORRELATION AND CROSS-CORRELATION COMPUTATION
c1=corr(A1,nl);
c2=corr(A1,A2,nl);
c3=corr(A2,A1,nl);
c4=corr(A2,nl);
//
cov=zeros(2,dnl);
ii=[1:2:(dnl-1)];ii1=ii+1;
//ARRANGING THE CORRELATIONS
cov(1,ii)=c1;
cov(1,ii1)=c2;
cov(2,ii)=c3;
cov(2,ii1)=c4;
cov=cov*nt;
```

From this code we get a (2×256) array, which is the first two lines of the Hankel matrix. From these two lines, the global Hankel matrix can be immediately constructed: The third and fourth lines of the matrix are obtained by deleting the first (2×2) block of the first two lines, and in this way we can build a (128×128) Hankel matrix $\mathcal{H}_{N,N}$ (we choose here to get a square matrix with $N = 128$). If the previous computations need to be performed on very long observed data sequences, the overlap-and-add technique can be used with the function **corr** to obtain the correlation computations. This is accomplished by making use of the input parameter **updt** (see the on-line help for more details).

The SVD decomposition of the Hankel matrix as a function of its order is the next step of the algorithm. The empirical Hankel matrix need not be square, for instance $\mathcal{H}_{p,N}$. The SVD is computed for each order, and the n largest singular values are retained. The remaining singular values are taken to be zero. In this way we obtain a sequence of modes and modal shapes for each 3-tuple (p, N, n). In what follows we choose $p = N$ where N is the number of lags used in computing the correlations. Also, we compute the modal characteristics for a large number of n.

An important point here is that we will later compare the same modal shapes identified for different orders (if they exist at each order). For this purpose we need always to have the same reference basis for all orders. Consequently, the SVD has to be performed while the different transformations are being stored to get back to the common initial basis for the eigenvectors.

The crucial point is then how to select a model order with the investigation of the stability of frequencies and mode shapes for increasing model order (the damping may be used too, even if its computation is very sensitive). This is the starting point of the selection (presented in Section 12.4.1 for the instrumental variable method): We need to take the good values to be the centers of the "frequency windows" used for the alignments. This is a natural and easy choice to be done by a visual examination but rather complicated to implement in an automatic way.

Interpretation of Results

The previous raw results could easily be obtained with Fortran or C routines. But Scilab facilitates the interpretation of the results. The interpretation is greatly aided by the use of various graphics tools.

First, the identified poles corresponding to consecutive model orders are compared. Given a pole at order i and a pole at order $i + 1$, if the corresponding natural frequencies are close, the pole is said to be stable in frequency. The damping ratios may also be compared, but it is well known that they are more difficult to determine. Thus, for damping, the comparison is not as conclusive. The third element for the comparison is the modal shapes. The stability of the pole is confirmed if the vectors are nearly co-linear. This can be a classical modal assurance criterion (MAC), which is typically a normalized scalar product.

These above-mentioned characteristics can be represented by a plot called the stabilization diagram. On the x-axis we have the increasing SVD model order and on the y-axis the natural frequencies. For a given order we retain only the largest eigenvalues. Our strategy is to keep a large number of eigenvalues, since the following procedure helps to eliminate the inconsequentially small values.

We then apply the selection procedure described above as a stability-in-frequency criterion. That is, when there is strong alignment in a y value as a function of x, it is retained as a legitimate eigenfrequency. For this procedure we examine all the identified values at the last order, we choose an interval in frequency around each of these values, and we count the number of occurrences in the interval for all the orders. In the example we compute values for 40 different orders, the interval is 0.2 Hz, and an eigenvalue is kept if the number of occurrences in the interval is greater than 15. This result is then improved with the plot of MAC values for modal shapes corresponding to the retained frequency.

12.5.2   Some Plots of Results

Here a classical approach for a single record is described. The beginning of a recorded signal is plotted in Figure 12.4. The spectral estimate of the signal is computed based on averaging blocks of 512 sample values and is plotted in Figure 12.5.

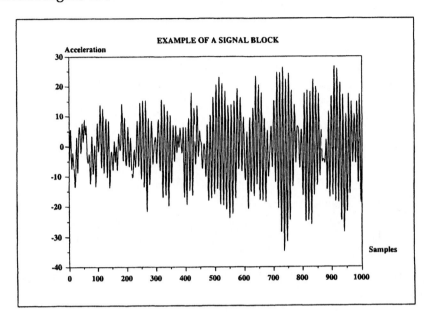

Figure 12.4: *Output of a sensor*

The peaks of the spectrum are then located, which is accomplished with the help of the **2D zoom** button on the graphics window and the **locate** command (which returns values clicked with the mouse in the graphics window). The corresponding frequencies are identified, and the result is the following:

```
-->xy=locate(6)
  xy   =

        column 1 to 4

!    9.0666667      20.977778     46.044444     49.955556 !
!    0.483          0.18          0.39          0.855     !

        column 5 to 6
```

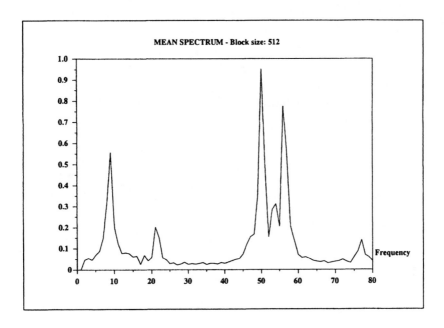

Figure 12.5: *Spectrum of previous signal*

```
!    76.977778       56.177778 !
!    0.084           0.324     !

-->xa=round(xy(1,:)+0.3)
 xa  =
!   9.      21.      46.      50.      56.      77. !


-->fre=(xa-1)'*410.004/511
 fre  =
!    6.4188493 !
!    16.047123 !
!    36.106027 !
!    39.315452 !
!    44.129589 !
!    60.979068 !
```

In the developed application that implements the algorithms discussed here many parameters can be chosen by using dialogue windows. Furthermore, significant numerical values are displayed for the specific cases where some sorting may be difficult to perform. A typical case would be the mixing of 2 very close frequencies. Displaying only a tight frequency interval and using simultaneously the MAC information for the 2 modes allows the separation of all the characteristics of these 2 modes.

We present here only the standard plots in a simple case and for a simple identified mode. Figure 12.6 is the stabilization diagram, and the selected frequencies are shown in Figure 12.7 for a window size of 0.2 Hz, with 12 occurrences of a frequency being necessary to retain it. Forty-one orders were computed.

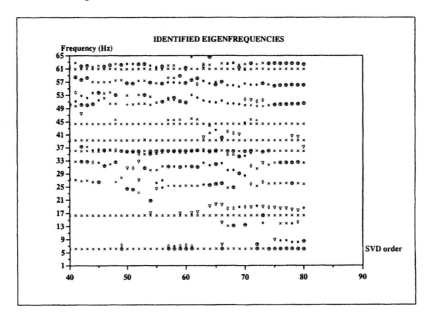

Figure 12.6: *Stabilization diagram*

The numerical values associated to the main 6 frequencies are

```
Number of modes :
     6
*************************************
FREQUENCIES
Frequencies            Damping coef(%)
     60.8923           0.5421
     44.3648           0.5120
     39.2900           0.5093
     35.9581           0.7881
     16.3577.          0.6072
      6.1400           1.2900
*************************************
Frequency means        Damping means
*************************************
```

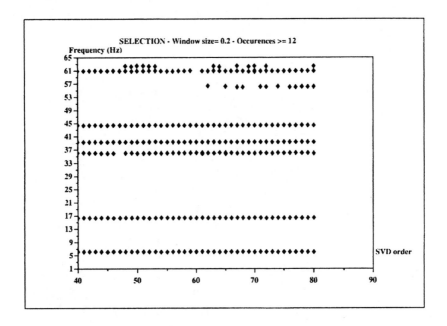

Figure 12.7: *Selected eigenfrequencies*

```
POLES
Real parts              Imaginary parts
     0.596089472             0.804565102
     0.778621815             0.629482810
     0.825168283             0.567084630
     0.853616257             0.524586748
     0.970180289             0.248427836
     0.998713760             0.094250701
************************************
```

The major point here is that we can manage how many simultaneous sensors we want, and we can get the corresponding damping coefficients and modal shapes of the mechanical structure.

The next few plots concern the eigenfrequency at 60.89 Hz. This value is a mean value. At the different orders we get values around this value (the role of the selection window is to retain values with small variations). Figure 12.8 shows the variation of this value with respect to the SVD order.

The ensuing two figures give the evolution of the MAC criterion and that of the damping coefficient. This is a favorable case with values close to 1 for the MAC criterion.

The MAC criterion is widely used for modal analysis. There are some slightly different definitions; it is basically the normalized scalar product

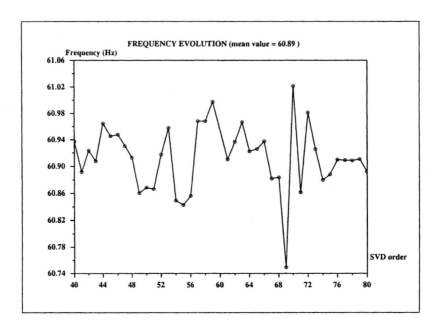

Figure 12.8: *Evolution of a selected eigenfrequency*

of two eigenvectors (this is strictly the case here). For complex structures, eigenvectors can be considered the same if the MAC criterion values are greater than or equal to 0.7.

It is common to represent the matrix of the values of the MAC criterion for all the couples of the identified eigenvectors. In our case we have simply taken a particular eigenvector v_{31} (number 31 value), and we represent the MAC values for all the pairs (v_i, v_{31}) in Figure 12.9.

For the same eigenfrequency we have represented in Figure 12.10 the variation of the values of the damping coefficient corresponding to the values of the eigenfrequencies of Figure 12.8. This figure confirms the fact that the identification of damping coefficients is more difficult than the identification of eigenfrequencies. The variations of the values of the damping coefficient are of the same magnitude as the mean value. The plot is important to get an acceptable "visual" estimation: In our example we can conclude that the damping coefficient is around 0.6%.

As a final example, the MAC criterion is computed for the eigenfrequency of 6.1399884 in Figure 12.11. This plot shows an excellent stability of the computed eigenfrequency.

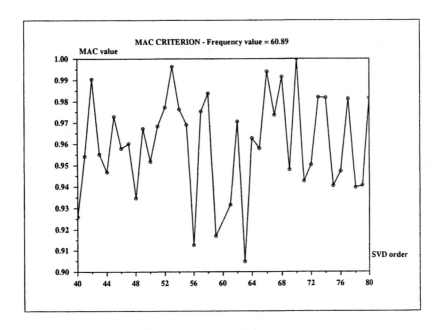

Figure 12.9: *MAC criterion for the previous eigenfrequency*

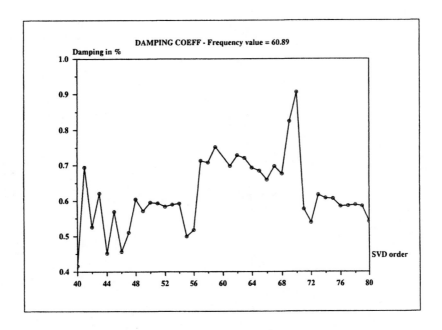

Figure 12.10: *Evolution of the damping coefficient*

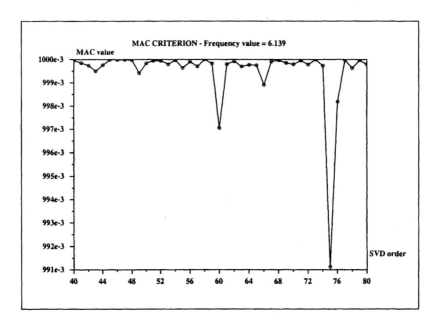

Figure 12.11: *MAC criterion*

Chapter 13

Control of Hydraulic Equipment in a River Valley

13.1 Introduction

This chapter concerns a real industrial application built with Scilab for the E.D.F., the French electrical utility. This application is the result of cooperation with E.D.F.-L.N.H[1] and is based on the Ph.D. thesis of H. Jreij [38]. Here Scilab is used to support a menu-driven integrated tool named Metalido for the analysis and automatic control of hydraulic equipment along a river valley.

Automatic control systems of hydraulic equipment in a river valley have many objectives. One concerns the optimization and control of electricity production in power plants. Also crucial is the need to protect people and property. In addition there are constraints due to navigation of ships and those due to the needs of irrigation (i.e., water levels at certain times must be kept within narrow tolerances).

The physical complexity of the system was a major challenge in solving this problem. Furthermore, the amount and variety of data, the industrial requirements concerning safety, and the compatibility of solutions with the evolution of future software developments were all important issues.

This application makes extensive use of many of Scilab's tools. Control and simulation play major roles in this problem. However, the example also illustrates the use of application-specific abstract data types, interactive dialogue and graphics functions, and the use of high-level Scilab tools such as Scicos (see Chapter 9) and Metanet (see Chapter 11).

---

[1]Électricité de France, Laboratoire National d'Hydraulique
6 quai Watier, 78401 Chatou CEDEX, France.

13.2   Description of a Managed River Valley

The aim of Metalido is to model river valleys controlled by river management equipment. A natural river valley is, in general, formed by a long depression on the earth's surface. Most of the time the water flows in a shallow part of this depression. However, in times of flood the water levels can change significantly. This is complicated by the fact that a river is normally a collection of flows coming from an ensemble of tributary rivers.

13.2.1   Hydraulic Equipment in a River Valley

A managed river valley is, in general, divided into a sequence of races (river portions between dams), as shown in Figure 13.1. Positions of dams are chosen, when possible, to optimize power generation. The mean distance between two consecutive dams is typically 10 km. Often, artificial deviations of water flow are created to provide water to power plants or to create special passages for navigation. These activities transform the one-dimensional topology of natural river flow to one that is much more complex.

13.2.2   Power Production

Each plant may work according to several modes.

- The first mode maintains the natural flow of the river with only small flow variations, which are performed to accommodate power modulation signals.

- The second mode is used to satisfy peak power demand. In this case races with no "hard level constraints" are used to increase (or decrease) significantly their turbined output flow. Under these conditions these races and all downstream races produce much more power for a short period of time.

- The third mode is for flood conditions. If a model for the flow of flood conditions exists, it is possible to decrease the water levels of downstream races in advance in order to get more storage capacity when flood peak flows arise.

13.2.3   Structural Analysis

Due to the nature of water flow, each upstream race acts mainly on the next downstream race. However, a slight downstream-to-upstream race coup-

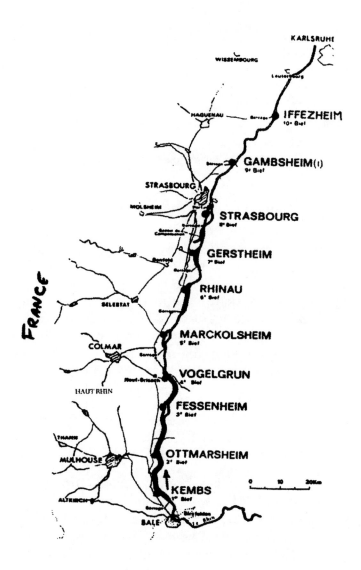

Figure 13.1: *Plants along the Rhine River*

ling exists due to the variation of water fall height. As shown in Figure 13.2, a cascade diagram may be used to represent the sequence of races.

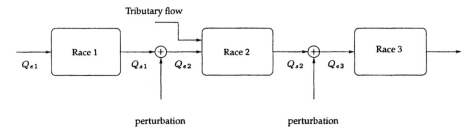

Figure 13.2: *Race sequence*

Perturbations can represent unmodeled tributary flows, error in controlled flow, etc.

13.2.4   Controller Structure

River valleys form large nonlinear systems with many possible working configurations and large propagation delays. Because of the numerous combinations of race working configurations, a global controller would probably be too complex. Utilizing up-to-down information flow (a race acts mainly on the next downstream race), a "natural" multilevel graded system may be used. It consists of a central hydraulic supervision station and a local controller associated with each race.

13.2.5   Central Hydraulic Supervision Station

The central hydraulic supervision station is in charge of power production and long-time scale actions. It receives power demand and synthetic information (working mode, support flow) concerning all plants and sends reference levels, power modulation signals, flow feed-forward (parallel anticipation) to each local controller. An example of a central hydraulic supervision structure is shown in Figure 13.3.

Parallel anticipations take into account input flow of the first river race and other *measured input flow* dynamics to compute long-time scale trends for each race output flow. Each of these trends is generated by parallel anticipation filters (**P**). The anticipation actions take advantage of water propagation time from one race to another to anticipate decisions for downstream races. For example, if a flood occurs at the first race of the river,

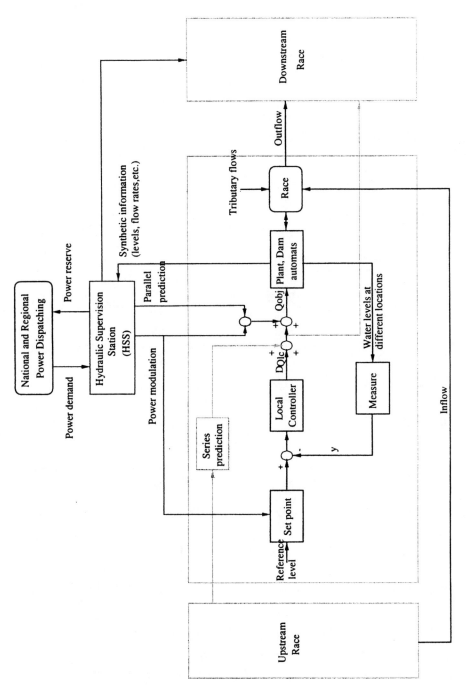

Figure 13.3: *Example of central hydraulic supervision structure*

downstream races may, if possible, decrease their levels to be able to store a part of the oncoming flood waters.

13.2.6   Local Controllers

Local controllers (one for each race) have only to maintain levels at their reference values when open-loop centrally specified levels are insufficient. The control signal for each local controller is the combined turbine and dam sluice gate regulated outgoing flow, as shown by Figure 13.4.

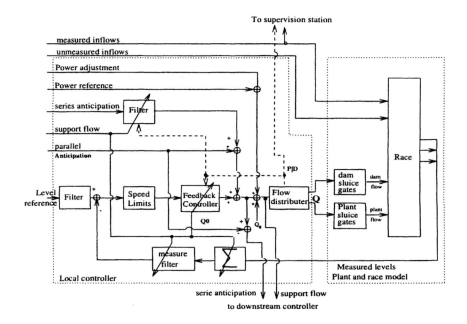

Figure 13.4: *Local controller typical structure*

The series anticipation signal is the regulated upstream race flow used to perform local control whenever possible through the anticipation filter (**Fas**). Series anticipation takes advantage of wave propagation delay on the race water surface to destroy the descending wave with an ascending one. The feedback controller uses measured levels to achieve the regulation objectives.

To take into account the large range of operating regimes and changes in race dynamics, the feedback controller, series anticipation filters, and measure filters are adjustable with the support flow. These filters also must change according to the feedback regulation actions on plant and dam sluice controls.

13.3 Race Modeling

13.3.1 Physical Description

A typical race is formed by two upstream and two downstream branches separated by dams, as shown in Figure 13.5(a). Figure 13.5(b) shows artificial deviations of water flow created to provide water to power plants or to create special passages for navigation.

However, various other configurations may occur. In particular, a power plant and the main dam may be on the same branch. In many cases the flood areas must also be taken into account for an accurate race modeling.

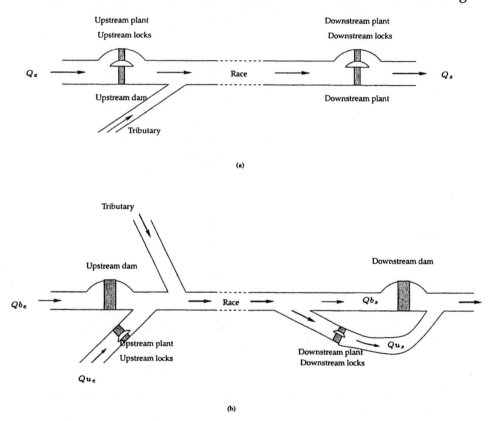

Figure 13.5: *Typical race diagrams*

Various operating conditions may arise. For low flows, water passes through the power plants. In case of floods, excess water passes through dam sluices. If a power plant fails, overflow weirs are opened. For navigable rivers, lock operations may cause strong stream impulses, which must be taken into account.

13.3.2 Mathematical Model

Development of a control system requires a model of the dynamics of each race that represents the wave propagation and the volume variation for a wide range of flow rates. Here each race is first split into a set of interconnected branches, as shown in Figure 13.6. Each branch is assumed to be

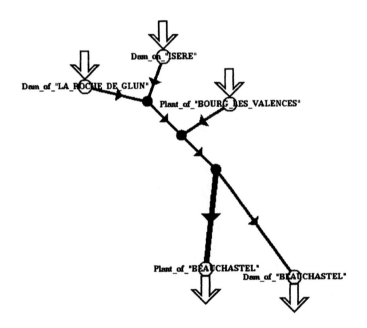

Beauchastel race on the Rhône River is formed by 7 branches.

Figure 13.6: *Branches interconnection representation using Metanet*

governed by the one-dimensional Saint Venant equations [41], [20] (in some cases turbulence models might be necessary):

$$\frac{\partial Q}{\partial x} + \frac{\partial S}{\partial Z}\frac{\partial Z}{\partial t} - q = 0,$$

$$\frac{\partial Q}{\partial t} + \frac{\partial (Q^2/S)}{\partial x} + gS(\frac{\partial Z}{\partial x} + \frac{Q^2}{K^2 S^2 R^{4/3}}) - qv_l - gS\theta = 0,$$

where

t :	Time,
x :	Distance along the flow axis,
$Q(x,t)$:	Mean flow,
$S(x,t)$:	Cross-sectional area,
$Z(x,t)$:	Free surface altitude,
$K(x,t)$:	Sliding Strickler coefficient,
$R(x,t)$:	Hydraulic radius,
$q(x,t)$:	Tributary flow,
$v_l(x,t)$:	Tributary flow speed,
g :	Acceleration due to gravity,
$\theta(x)$:	The slope of the bottom of the branch,
x_b :	coordinates of the boundaries of the branch,

and the boundary conditions are given by the evolution of $Q(x_b, t)$ or $Z(x_b, t)$ over time. The functions $Z(x,t)$ and $Q(x,t)$ are to be determined. The functions $S(x,t)$ and $R(x,t)$ are given by the geometric characteristics of the race.

Branch connections are modeled by two equations that ensure the conservation of flow and the equality of free surface altitude at the common boundaries. Plant and dam sluices also have to be modeled to describe the relationship between sluice aperture and flow. These models are, in general, provided by table lookup, where flow is given as a function of sluice aperture, water level, and fall height. Sluice opening dynamics are often modeled by linear first-order systems.

13.3.3 Race Numerical Simulation

Simulator Design

A numerical simulator for multibranch races with singularities and flooding areas has been implemented in Fortran. This program, called LIDO, was designed by the Laboratoire National d'Hydraulique E.D.F. [40]. For this application, LIDO has been interfaced with Scilab using `intersci` (see Section 5.5) .

A given race is described by cross sections measured along each branch of the race (see Figure 13.7), singularity positions and types, and, finally, by the values of the Strickler coefficients along the race. The Saint Venant equations are then solved using an implicit finite difference method using

Wendroff sampling. The Strickler coefficients are calibrated so that simulations match real static and dynamic tests carried out on the race.

Race of Beauchastel

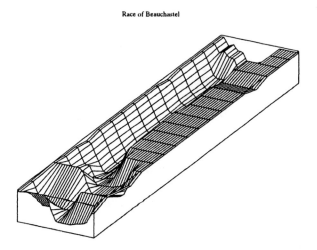

Branch to Beauchastel plant (highlighted arc in Figure 13.6)

Figure 13.7: *3-D view of a branch*

Race description data (measured cross sections, singularity descriptions, branch connections, etc.) are read from ASCII files and stored in a specialized Scilab data structure, `lido_dat`, using the function `leclido`. The `lido_dat` structure contains all the physical descriptions of the race and its sampling intervals along the x and z coordinates. For the sake of efficiency it stores data as a copy of the LIDO data Fortran COMMONs. Each component of a `lido_dat` data structure may be extracted or modified through special Scilab functions called `getlido` and `setlido`. For example, `getlido('DXP')` and `getlido('DYP')` get the coordinates of the race cross sections.

The simulator function `runlido` can be used to compute the values of $Z(x, t)$ and $Q(x, t)$ for specified values of $t > t_0$ given the initial state $(Z(x, t_0), Q(x, t_0))$ and the time evolution of all boundary values of the flows $Q(., t)$ and the levels $Z(., t)$. The evolution of boundary values over time is described in lookup tables. An example of the calling syntax to `runlido` is

```
[Q,Z]=runlido(Q0,Z0,x,t_desc,boundaries,pflags,lido_dat)
```

where `t_desc` gives the initial, final, and incremental values of time.

Steady-State Qualitative Behavior

A race steady state corresponds to constant input and output flows whose sum equals zero. These steady states are computed in Scilab by repeated calls to **runlido** until water surfaces stabilize. Due to race dynamics, to establish real steady state of a race can take up to 24 hours.

For different values of water volume in the race but with identical flows, the water surface of the race is nearly translated as indicated in Figure 13.8, which represents water surface levels along the race flow axis corresponding to three different stored volumes. Note also that since the race is carrying flow, the race surface is not horizontal.

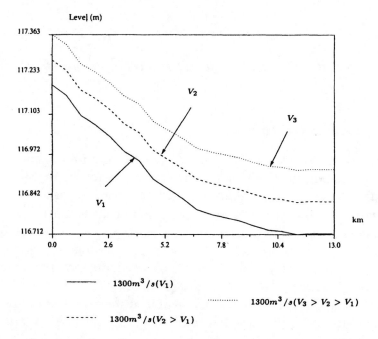

Figure 13.8: *Steady-state water level for various volumes V_1, V_2, V_3*

On the other hand, steady-state water levels corresponding to different stationary flows but identical stored volumes show a narrow zone where the level remains constant. This zone is often reduced to a single point, as shown in Figure 13.9, which represents water surface levels along the race flow axis corresponding to four different flows. Water surface slope increases with the flow rotating around this point.

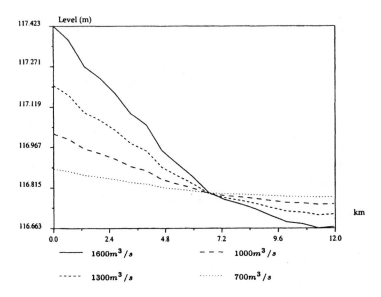

Figure 13.9: *Steady-state water level for various flows*

Dynamic Qualitative Behavior

If downstream flow increases, waves are created and propagated along the water surface. Figure 13.10 shows the water surface level along the race flow axis at several times after the downstream flow increases.

One may also observe the time evolution of the water level at a given point of the race after a downstream flow increase. As shown in Figure 13.11, a transient phase corresponding to wave propagation and reflection is observed after a time delay (due to the rate of wave propagation at the surface of the race). Finally one observes a nearly linear decrease of the level. This corresponds to the exhaustion of the water in the race and is linear if the free water surface of the race does not vary (race banks are vertical in the working region).

13.4   Choice of Observation

Two possibilities can be considered for the choice of input signal to the local controller for each race. The first is the level of a critical point that must be particularly supervised. The second is the volume of water stored in the race. In particular, this last objective is useful for cumulative extra flow regulation. For both cases, the measurement of water levels at only a few points

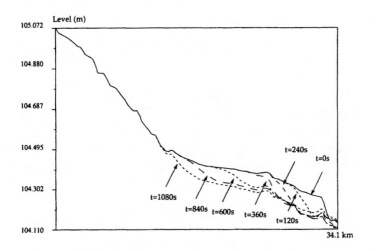

Figure 13.10: *Transient behavior*

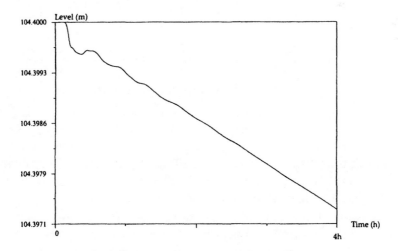

Figure 13.11: *Step response at a given point*

in the race are available. In fact, normally only two measurement points, located on upstream and downstream dams, are available. However, a few other points may be added if necessary.

13.4.1 Volume Observer

Control of the volume is particularly useful for satisfying peak power demand. In this case, races with no "hard level constraints" (i.e., races with no navigation or safety constraints) are made to increase (or decrease) significantly their turbined flow compared to natural river flow (see Figure 13.12). This extra flow allows extra power production for this race and all downstream races.

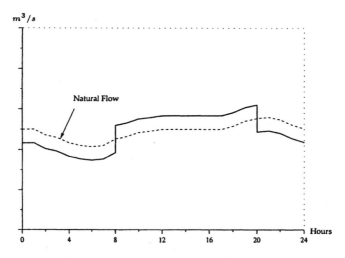

Figure 13.12: *Daily flow modulation*

For races with simple geometries, steady-state constant-volume water surfaces rotate around an equilibrium point. Under these circumstances the volume stored is a simple function of level at the equilibrium point (see Figures 13.9 and 13.13). Note that although the dynamics of the race depend on nonlinear partial differential equations, the volume stored is just the integral of the algebraic sum of input and output flows.

For complex races, it may be shown (even for small n) that the stored volume is well approximated by

$$y = \sum_{i=1}^{n} C_i h_i,$$

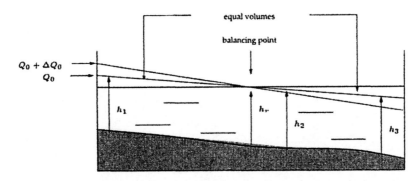

Figure 13.13: *Water surfaces for different steady states with constant volume*

where the h_i are water levels at measurement points and the C_i are positive weighting factors satisfying $\sum_{i=1}^{n} C_i = 1$. The C_i are computed to minimize

$$J = \| \sum_{Q,t_k} \sum_{i=1}^{n} C_i h_i(Q, t_k) - h_r \|_2$$

for a given reference level h_r and for an *equivolume* dynamical test performed on a set of steady states relative to different support flows Q for the same reference volume. Here $h_i(Q, t_k)$ is the water surface level at time t_k during the dynamical test associated with support flow Q.

The minimization of the above equation is done using Scilab's linear least-square solver. In this way an approximation of the volume over the entire working domain is obtained with only a few measurement points, as shown in Figure 13.14, which represents the evolution of the measured level h_i at each measurement point as a function of steady-state flow and shows that the composite measure, C_m, is nearly independent of the flow.

If it is necessary to add measurement points, their locations are chosen to minimize J at all possible locations. Figure 13.15 shows J as a function of position of a new measurement point along the cross sections of a branch.

13.4.2 Level Observer

Control of the level is generally used to assign a critical point when the race turbine flow is the natural river flow. The critical point level may not be measured. In this case a critical level observer is needed.

Water surfaces are computed for different steady flows with *constant level h_r at the fixed point*. The computation of these water surfaces is performed using a simple regulator to adjust output flow to the level at the

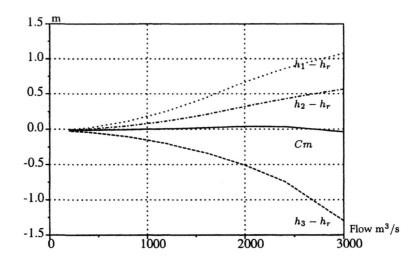

Figure 13.14: *Individual ($h_i - h_r$) and observer ($C_m = y - h_r$) evolution vs. steady-state flow*

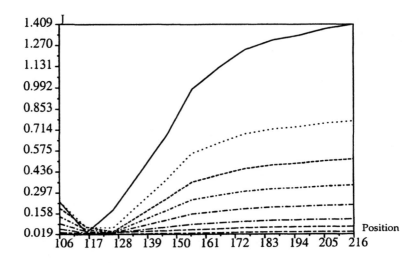

Figure 13.15: *J vs. position and various steady-state flows*

fixed point. Note that these water surfaces correspond to different stored volumes, as illustrated in Figure 13.16.

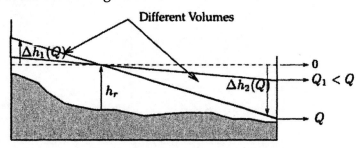

Figure 13.16: *Water surfaces for different steady flows with constant level*

The level at the fixed point may then be estimated by

$$y = \sum_{i=1}^{n} C_i(h_i - \Delta h_i(Q)),$$

where the $\Delta h_i(Q)$ are tabulated from isolevel steady-state surfaces, and the constants C_i are computed so as to minimize

$$J = \sum_{Q,t_k} \sum_{i=1}^{n} (h_r - C_i(h_i(Q, t_k) - \Delta h_i(Q)))^2$$

on a set of dynamical tests performed on steady states relative to different support flows Q for the same reference level h_r.

13.5   Control of a Race

As shown in Figure 13.17, from the control point of view a race may be seen as a dynamical system whose inputs are the input and whose output flows and whose output is the observer built with the water level measurements at a few points.

The controlled input is the variation ΔQ_s of the flow Q_s passing through downstream plant or dam sluices.

13.5.1   Race Dynamics Identification

To improve the regulator design we need more tractable models to describe the dynamics of the input/output flows than what the Saint Venant equations provide. As the dynamics of input flows are generally slower than

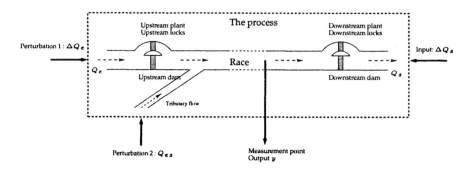

Figure 13.17: *Control diagram of a race*

those for the race, models linearized around a set of steady-state regimes are sufficient to regulate races.

A set of linear models is computed for each race using computed step responses relative to main input or output flows for operating regimes spread from low flow to flood (see Figures 13.18 and 13.19). Due to the physics of the race, the transfer from output flow to level measurements and input flow to level measurements are essentially formed by a time delay (due to wave propagation), an integrator (for storage of water), and a pure dynamical part.

The transfer functions have the following structure:

$$T(p, Q) = e^{-\tau(Q)p}\left(\frac{\pm 1}{S(Q)p} + G(p, Q)\right),$$

where

$$G = \frac{\sum_{i=0}^{m} \beta_i(Q)p^i}{\sum_{i=0}^{m} \alpha_i(Q)p^i}.$$

This can be approximated by a discretized model as

$$T(z, Q) = z^{-\tau(Q)/\Delta t}\left(\frac{\Delta t}{S(Q)(1 - z^{-1})} + G(z, Q)\right),$$

where

$$G = \frac{\sum_{i=0}^{m} b_i(Q)z^i}{\sum_{i=0}^{m} a_i(Q)z^i}.$$

These transfer functions may be computed using step responses for the race. The propagation delay τ can easily be determined from each step response (see Figure 13.18). Note large delays and dispersion in the step responses relative to different flows.

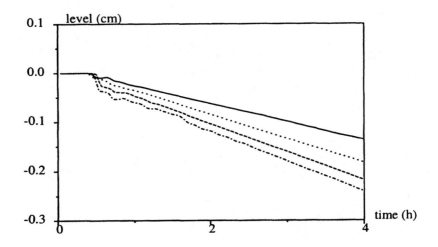

Figure 13.18: *Step responses with one measurement point in the middle of the race for various flows*

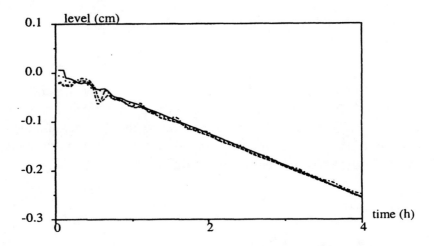

Figure 13.19: *Typical step responses for three-point composite measure*

To identify G one must use traditional least squares identification methods applied to the step response. However, for low flows least squares often fails to fit the oscillations due to wave reflections. Under these circumstances least squares can also give rise to an unstable model. In this case L_2 approximation methods applied to the derivative of the impulse response (see Figure 13.20) such as those proposed by [6] and [5] (corresponding to the Scilab function **arl2**, see Section 6.10) are more efficient (see Figure 13.21).

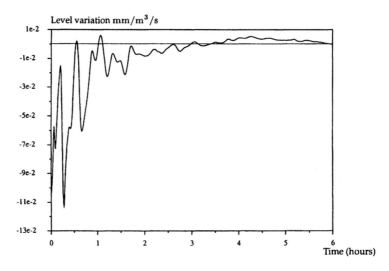

Figure 13.20: *Impulse response*

High-order models are necessary to fit the data with good precision, but note that the most important part of the model, from a control theory point of view, is given by the integral part and the propagation delay of G. A second- or third-order model of G is often sufficient, and modeling errors may be handled by a robust controller design.

13.5.2   Local Control Synthesis

The two major parts of the local controller are the series anticipation feedforward, which adjusts parallel anticipation using previous race-regulated flow, and the feedback controller, which ensures precise level control.

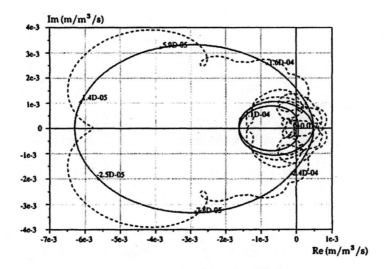

Nyquist plot of the initial impulse response (dashed line) and of the fourth-order model (solid line)

Figure 13.21: *Fourth-order identification of the impulse response*

13.5.3   Series Anticipations Design

Main input flow of a race is often known quite precisely (it is the regulated output flow of the previous race). So it is possible to take advantage of the water propagation delay from input to measurement to obtain an efficient action. Series anticipation actions differ according to regulation objectives. For the control of the stored volume the action is to directly "copy" the input flow variations on the output flow because the volume is just the integral of all flows coming in or out the race. For control of the level, series anticipation actions have to change race output flow to generate an ascending wave designed to kill (at the critical point, as shown in Figure 13.22) descending waves created by input flow.

Let T_1 and T_2 be the discrete impulse response transfer from input flow to measurement and output flow to measurement, respectively, and S the discrete anticipation transfer impulse response from input to output flow. Then

$$\frac{\Delta h}{\Delta Q_{in}} = T_1 + S * T_2.$$

Given this equation, the first inclination is simply to invert T_2 to get S by solving $T_1 + S * T_2 = 0$, but this may produce an unstable series anticip-

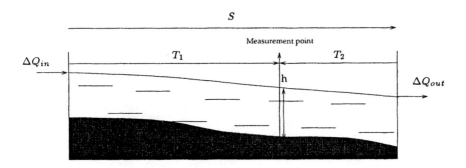

Figure 13.22: *Series anticipation diagram*

ation filter. Moreover, it is not advisable to compute such an exact solution, because it may produce high-frequency changes in the output flow and, consequently, in the input of the next race.

The form of T_1 and T_2 (see page 458) suggests looking for an S of the form $S = z^{-k}(1 + (1 - z^{-1})\overline{S})$, where $\overline{S}$ is a low-order transfer function. It is required that S be such that the observation steady-state error is zero for impulse and step variation ΔQ_e. This condition gives rise to linear equality constraints on $\overline{S}$. The coefficients of $\overline{S}$ are then optimized to minimize

$$J = \|\frac{\Delta h}{\Delta Q_e} - T_1 + S * T_2\|_2.$$

The above computation can be made using Scilab's polynomial and optimization toolboxes. Optimization with boundary constraints is used here to ensure filter stability. Furthermore, the Maple to Scilab toolbox (see Chapter 10) is used to solve equality constraints on $\overline{S}$ and to generate Fortran code, which is used to evaluate the cost function, J, and its gradient with respect to the parameters of S. This function is subsequently linked (see Section 5.6) to Scilab and used as a functional argument to **optim** (see Section 8.4), which searches for a local minimum of J.

Figure 13.23 shows the action of a second degree optimized series anticipation filter for a $100\,\mathrm{m}^3/\mathrm{s}$ step input flow. The top plot shows the input and output flow evolution and the bottom one the measured level evolution. For a $100\,\mathrm{m}^3/\mathrm{s}$ input flow step the critical point level variation is less than one centimeter. The unexpected residual steady-state error is due to the unmodeled delay fraction smaller than the time step.

Note that T_1 and T_2 depend on the support flow Q, as must S. To take this dependency into account we compute a set of transfer functions, $S(Q_k)$, around steady-state support flow Q_k. The global S results from pole and

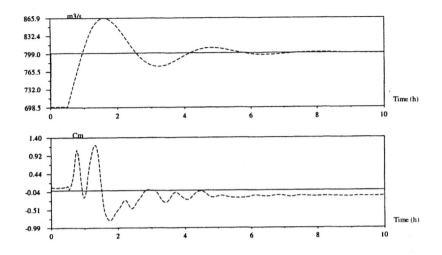

Figure 13.23: *Series anticipation simulation*

zero interpolation along with steady-state flow Q. Figure 13.24 shows the simulation results for such an adaptation with an input flow increase of $500 \, \text{m}^3/\text{s}$ (top plot solid line).

13.5.4 Parallel Anticipation Design

For series anticipation, long propagation delays along the river can be used to compute open-loop actions (race output flow variations) to preventively increase or decrease the volume stored in each downstream race. Each measured or predicted tributary input flow along the river may be used to compute a parallel anticipation action. An anticipation filter (P_i) (see page 444) is designed for each measured input flow and for each downstream race. Figure 13.25 shows a diagram of series (S_i) and parallel (P_i) feedforward actions on a three-race sequence with only one input flow.

The level variation of the ith race for an upstream flow variation ΔQ_{in} is given by

$$\Delta h_i = (P_{i-1} T_1^i + P_i T_2^i) \Delta Q_{in}.$$

So if one knows P_{i-1}, it is possible to compute P_i in a way similar to series anticipation.

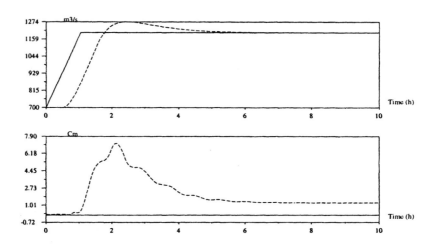

Figure 13.24: *Series anticipation simulation for big flow changes*

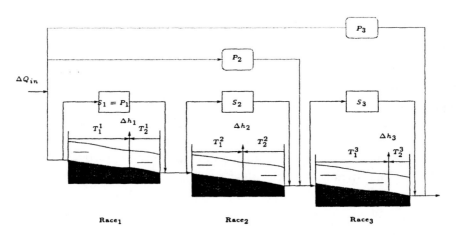

Figure 13.25: *Diagram of parallel anticipation actions due to top input flow*

13.5.5 Feedback Controller Design

In most cases anticipation actions satisfy most regulation objectives. Thus, simple discrete proportional integral (PI) controllers designed for low flow variations and only taking into account the integrator term $(1/(S(Q)p))$ are often sufficient for simple races. This may not be true if delays are present (control of the volume).

When output flow to observation transfer functions presents important delays, the problem is more difficult. Many of the classical methods in Scilab's control toolbox have been tested for this feedback controller (LQG, LQG/LTR [44] [14], adaptive model matching [22] [23], etc.), but none of these permits the design of a robust controller, with a proven stability, for all the support flow range. The major difficulties are the large wave propagation delay and the robustness to model errors and unmeasured flows. H. Jreij [38] has implemented a set of new controller design functions in Scilab (based on Smith [54] [55], Watanabe [57], Morari and Zafiriou [45], and Åström [4]) that produce more effective feedback controller designs than the classical approaches. All these new methods require polynomial computations. The most efficient method in this case was the so-called H_2 optimal, two degree of freedom, internal model controller (I.M.C.) [4] (see Figure 13.26).

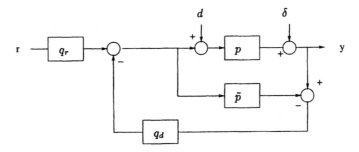

Figure 13.26: *Diagram of two degrees of freedom I.M.C.*

In Figure 13.26, p is the race dynamic and $\tilde{p}$ is the race model. The filter q_d is computed to reject perturbations d and δ (this filter is required because p contains an integrator), and q_r is a filter for good reference following. Both filters may be tuned separately to achieve robust performance objectives.

Figure 13.27 shows the evolution of the observation for an initial step of 10 cm of the level reference and a step of $-48\,\mathrm{m^3/s}$ at $t = 10\,\mathrm{h}$ on the perturbation d. The figure shows both the model matching controller (solid line) and I.M.C. controller (dashed line). As seen in the figure, the I.M.C.

controller yields good tracking and better disturbance rejection than the model matching controller.

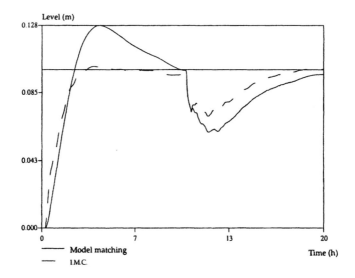

Figure 13.27: *Two degree of freedom I.M.C. action*

13.6 Metalido Overview

All the computations described above have been implemented in the Metalido application package written mainly in the Scilab programming language (the package consists of more than 30,000 lines of Scilab code).

13.6.1 Graphical User Interface

Metalido is a menu-based system. All the modeling, conception, and validation work is driven by menus. At the beginning the user may choose to work on a single race or on the entire river. When a race is chosen, Metalido first looks for Metalido saved files in a directory associated with the chosen race. If no files exist, it looks for LIDO-specific files (see page 449) and automatically creates the main data structure, **race** (see page 478). At this point the user can begin design work. The main menu (see Figure 13.28) consists of a palette of submenus.

The **Hydraulic model** submenu gives access to visualization and graphical editing of the branch cross sections.

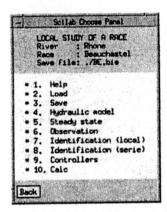

Figure 13.28: *Metalido main menu*

The observations (see Section 13.4) are specified using the submenu **Observation** (Figure 13.29). It is possible, here, to choose between level or volume observations and to compute the optimal location of the measurement points and measurement weightings.

It is then possible to define and compute a set of steady states (see Section 13.3.3 and Figure 13.30) around which the race dynamics are linearized using the **Identification (local)** and **Identification (serie)** submenus. These submenus (see Figure 13.31) allow the computation of the step responses (whose parameters may be entered in the Scilab dialogue box shown in Figure 13.32). It also allows the computation and validation of chosen linear models using various identification methods.

The final step of analysis is controller design and simulation. This part is accomplished using Scilab's interactive graphical system builder, Scicos. This is discussed in the next section.

13.6.2   Scicos

A complete description of the local controller is quite complex. However, for design purposes it is desirable to have a flexible specification in order to allow iterative adaptation of the design. Furthermore, event handling, management of sampling rates, failure management, and mode switches are all important aspects of the river valley design problem. Scicos (see Chapter 9) has been extensively used by E.D.F. to graphically describe the simulation

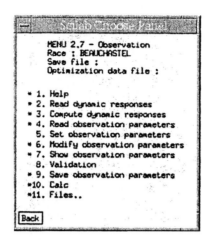

Figure 13.29: *Metalido observation determination menu*

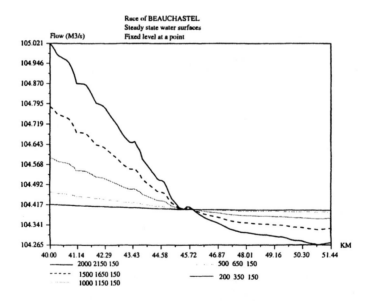

Figure 13.30: *Steady-state water surface in Beauchastel main branch*

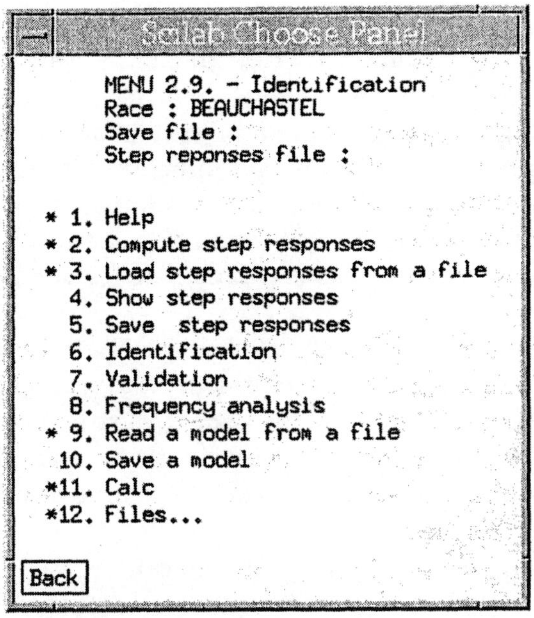

```
┌──────────────────────────────────────────┐
│ ─      Scilab Choose Panel                │
├──────────────────────────────────────────┤
│         MENU 2.9. - Identification        │
│         Race : BEAUCHASTEL                 │
│         Save file :                        │
│         Step reponses file :               │
│                                            │
│       * 1. Help                            │
│       * 2. Compute step responses          │
│       * 3. Load step responses from a file │
│         4. Show step responses             │
│         5. Save  step responses            │
│         6. Identification                  │
│         7. Validation                      │
│         8. Frequency analysis              │
│       * 9. Read a model from a file        │
│        10. Save a model                    │
│       *11. Calc                            │
│       *12. Files...                        │
│                                            │
│  ┌──────┐                                  │
│  │ Back │                                  │
│  └──────┘                                  │
└──────────────────────────────────────────┘
```

Figure 13.31: *Identification submenu*

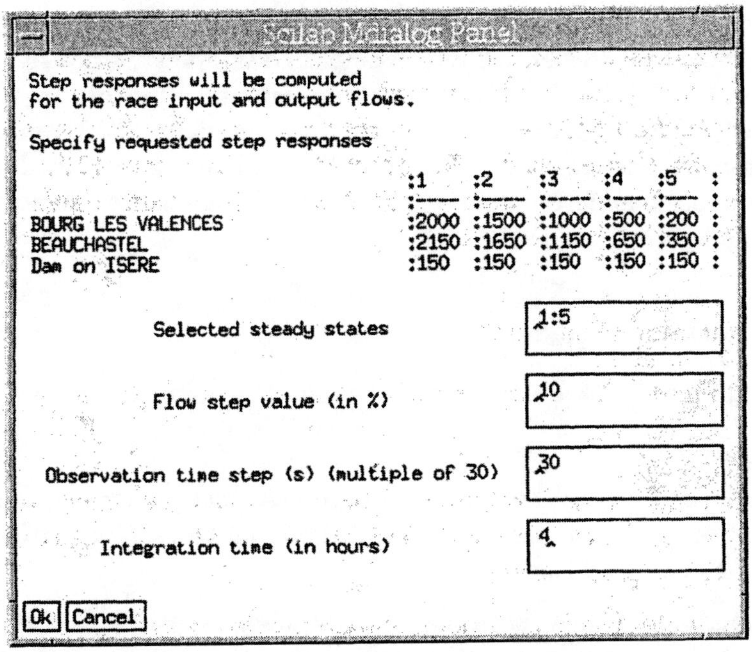

```
┌──────────────────────────────────────────────────┐
│ ─         Scilab Metalog Panel                     │
├──────────────────────────────────────────────────┤
│ Step responses will be computed                    │
│ for the race input and output flows.               │
│                                                    │
│ Specify requested step responses                   │
│                          :1   :2   :3   :4   :5  : │
│                          :─── :─── :─── :─── :─── :│
│ BOURG LES VALENCES       :2000 :1500 :1000 :500 :200 : │
│ BEAUCHASTEL              :2150 :1650 :1150 :650 :350 : │
│ Dam on ISERE             :150  :150  :150  :150 :150 : │
│                                                    │
│        Selected steady states        ┌─────────┐  │
│                                       │ 1:5     │  │
│                                       └─────────┘  │
│                                                    │
│        Flow step value (in %)         ┌─────────┐  │
│                                       │ 10      │  │
│                                       └─────────┘  │
│                                       ┌─────────┐  │
│ Observation time step (s) (multiple of 30)│ 30  │ │
│                                       └─────────┘  │
│                                       ┌─────────┐  │
│     Integration time (in hours)       │ 4       │  │
│                                       └─────────┘  │
│ ┌────┐┌────────┐                                   │
│ │ Ok ││ Cancel │                                   │
│ └────┘└────────┘                                   │
└──────────────────────────────────────────────────┘
```

Figure 13.32: *An example of Metalido dialogues*

diagram and is eminently capable of handling the above-mentioned design problems. (Note: The ensuing Scicos diagrams are a courtesy of E.D.F.-L.N.H.).

The Scicos diagram description of the system allows for the interactive update of regulation and simulation structures for each race. It also allows for easy testing on each race (some examples include fault simulation, silt-up simulation, etc.). In fact, Scicos is a very useful working tool for the entire implementation process (simulation, testing, implementation, and validation).

For the river valley problem a palette of basic Scicos blocks was added. This was due to the fact that these blocks need to use data computed in previous design steps and stored in the **race** data structure (see page 477). The blocks were written either in Scilab or Fortran code. The **lido** block, which interfaces the **runlido** function, is an example of one of the Scicos blocks created for this problem.

In the following series of figures hierarchical diagram descriptions using Scicos **Superblocks** have been extensively used. It should be noted that for simulation auditing major blocks generate their own status output. Furthermore, a global Scicos **Supervisor** block (see Figure 13.34) periodically generates a report of the system's state.

Figure 13.33 shows a simplified version of the local controller implemented for the race "Sisteron," which is located on the Durance river. The diagram in this figure has been used to simulate the effect of race silt-up. The following figures show a part of the Scicos simulation diagram for the controller of the race named "Beauchastel" on the Rhône River. It is currently being used as part of a study on how to change the regulation of all the races along the Rhône Valley.

Main Simulation Diagram

Shown in Figure 13.34 is the main simulation diagram, which consists of the following blocks:

- a super-block representing the "Bourg Les Valence" race, which produces outputs of the upstream plant, dams, and locks (located at Bourg Les Valences and Isere),

- a super-block called "Supervision Controller," which simulates the actions of the central supervision controller such as daily power production evolution, modulation to follow actual power demand, reference, and working mode (see Section 13.2.5 for this race),

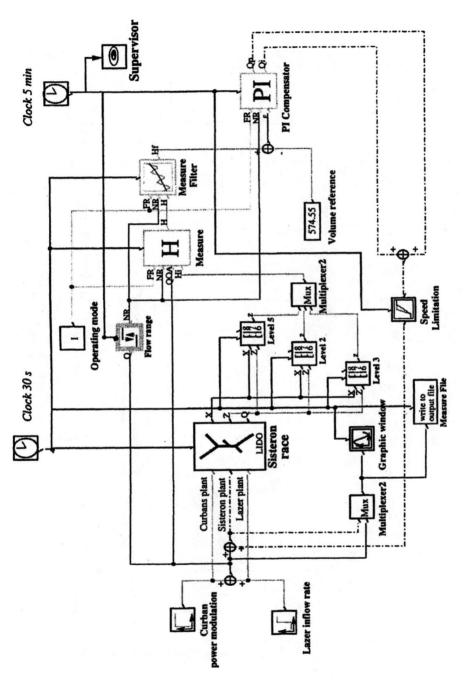

Figure 13.33: *Simple Scicos simulation diagram*

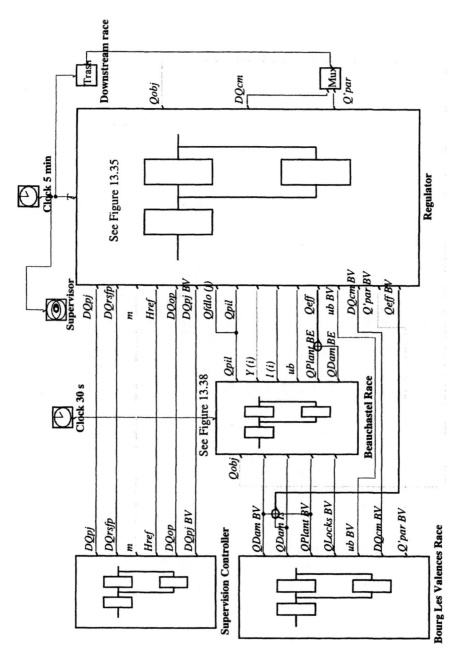

Figure 13.34: *Scicos main diagram*

- the "Beauchastel" race block, which constitutes the physical model of the race, its sluices, and the measurements described in Figure 13.38,

- the local controller block (labeled "Regulator") described in Figure 13.35,

- the supervisor block used to generate a report of the system's state, and

- two activation clocks with different time periods.

Local Controller

Figure 13.35 is a high-level representation of the local controller. Actions made by the series and parallel anticipation filters and by the regulator (see Figure 13.36) are used to compute the flow objective (i.e., the desired output flow).

Regulation Error

Figure 13.36 represents how the observation, Y, and the reference, Y_c, are obtained and shows how they are used to compute the compensator error input (see Figure 13.37). The "Observation" block computes the observation Y from the physical measurements $Y(i)$, working mode m, and information on dam saturation ub. The "Reference" block computes the reference value Yc for the observation Y from information transmitted by the central supervision controller.

Compensator

Figure 13.37 represents the adaptive PI compensator used for this race. Parameters are continuously adjusted with the support flow Q_{pil}, and they switch with the working modes m and ub.

Race Model

Figure 13.38 shows the model diagram of the physical behavior of the race. It consists of:

- water dynamics within the race represented by the $LIDO$ block,

- flow dispatcher model, which opens the dam sluices when the plant is saturated. The ub value indicates whether or not the plant is saturated. Changes of this indicator correspond to changes in system dynamics and so have to be transmitted to regulator blocks,

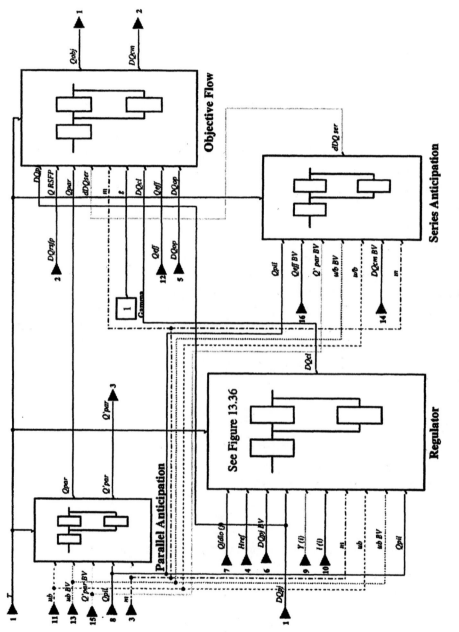

Figure 13.35: Scicos regulator main diagram

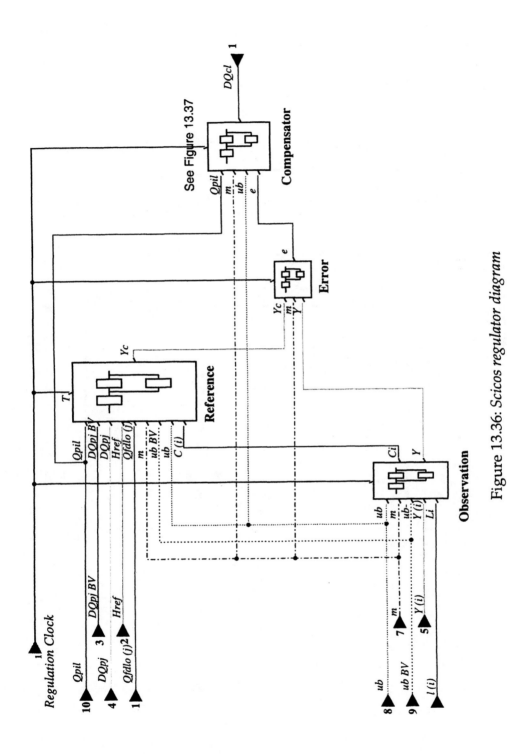

Figure 13.36: *Scicos regulator diagram*

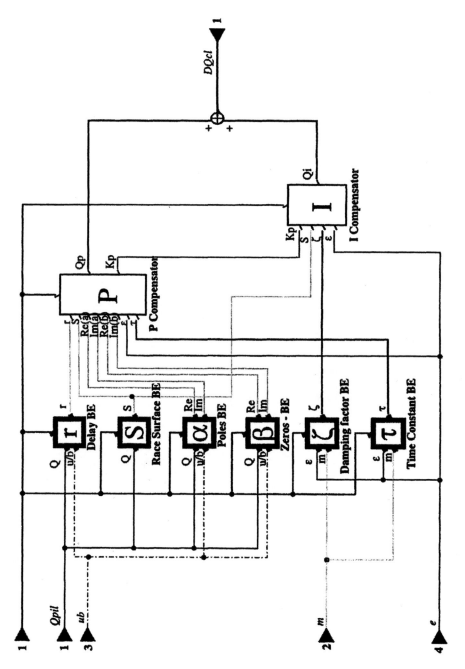

Figure 13.37: *Scicos compensator diagram*

- sluice models that form real output flow passing through Q_{eff}Plant plant or Q_{eff}Dam dam given the flow objective, Q_{obj}, and

- a model of level measurements, $Y(i)$, and support flow determination, Q_{sup} (see Figure 13.39).

13.6.3 Data Structures

The `lido_dat` data structure (see page 450) works with the data structure `race`, which maintains all the data computed with the Metalido race menus. This data structure is a Scilab `tlist` data type (see Section 2.2.8), which contains the following fields associated with the main Metalido submenus:

- **GEOMETRY**: the hydraulic model data structure consisting of cross sections, Strickler coefficients, branch interconnection data structures, and hydraulic singularities,

- **LINES**: the computed steady states data structure,

- **OBSERVATION**: the observation data structure, position of measurement points, and weighting coefficients,

- **MOD_LIN**: a data structure of linear models for the race dynamics,

- **MOD_SER**: the series anticipation filter data structure,

- **CONTROLLERS**: the regulator data structure.

Finally, for a particular race it is possible to define and save sets of data structures each set associated to different user choices. Each submenu allows one to save its corresponding subsidiary data structure and some text comments in a file.

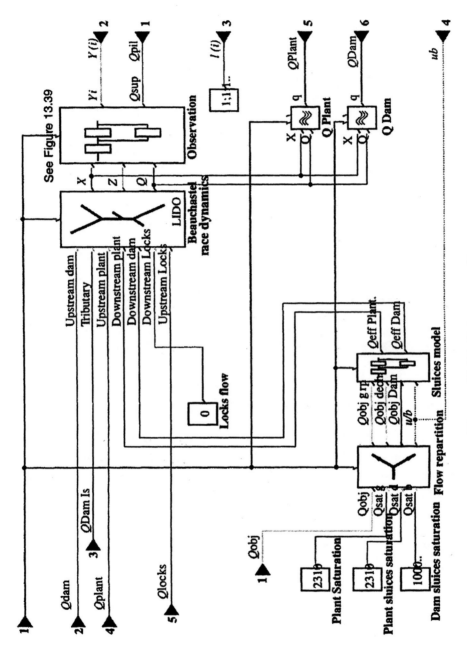

Figure 13.38: *Scicos race diagram*

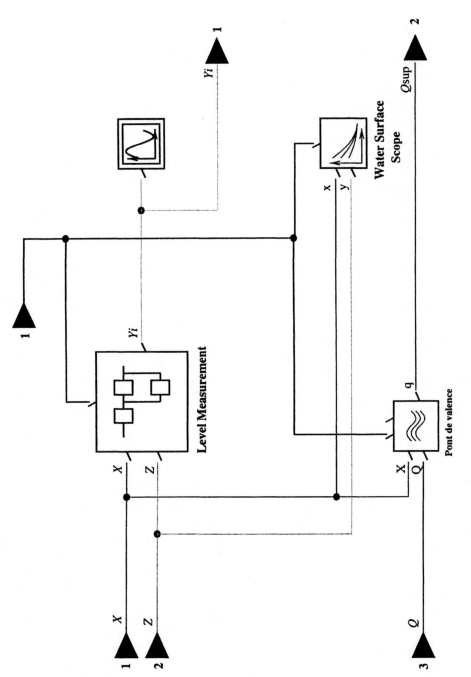

Figure 13.39: *Scicos observation*

Bibliography

[1] M. Abdelghani, M. Basseville, and A. Benveniste. Subspace-based fault detection and isolation methods. Research Report IRISA 1143 - INRIA 3299, INRIA-IRISA, 1997.

[2] C. Allal, S. Abdou, P. Daviet, and M. Parent. Tactical and technical routing of empty vehicles in the PRAXITELE project. In *4th IEEE Symposium on Control and Automation*, Crete, Greece, June 10–14, 1996.

[3] C. Allal, M. Goursat, and J.P. Quadrat. Dynamic models for the allocation and routing of empty vehicles in the PRAXITELE project. In *IFORS Conference*, Vancouver, Canada, July 8–12, 1996.

[4] K.J. Åström, C.C. Hang, and B.C. Lim. A new Smith predictor for controlling a process with an integrator and long dead-time. *IEEE Trans. on Automatic Control*, 39(2):343–345, 1994.

[5] L. Baratchart. Existence and generic properties for l^2 approximant of linear constant systems. *I.M.A. Journal of Math. Control and Identification*, 3:89–101, 1986.

[6] L. Baratchart, M. Cardelli, and M. Olivi. Identification and rational l^2 approximation: a gradient algorithm. *Automatica*, 27(2):413–418, 1991.

[7] O. Baudon and J.M. Laborde. Cabri-graph, a sketchpad for graph theory. In G. Jacob, N.E. Oussous, and S. Steinberg, editors, *IMACS Symposium on Symbolic Computation, New Trends and Developments*, pages 50–55, 1993.

[8] Th. Beelen and P. Van Dooren. An improved algorithm for the computation of Kronecker's canonical form of a singular pencil. *Linear Algebra and Applications*, 105:9–65, 1988.

[9] A. Benveniste. Compositional and uniform modeling of hybrid systems. *IEEE Trans. Automat. Control*, AC-43, 4 1998.

481

[10] A. Benveniste and J.J. Fuchs. Single sample modal identification of a non-stationary stochastic process. *IEEE Trans. Aut. Control*, AC-30, 1985.

[11] A. Benveniste and P. Le Guernic. Hybrid dynamical systems theory and the SIGNAL language. *IEEE Trans. Automat. Control*, AC-35:536–546, 5, 1990.

[12] D.P. Bertsekas and P. Tseng. Relaxation methods for minimum cost ordinary and generalized network flow problems. *Operations Research J.*, 36:93–114, 1988.

[13] J.F. Bonnans, J.C. Gilbert, C. Lemaréchal, and C. Sagastizàbal. *Optimisation numérique*. Number 27 in Mathématiques et Applications. Springer, 1997.

[14] H. Bourles and E. Irving. La méthode lqg/ltr : une interprétation polynomiale temps continu/temps discret. *RAIRO APII*, 25(6):545–592, 1991.

[15] S. Boyd, L. El Ghaoui, E. Feron, and V. Balakrishnan. *Linear Matrix Inequalities in System and Control Theory*. Studies in Applied Mathematics. SIAM, 1994.

[16] B.W. Brow and J. Lovato. Randlib: Library of Fortran routines for random number generation. Technical report, The University of Texas, 1994.

[17] B.W. Brow, J. Lovato, and K. Russell. Dcdflib: Library of Fortran routines for cumulative distribution functions, inverses and other parameters. Technical report, The University of Texas, 1994.

[18] P. Capolsini. MacroC, C code generation within Maple. Rapport technique 151, INRIA, February 1993.

[19] E.W. Cheney. *Introduction to Approximation Theory*. McGraw-Hill, 1966.

[20] J.A. Cunge, F.M. Holly Jr., and A. Vervey. *Practical Aspects of Computational River Hydraulics*. Pitman, 1980.

[21] S. Dalmas, M. Gaëtano, and S. Watt. An OpenMath 1.0 implementation. In *Proceedings of ISSAC 97*, pages 241–248, 1997.

[22] H. Dang Van Mien. Étude pour la conduite automatisée et centralisée des aménagements hydrauliques du Neckar. Technical Report HE-43/93/38/A, Électricité de France (DER), 1993.

[23] H. Dang Van Mien and K. N'Guyen. Centralized and automatic control of hydraulic plants. In *Proc. sixth international conference, Swansea*, volume 6 of *Numerical Methods in Laminar and Turbulent Flow*, 1989.

[24] F. Delebecque. State-space algorithms for manipulating rational matrices. In *Proceedings IEEE/IFAC Joint Symposium on CACSD, Tucson, Arizona*, pages 475–480, March 7–9, 1994.

[25] A. Emami-Naeini and P. Van Dooren. Computation of zeros of linear multivariable systems. *Automatica*, 18:415–430, 1982.

[26] E. Fehlberg. Low-order classical Runge-Kutta formulas with step size control and their application to some heat transfer problem. *Computing*, 6:61–71, 1970.

[27] G.F. Franklin, J.D. Powell, and R.W. Schafer. *Feedback Control of Dynamic Systems*. Addison-Wesley, 1990.

[28] B. Gach-Devauchelle. *Diagnostic mécanique sur les structures soumises à des vibrations en ambiance de travail*. PhD thesis, Université Paris-IX Dauphine, Décembre 1991.

[29] G. Golub and C. van Loan. *Matrix Computations*. Johns Hopkins Univ. Press, 1989.

[30] C. Gomez. Macrofort: a Fortran code generator in Maple. Rapport technique 119, INRIA, Mai 1990.

[31] C. Gomez and P. Capolsini. MacroC and Macrofort, C and Fortran code generation within Maple. *Maple Technical Newsletter*, 3(1):14–19, 1996.

[32] M. Gondran and M. Minoux. *Graphes et algorithmes*. Eyrolles, 1985.

[33] E. Hairer and G. Wanner. *Solving Ordinary Differential Equations II, Stiff and Differential-Algebraic Problems*. Number 14 in Series in Computational Mathematics. Springer, second edition, 1996.

[34] M. Himsolt. GraphEd: an interactive graph editor. In Springer-Verlag, editor, *Proc. STACS 89*, volume 349 of *Lecture Notes in Computer Science*, pages 532–533. Springer-Verlag, 1989.

[35] A.C. Hindmarsh. Odepack, a systematized collection of ode solvers. In R.S. Stepleman et al., editor, *IMACS Transactions on Scientific Computing*, pages 55–64. North-Holland, 1983.

[36] J.B. Hiriart-Urruty and C. Lemaréchal. *Convex Analysis and Minimization Algorithms*. A Series of Comprehensive Studies in Mathematics. Springer, 1993.

[37] M.A. Jenkins. An algorithm for finding all the zeros of a real polynomial. *ACM Transactions on Mathematical Software*, 1:178–189, 1975.

[38] H. Jreij. *Sur la régulation des cours d'eau aménagés*. PhD thesis, Université Paris-IX Dauphine, April 1997.

[39] K. Kundert. *Circuit Analysis, Simulation and Design*, chapter Sparse Matrix Techniques. North-Holland, 1986.

[40] A. Lebosse. Codes de calcul d'écoulement à surface libre filiaire LIDO, SARA et REZO. Technical Report HE-43/92-4, Électricité de France (DER), 1993.

[41] A. Lebosse, J. Allard, and M. Uan. Principe de résolution de modélisations filiaires des écoulements à surface libres LIDO et SARA. Technical Report HE/43-88-45, Électricité de France (DER), 1988.

[42] P. L'Écuyer and S. Côté. Implementing a random number package with splitting facilities. *ACM Transactions on Mathematical Software*, 17(1):98–111, 1991.

[43] D.G. Luenberger. Time invariant descriptor systems. *Automatica*, 14:473–480, 1978.

[44] J.M. Maciejowsky. Asymptotic recovery for discrete-time systems. *IEEE Trans. on Automatic Control*, AC-30(6):602–605, 1985.

[45] M. Morari and E. Zafiriou. *Robust Process Control*. Prentice-Hall International, Englewood Cliffs, NJ 07632, 1989.

[46] A.V. Oppenheim and R.W. Schafer. *Digital Signal Processing*. Prentice Hall, 1975.

[47] L.R. Petzold. A description of dassl: a differential/algebraic solver. In R.S. Stepleman et al., editor, *IMACS Transactions on Scientific Computing*, volume 1, pages 65–68. North-Holland, 1983.

[48] W.H. Press, B.P. Flannery, S.A. Teukolsky, and W.T. Vetterling. *Numerical Recipes, the Art of Scientific Computing*. Cambridge University Press, 1986.

[49] L. Rabiner and B. Gold. *Theory and Application of Digital Signal Processing*. Prentice Hall, 1975.

[50] L.R. Rabiner, R.W. Schafer, and C.M. Rader. The chirp z-transform algorithm and its applications. *Bell Sys. Tech. J.*, 48:1249–1292, May 1969.

[51] N. Rouillon. Gestionnaire de communications interactif. In *Journée sur les interfaces graphiques et mathématiques discrètes, Bordeaux, France*. LaBRI, Septembre 1992.

[52] R.W. Schafer and L.R. Rabiner. A digital signal processing approach. *Proc. IEEE*, 61(6):692–702, June 1973.

[53] L.M. Silverman. Inversion of multivariable linear systems. *IEEE Trans. Aut. Control*, AC-14, 1969.

[54] O.J.M. Smith. Closer control of loops with dead time. *Chemical Engineering Progress*, 53:217, 1957.

[55] O.J.M. Smith. A controller to overcome dead time. *ISA Journal*, 6:28–33, 1959.

[56] L. Vandenberghe and S. Boyd. Sp, software for semi-definite programming. Available at ftp://isl.stanford.edu/pub/boyd/semidef_prog.

[57] K. Watanabe and M. Ito. A process-model control for linear system with delay. *IEEE Trans. on Automatic Control*, 26(6):1261–1266, 1981.

[58] Working Group on Software. The control and system library SLICOT. Available at http://www.win.tue.nl/wgs/slicot.html.

Index

A

Advanced programming

    `addinter` 124, 125, 146

    `c_link` 145

    `call` 121

    `CheckLhs` 126

    `CheckRhs` 125

    `CreateVar` 126

    `CreateVarFromPtr` 126

    `Error` 127

    `fort` 121

    `FreePtr` 126

    `GetMatrixptr` 130

    `GetRhsVar` 126

    `intersci` 135

    `IsOpt` 127

    `link` 125, 145

    `NumOpt` 127

    `ReadMatrix` 129

    `ReadString` 130

    `ulink` 145

    `WriteMatrix` 130

    `WriteString` 130

C

Commands

    `abort` 44

    `apropos` 43

    `clear` 43

    `exit` 44

    `help` 43

    `pause` 44

    `resume` 44

    `return` 44

    `who` 44

Control

    `augment` 189

    `balreal` 202

    `black` 183

    `bode` 183, 219

    `contr` 168

    `dscr` 159, 172

    `dt_ility` 171

    `evans` 182

    `factors` 205

    `gspec` 167

    `h_inf` 198

    `horner` 153

    `kroneck` 167

    `lft` 159, 191

    `lmisolver` 208

    `lmitool` 208

    `lqe` 171

    `lqg` 193

    `lqg2stan` 192

    `lqr` 170

    `nyquist` 183

    `obscont` 194

    `observer` 171

    `ppol` 169

    `repfreq` 183

    `roots` 167

    `st_ility` 169

    `stabil` 169

    system transformations

        `arl2` 203

bilin 203
des2ss 164
des2tf 165
frep2tf 204
imrep2ss 203
ltitr 173
ss2des 164
ss2ss 166
ss2tf ... 153, 155, 165, 217
syslin 152, 216, 217
tf2des 165
tf2ss 155, 165, 217
trzeros 167
unobs 171

D
Decimation 210

E
Eigenvalues 99

F
FFT 210
Files
    callinterf.h 147
    fundef 147
    matusr.f 148
Flow control
    break 36
    case 33
    for 34
    if 33
    select 33
    while 36

G
Global variables 41
Graphics
    black 78
    bode 78
    champ 71
    chart 78

contour 83
displaying text 76
driver 59
edit_curv 79
elliptical plotting 76
errbar 72
eval3d 83
eval3dp 83
evans 78
fchamp 71
fcontour 83
fgrayplot 72
fplot2d 71
fplot3d{1} 81
gainplot 78
genfac3d 83
geom3d 84
gr_menu 79
graduate 73
grayplot 72, 83
hist3d 83
histplot 72
isoview 63
locate 79
m_circle 78
nyquist 78
param3d 83
plot 64
plot2d{1,2,3,4} 64–70
plot3d{1} 81
plotframe 73
plzr 78
polygon plotting 75
rotate 64
scaling 64
secto3d 83
square 63
titlepage 73
winsid 60
xbasc 59
xbasr 59

xclear 59
xclick 59
xclip 64
xdel 59
xend 59
xgetmouse 59
xgrid 73
xinfo 60
xinit 59
xload 59
xpause 59
xs2fig 92
xsave 59
xselect 59
xset 61–63
xsetech 64
xtitle 73
zgrid 78

I
Index of DAE 267
Interpolation 210

K
Kirchhoff law 397

L
Least squares 96
Linear state-space system 321
Lists 21, 29–32

M
MacroC 344
Macrofort 344
Maple 340
Maple interface
  maple2scilab 350
  n-link pendulum 360, 363
  rolling wheel 354
Metanet
  adj_lists 372
  adjacency lists 371

arc 368
bandwr 375
chain_struct 377
chained lists 375
current Metanet window 381
edge 368
edge label 371
edge name 371
flow 391
gen_net 385
graph 368
graph file 380
graph list 370
graph_2_mat 373, 374
head 368
internal number 371
Kirchhoff law 392
load_graph 380
make_graph 369, 378, 380
mat_2_graph 373, 374
max_cap_path 390
max_flow 392
metanet 381
Metanet window ...378, 380
minimum cost flow
  min_lcost_cflow 394
  min_lcost_flow1 394
  min_lcost_flow2 394
  min_qcost_flow .283, 397
netwindow 384
netwindows 384
node 368
node label 371
node name 371
node-arc matrix 373
node-node matrix 374
path 382
pipe_network 398
plot_graph 381
save_graph 380
show_arcs 383

show_graph 381
show_nodes 383
sink node 386
source node 386
tail 368
tail/head representation .369

N
n-link pendulum 360, 363

O
Operators
   arithmetic 24
   logical 27
   matrix 27
   relational 26
   string 27
Optimization
   fsolve 291
   linpro 248
   optim 248, 284, 286, 290
   quapro 248, 282, 284
Other functions
   bdiag 101
   cdf{bet,bin, ... } 117
   clean 108
   code2str 46
   coeff 45
   coff 102
   coffg 106
   colcomp 104
   derivat 339
   det 106, 109
   expm 102
   factors 108
   fsolve 171
   full 46
   grand 114
   gschur 109
   gspec 109
   im-inv 104

inv 106, 109
kroneck 109
linpro 395
list 21, 29–32, 46
lufact 111
luget 112
lusolve 111
pbig 102
pencan 109
penlaur 109
poly 44
projspec 102, 109
psmall 102
qr 96
rand 114
roots 106
rowcomp 104
schur 99
simp 108
spaninter 105
spanplus 104
spantwo 105
sparse 45, 111
spec 99
spget 111
str2code 46
svd 98
tangent 171
tlist 21, 29–32, 46
trfmod 108
type 47
typeof 47

P
Pendulum 271
Programs
   BEpsf 92
   Blatexpr{2,s} 89
   Blpr 88

R

RLC circuit 251
Rolling wheel 354

S

Saint Venant function 336
Sampled systems 154
Scicos
    activation code 307, 323
    activation time 294
    basic blocks 294
    C block 315
    computational Function . 323
    computational function . 313, 324
    context 301, 303, 315, 331
    event 294
    Fortran block 315
    inheritance 311
    initial firing 308
    interfacing function 313
    `ode` 294
    palettes ... 293, 295, 297, 313, 314, 325, 332
    RBB 307
    regular basic block 307
    SBB 311
    `scicos` 295
    `scicosim` 306
    super block 303, 313, 314
    symbolic parameters 301, 318, 331
    synchro basic block 311
    synchronization 312
    time dependence 311, 319
    ZBB 310
    zero crossing basic block 310
Signal processing
    `convol` 224
    `cspect` 244
    `czt` 224

    `eqfir` 230
    `eqiir` 236
    `fft` 212
    `flts` 225
    `frmag` 220
    `group` 221
    `iir` 234
    `intdec` 210
    `mfft` 212
    `pspect` 242
    `remezb` 233
    `wfir` 227
Simulation 293, 299, 301, 308, 313, 315, 323, 331
    `csim` 172, 174
    `dasrt` 248, 280
    `dassl` 248, 270–279
    `dsimul` 173
    `flts` 174
    `impl` 248, 265–267
    index of DAE 267
    `ode` 248–263
    `odedc` 176, 264
    pendulum 271
    RLC circuit 251
    `rtitr` 173
Singular values 98
Special constants 18

T

Typed lists 21, 29–32